AF443209

Asymmetric Synthesis of 3,3-Disubstituted Oxindoles

Asymmetric Synthesis of 3,3-Disubstituted Oxindoles

Renato Dalpozzo

University of Calabria, Italy

Hélène Pellissier

Aix-Marseille University, France

Swapandeep Singh Chimni

Guru Nanak Dev University, India

Ramon Rios Torres

University of Southampton, UK

World Scientific

NEW JERSEY · LONDON · SINGAPORE · BEIJING · SHANGHAI · HONG KONG · TAIPEI · CHENNAI · TOKYO

Published by

World Scientific Publishing Europe Ltd.

57 Shelton Street, Covent Garden, London WC2H 9HE

Head office: 5 Toh Tuck Link, Singapore 596224

USA office: 27 Warren Street, Suite 401-402, Hackensack, NJ 07601

Library of Congress Cataloging-in-Publication Data

Names: Dalpozzo, Renato, author.
Title: Asymmetric synthesis of 3, 3-disubstituted oxindoles / by Renato Dalpozzo
   (University of Calabria, Italy), Hélène Pellissier (Aix-Marseille University, France),
   Swapandeep Singh Chimni (Guru Nanak Dev University, India) and
   Ramon Rios Torres (University of Southampton, UK).
Description: New Jersey : World Scientific, 2019. | Includes bibliographical references.
Identifiers: LCCN 2019013757 | ISBN 9781786347299 (hc)
Subjects: LCSH: Heterocyclic chemistry. | Oxindoles. | Indole. | Asymmetric synthesis. |
   Isomerism.
Classification: LCC QD400 .A89 2019 | DDC 547/.59--dc23
LC record available at https://lccn.loc.gov/2019013757

British Library Cataloguing-in-Publication Data

A catalogue record for this book is available from the British Library.

For any available supplementary material, please visit
https://www.worldscientific.com/worldscibooks/10.1142/Q0216#t=suppl

Desk Editors: Dipasri Sardar/Jennifer Brough/Shi Ying Koe

Typeset by Stallion Press
Email: enquiries@stallionpress.com

Preface

Oxindole is one of the most abundant heterocyclic compound available in nature, and it is often present in the enantiopure form. Therefore, its synthesis represents one of the most challenging asymmetric preparations in organic chemistry. Many scientists have ventured into this synthesis. This book aims to give a comprehensive literature survey on the preparation of enantioenriched 3,3-disubstituted oxindoles, the most intriguing oxindole derivative. Owing to the pharmacological activities of oxindole derivatives (show anticancer, anti-HIV, antidiabetic, antibacterial, antioxidant, kinase inhibitory, AChE inhibitory, antileishmanial, $\beta 3$ adrenergic receptor agonistic, phosphatase inhibitory, analgesic, and spermicidal properties; act as vasopressin and progesterone antagonists; possess neuroprotective characteristics; and have NMDA blocker activities) and the number of approved drugs as well as natural products that belong to this family, we hope that this book will become an important reference material for organic chemists both in the academic laboratories and in the R&S industrial laboratories.

About the Authors

Renato Dalpozzo was born in 1957. He graduated from the University of Bologna in 1981, with a laurea in industrial chemistry. He was Researcher of Organic Chemistry at the University of Bologna from 1983 to 1992. In 1992, he was appointed as Associate Professor of Organic Chemistry at the University of Calabria. In 2002, he became a Full Professor of Environmental and Cultural Heritage Chemistry and then of Organic Chemistry again, at the University of Calabria, where he is teaching currently.

His research interests started with the reactivity of organometallic compounds with nitroarenes, the use of dianions derived from enamino carbonyl compounds, and the stereoselective reduction of various classes of ketones. More recently, he was interested in the development of new Lewis acid systems, and presently, he is working on enantioselective organocatalysis, especially on heterocyclic compounds.

Ramon Rios Torres was born in 1974 in Barcelona. In 1996, he got his Bachelor's Degree from the University of Barcelona. In 2000, he obtained his Ph.D. degree under the supervision of Prof. Albert Moyano, working on the study of the Pauson Khand reaction with electron-deficient enynes. After several postdocs and research stages with Prof. Greg C. Fu (MIT, Boston, 1999), Prof. Patrick Walsh (University of Pennsylvania, Philadelphia, 2001), Prof. Benjamin List (Max Plank, Muelheim an der Ruhr, 2004–2006), Prof. Armando Córdova (University of Stockholm, Stockholm, 2006–2007), and Prof. Jose Mᵃ Alvarez-Pez (University of Granada, Granada, 2007) and an industrial experience (J.C. Uriach S.A., Barcelona, 2002–2004), he was awarded with an ICREA Junior Academia in 2008. He began his independent career as an ICREA Researcher in 2008 at the University of Barcelona. In August 2012, he became Associate Professor in Organic Chemistry in the University of Southampton. He has been Visiting Professor in Nagoya Institute of Technology (Japan) and Sungkyunkwan University (SKKU) (South Korea). He is the author of more than 125 papers and 15 chapters in chemical books and has obtained one patent. He has edited three books — one on the Pauson–Khand reaction, one on organocatalysis, and a recent one on Spiro compounds. In all, he has more than 6300 citations and Idh = 45.

He is currently the Associate Editor of *Scientific Reports* (Nature publishing group) and *Frontiers in Organic Chemistry*. In 2018, he became the Senior Fellow of the Higher Education Academy.

His research interests are in the development of new asymmetric methodologies based on organocatalysis and/or organometallic chemistry, synergistic catalysis, photocatalysis, heterogenous catalysis, and biomass valorization.

Marta Meazza was born in 1987 in Saronno (Italy). She graduated from the University of Milan in 2012 with a Master's Degree in Chemistry and Pharmaceutical Technology, under the supervision of Prof. Egle Beccalli. She obtained her Ph.D. under the supervision of Dr. Ramon Rios at the University of Southampton (UK) in 2016. After a postdoctoral position in the group of Prof. Karl Anker Jørgensen at Aarhus University (DK), she moved to the University of Southampton as Research Fellow in 2017. She is the author of 30 papers, with 300 citations and Idh = 10. Her research interests include synergistic catalysis, organocatalysis, organometallic chemistry, organocatalytic photoactivation of organic compounds, and heterogeneous catalysis.

Hélène Pellissier is currently a researcher at the National Center for Scientific Research at Aix-Marseille University (France). She carried out her Ph.D. under the supervision of Dr. G. Gil in Marseille in 1987. After a postdoctoral period in Prof. K.P.C. Vollhardt's group at the University of California, Berkeley, she joined the group of Prof. M. Santelli in Marseille in 1992, where she developed novel very short total syntheses of steroids starting from 1,3-butadiene and benzocyclobutenes. She is the author of 110 papers including reviews in international journals, eight books, and seven book chapters.

Swapandeep Singh Chimni was born in 1962 at Amritsar, India. He received his M.Sc. (Hons. Sch.) in Chemistry in 1985 and Ph.D. in 1991 from Guru Nanak Dev University, Amritsar. After 2 years as a lecturer at the Regional Engineering College (now NIT) in Jalandhar, he joined the Department of Chemistry, Guru Nanak Dev University as Lecturer in 1992. He is presently working as a Professor in the same department. He has a research experience of 29 years and published over 125 publications. He works in the area of synthetic organic chemistry with emphasis on asymmetric organocatalysis, biocatalysis, and phase-transfer catalysis, as well as green chemistry.

Jasneet Kaur was born in 1989 in Vain Poin Village of Tarn Taran, Punjab, India. She obtained her B.Sc. (Hons. Sch.) in Chemistry and M.Sc. (Hons. Sch.) in Chemistry from Guru Nanak Dev University, Amritsar. She completed her Ph.D. under the supervision of Prof. Swapandeep Singh Chimni from Guru Nanak Dev University, Amritsar, India, in 2017. After that she worked as a Research Associate with the same research group until July 2018. Since August 2018, she is working as an Assistant Professor in the Department of Chemistry at Khalsa College, Amritsar. So far she has published 18 research papers and review articles. Her research interests include the synthesis of new chiral bifunctional organocatalysts and their applications for enantioselective carbon–carbon and carbon–heteroatom bond formations and domino reactions.

Contents

Chapter 1

Introduction

Oxindole (indolin-2-one) chemistry began to develop with the study of the dye indigo, which was easily converted to isatin and then to oxindole. When in 1866, Adolf von Baeyer reduced oxindole to indole, it was then able to propose its chemical formula [1]. It is an aromatic heterocyclic organic compound with a bicyclic structure, consisting of a six-membered benzene ring fused to a five-membered nitrogen-containing ring (indoline skeleton) substituted with a carbonyl group at the second position of the five-member ring (Figure 1.1). After their discovery, indole derivatives were prepared as important dyestuffs in analogy with indigo, and only since the 1930s, interest in indole was intensified as it became clear that indole skeleton was present in many important alkaloids. Now, the indole ring is recognized as the most common heterocycle present in nature [2]. For this reason, substituted indoles have been referred to as "privileged structures" and a number of approved drugs as well as natural products belong to this family. This biological activity and structural complexity have stimulated generations of synthetic chemists to design strategies for assembling these structures and this field of chemistry is growing drastically.

Among them, the therapeutic potential of oxindole nucleus has captured the interest of medicinal chemists to synthesize novel oxindole derivatives [3]. If Baeyer obtained it from indigo, oxindole was first extracted in nature from the cat's claw plant *Uncaria tomentosa* (Rubiaceae) found in the Amazon rainforest and other tropical areas of South and Central America [4]. In human biology instead, oxindole is

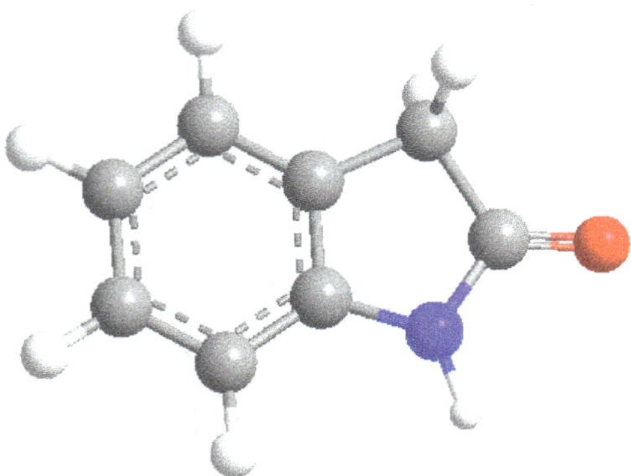

Figure 1.1.   Structure of oxindole.

formed by gut flora and is normally metabolized and detoxified from the body by the liver. Its excess can cause sedation, muscle weakness, hypotension, and coma. For instance, elevated serum oxindole levels were recorded in patients with hepatic encephalopathy [5].

The detailed pharmacological activities of oxindole derivatives include anticancer, anti-HIV, antidiabetic, antibacterial, antioxidant, kinase inhibitory, AChE inhibitory, antileishmanial, $\beta 3$ adrenergic receptor agonistic, phosphatase inhibitory, analgesic, spermicidal, vasopressin antagonists, progesterone antagonists, neuroprotection, and NMDA blocker activities [6]. Then structure–activity relationship studies have found that the substituent pattern at the C-3 position of oxindole greatly affects all these biological activities, thus, in recent years, the development of efficient synthetic methods for the enantioselective assembly of 3,3-disubstituted oxindoles has attracted the interest of the chemical community.

The aim of this book is to allow chemists to orientate in the forest of enantioselective syntheses appearing in the literature in recent years. Every chapter covers the syntheses of a particular class of disubstituted oxindoles and also provides a survey of the enantioselective metal complex and organo-catalysis. Asymmetric oxindole synthesis was demonstrated, in fact, as a training area for organic chemists in the fascinating field of enantioselective synthesis.

Starting materials for asymmetric syntheses on oxindoles are commercially available or can be easily prepared, thus allowing a wide accessibility

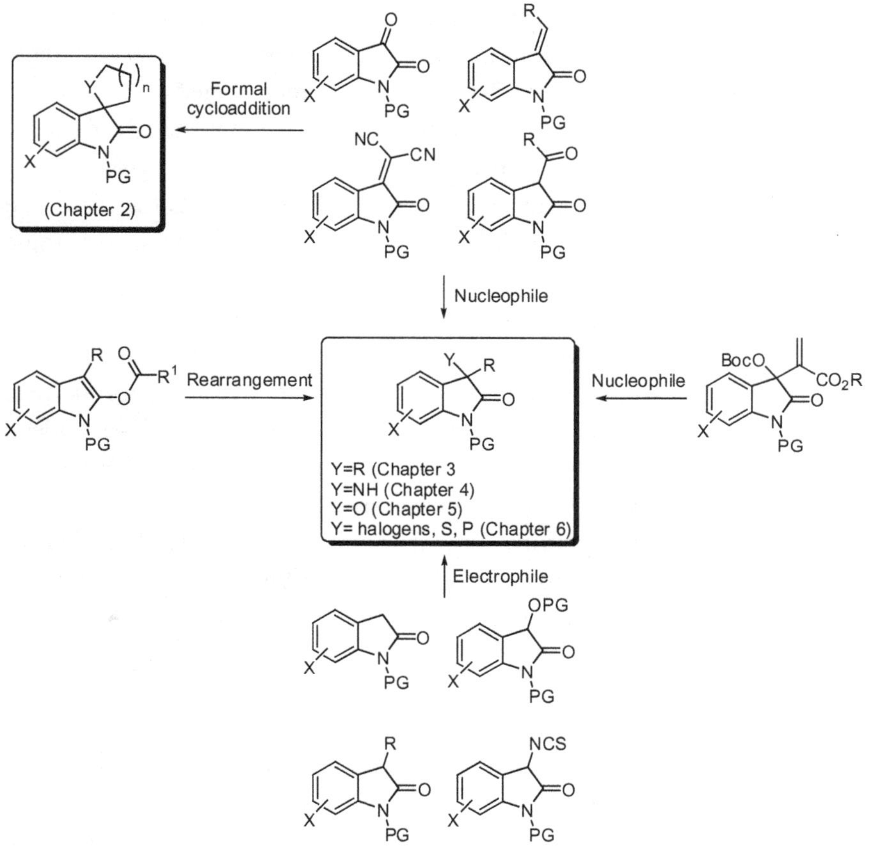

Scheme 1.1. Main reactions described in this book.

(Scheme 1.1). Alkylidene oxindole, isatin and their derivatives can undergo formal cycloadditions that actually are often cascade reactions. In Chapter 2, Prof. Ramon Rios explores these syntheses, which perhaps form the most fascinating topic of the book. These compounds can also be nucleophilically attacked. It should be noted that, in alkylidene oxindoles, which carry two different electron-withdrawing groups at both termini of the double bond, the negative charge will accommodate near to the out-of-ring stabilizing group. The C-3 position of oxindoles shows nucleophilic character, thus asymmetric catalysts often show basic character in order to form enol or enolate, which can undergo nucleophilic substitution, aldol or Mannich reactions. Moreover, oxindoles bearing leaving groups onto the C-3 can undergo nucleophilic substitution and 2-indolyl carbonate can rearrange. Among oxindole derivatives, 3-aryloxindoles represented an intriguing challenge because the aryl group increases (i) the acidity and (ii) the steric hindrance at the C-3 position (increasing the chance of a background reaction or diminishing the reactivity, respectively), and (iii) has the same size as the oxindole ring (thus, the differentiation between the two enantiotopic faces becomes more difficult).

All these reactions allow the synthesis of open chain compounds and Chapter 3 by Prof. Renato Dalpozzo describes how to prepare open-chain 3,3-dialkyloxindoles, compounds with manifold bioactive properties. Chapter 4 by Dr. Hélène Pellissier and Chapter 5 by Prof. Swapandeep Chimni deal with the syntheses of 3-substituted-3-aminooxindoles and 3-substituted-3-hydroxyoxindoles, respectively, the units largely encountered in alkaloids and pharmaceutical candidates. Finally, Prof. Renato Dalpozzo presents a survey on 3-hetero-3-substituted oxindoles, the less explored class of oxindole derivatives.

References

1. (a) A. Baeyer, *Annalen der Chemie und Pharmacie* **1866**, *140*, 296. (b) A. Baeyer, A. Emmerling, *Ber.* **1869**, *2*, 679.
2. (a) R. Sundberg, *The Chemistry of Indole*, Academic Press, Cambridge, **1970**. (b) R. Sundberg, *Indoles* (Eds.: A. Katritzky, O. Meth-Cohn, C. Rees), Academic Press, Cambridge, **1998**. (c) Rahman, *Indole Alkaloids*, CRC Press, Boca Raton, **1998**. (d) G. W. Gribble *Indole Ring Synthesis: From Natural Products to Drug Discovery*, Wiley, Chichester, **2016**.
3. J. Bergman, Oxindoles, in: *Advances in Heterocyclic Chemistry*, Vol. 117 (Eds.: E. F. V. Scriven, C. A. Ramsden), Elsevier, Amsterdam, **2015**, pp. 1–81.
4. S. R. Hemingway, J. D. Phillipson, *J. Pharm. Pharmacol.* **1974**, *26*, 113.
5. O. Riggio, G. Mannaioni, L. Ridola, S. Angeloni, M. Merli, V. Carlà, F. M. Salvatori, F. Moroni, *Am. J. Gastroenter.* **2010**, *105*, 1374.
6. (a) S. R. S. Rudrangi, V. K. Bontha, V. R. Manda, S. Bethi, *Asian J. Res. Chem.* **2011**, *4*, 335. (b) M. Kaur, M. Singh, N. Chadha, O. Silakari, *Eur. J. Med. Chem.* **2016**, *123*, 858. (c) N. H. Greig, X.-F. Pei, T. T. Soncrant, D. K. Ingram, A. Brossi, *Med. Res. Rev.* **1995**, *15*, 3.

Chapter 2

All-Carbon Spirooxindoles

2.1. Introduction

Spirooxindoles are one of the most important targets for synthetic organic chemists due to their inherent biological activity, as witnessed by the presence of spirooxindole motifs in many natural products and pharmaceuticals. Examples of natural products are Gelsemine (a strong CNS stimulant showing antihypertensive activity) whose synthesis has been reported by several groups, Welwitindolinone A (with antifungal activity) and (−)-Horsfiline, which present analgesic properties. As pharmaceuticals, we can highlight the anti-cancer agent MI-219 and the antimalarial compound NITD609 (Figure 2.1) [1].

In this chapter, we will summarize the most important methodologies developed for the enantioselective synthesis of spirooxindoles, focusing on spirooxindoles with all C in the spiroquaternary center.

In Section 2.2, we will present the enantioselective syntheses of spirocycles using organometallic catalysts. This section is divided according to the metal employed in the synthetic reaction rather than in the strategy employed, highlighting the most important advancements in this area. Section 2.3 is devoted to enantioselective organocatalytic methodologies for the synthesis of spirocycles. This section is organized on the basis of the nature of the synthesized spirocyclic ring.

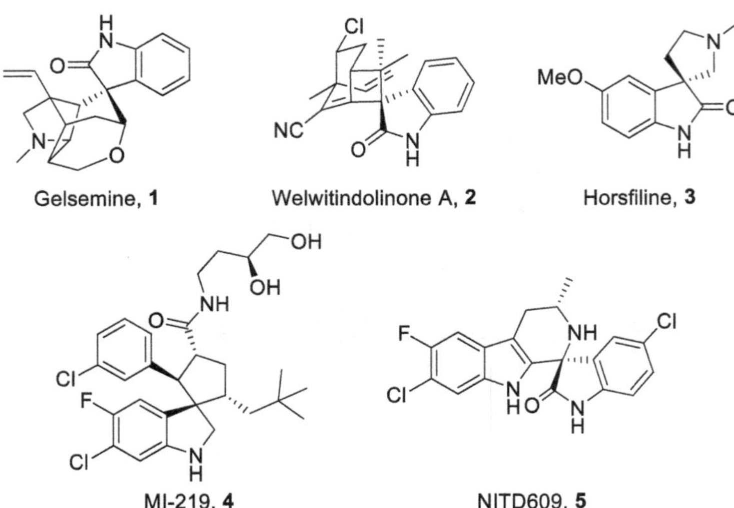

Figure 2.1.   Examples of natural products and pharmaceuticals containing spirooxindoles.

2.2. Organometallic Approaches

For decades, organometallic chemistry ruled organic synthesis. The combination of chiral ligands and transition metals such as Zn, Rh, Pd, and so forth have demonstrated a powerful combination for the development of enantioselective C–C bond forming reactions. In this section, we will focus on the enantioselective methodologies for the synthesis of spiro compounds developed over the past few years using organometallic reactions. As stated before, spirooxindoles have greatly benefitted from their use in medicinal chemistry as the intense quest for new oxindole scaffolds with rigid sructures has increased the interest to develop new strategies for the synthesis of spirooxindoles.

2.2.1. *Zinc-catalyzed methodologies*

One of the first zinc-catalyzed syntheses of 3,3-disubstituted spirooxindoles was reported by Feng *et al.* In this report, methyleneidolinones **6** react with Brassard's type dienes **7** in a Diels–Alder cycloaddition fashion

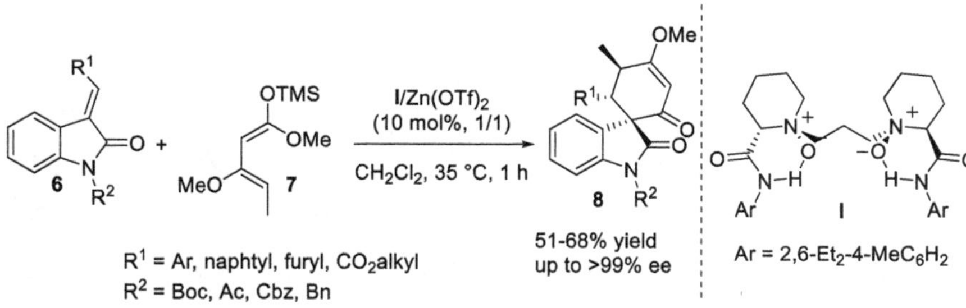

Scheme 2.1. Spirocyclization reported by Feng.

to furnish the spirooxindole **8** [2]. The reaction is efficiently catalyzed by Zn complexes being the best ligand *N*-oxide *L*-PiEt$_2$-Me (**I**) which delivers the oxindoles with good yields (58–68%) and excellent enantioselectivities (98–99% ee). The reaction scope is quite broad allowing various substituents on the *N*-Boc-protected indolinones, such as substituted phenyl, naphthyl, furyl, and alkylesters. The electron-donating group (EDG) in the phenyl moiety afforded better results over the electron-withdrawing group (EWG). Several protecting groups can be used on the nitrogen of the oxindole such as Boc, Ac, and Cbz with the only limitation of Bn protecting group. This behavior demonstrates that both carbonyl groups present in the indolinone exhibit bidentate coordination with the Zn catalyst favoring a more stable (and rigid) transition state. Moreover, a linear effect between the enantiomeric excess of the ligand and the product was observed, indicating that the monomeric catalyst could be the active species (Scheme 2.1).

Another approach for the synthesis of spirooxindoles was reported by Wang, based on a desymmetrization of diketone oxindole **9**. Trost's dinuclear zinc catalyst **Zn(II)** promotes the asymmetric intramolecular aldol reaction achieving reasonable yields (53–81%) with moderate to good enantioselectivities (63–95% ee) (Scheme 2.2) [3].

2.2.2. *Palladium-catalyzed methodologies*

One of the first methodologies for the synthesis of spirooxindoles was developed in 2007 by Trost, who used a palladium-catalyzed

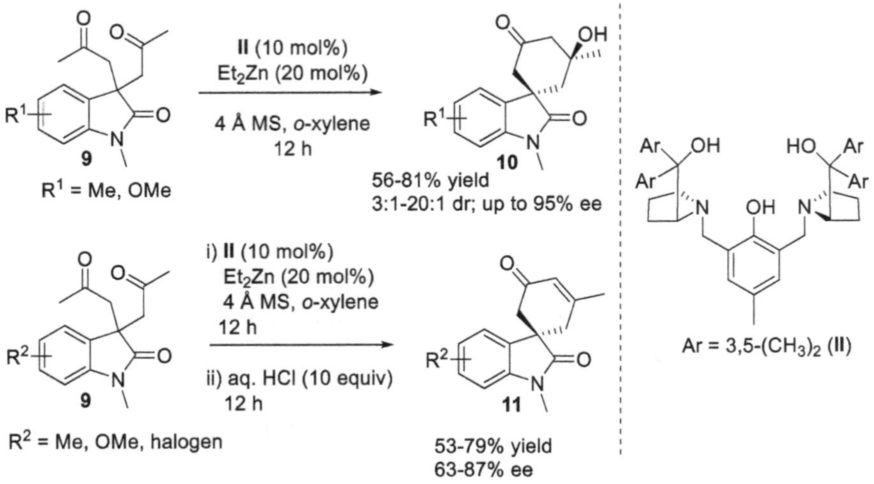

Scheme 2.2.   Spirocyclization reported by Wang.

trimethylenemethane [3+2] cycloaddition to afford spirocyclic oxindolic cyclopentanes [4]. 3-Acetoxy-2-trimethylsilylmethyl-1-propene derivative **13** catalyzed by Pd reacts with benzylidene oxindoles to produce spirooxindoles **14** and **15** in a highly chemoselective, regioselective, and stereoselective manner. The best ligand in terms of stereoselectivity was **III**, rendering the final spiro-cyclo oxindoles in high yields and with excellent diastereo- and enantio-selectivities (Scheme 2.3).

Lu *et al.* reported a [3+2] cycloaddition between vinyl aziridines **17** and alkylidene oxindoles **16** using Pd complexes as catalysts [5]. The reaction starts with the formation of the 1,3-dipole by ring opening of the vinylaziridine under Pd catalysis. Next, a Michael addition to methylene indoline takes place, followed by intramolecular asymmetric allylic alkylation to furnish the 3,3′-pyrrolidonyl spirooxindole and regenerate the Pd(0) catalyst. The best ligand is phosphoramidate **V** achieving the final compounds **18** in good yields (69–90%), good to excellent diastereoselectivities (up to 19:1 d.r.) and excellent enantioselectivities (82–99% ee). The scope of the reaction presents certain limitations like the need to use Ts-protected aziridines and electron-withdrawing substituents at the alkylidene position of the oxindole (Scheme 2.4).

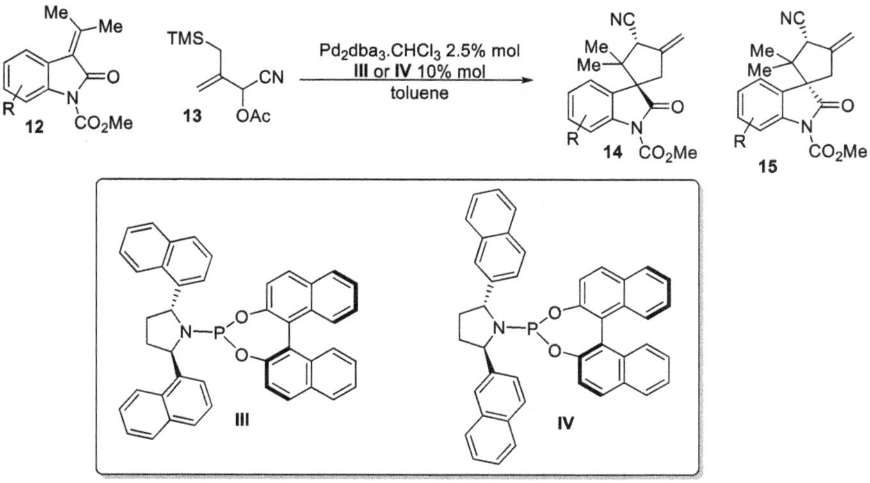

III R = H 97% yield, 1:6.2 (77a:78a) 96% ee, 99% ee
IV R = H 97% yield, 4.3:1 (77a:78a) 92% ee, 95% ee

Scheme 2.3.   [3+2] Cycloaddition reported by Trost.

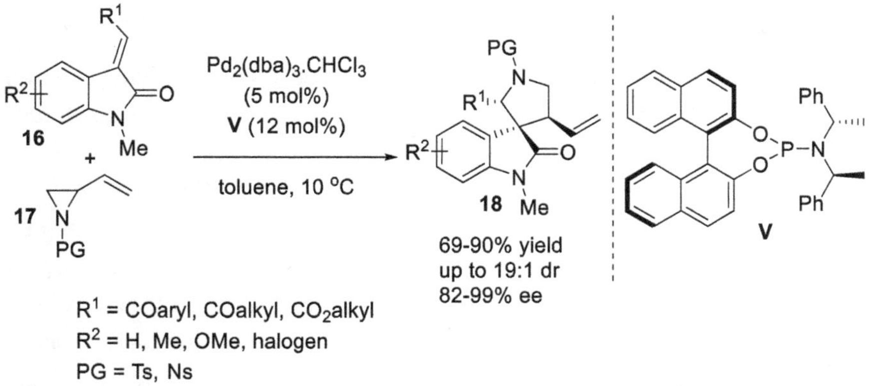

69-90% yield
up to 19:1 dr
82-99% ee

R^1 = COaryl, COalkyl, CO_2alkyl
R^2 = H, Me, OMe, halogen
PG = Ts, Ns

Scheme 2.4.   Spirocyclization reported by Lu.

2.2.3. *Scandium-catalyzed methodologies*

The first enantioselective [3+2] carboannulation of alkylidene oxindoles **19** with allylsilanes **20** was reported by Franz and co-workers in 2014. The reaction was simply catalyzed by a scandium(III)/(R,S)-indaPybox **VI** complex, using sodium tetrakis-[3,5-*bis*(trifluoromethyl)phenyl]borate

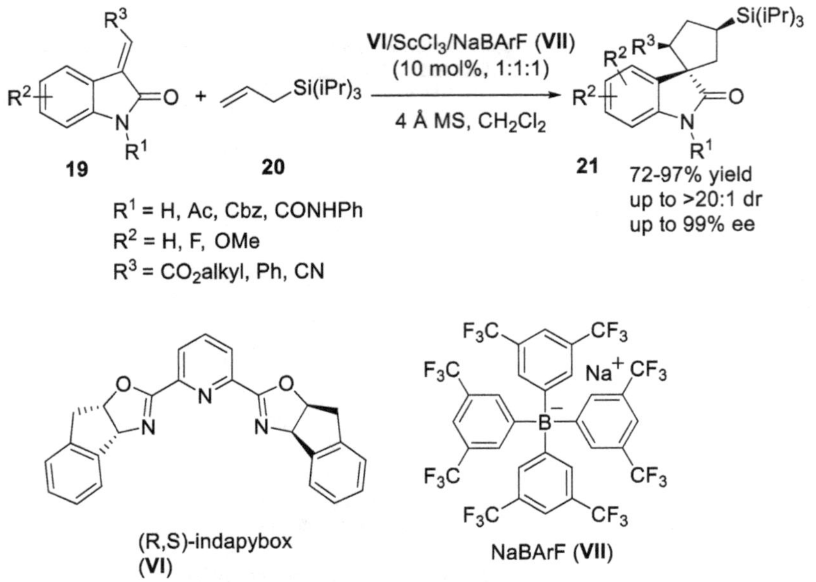

R^1 = H, Ac, Cbz, CONHPh
R^2 = H, F, OMe
R^3 = CO_2alkyl, Ph, CN

72-97% yield
up to >20:1 dr
up to 99% ee

(R,S)-indapybox
(**VI**)

NaBArF (**VII**)

Scheme 2.5.   Spirocyclization reported by Franz.

(NaBArF, **VII**) as additive to improve the stereoselectivity. The reaction allows access to spirooxindolecyclopentanes **21** with up to three stereocenters with excellent stereoselectivities (Scheme 2.5) [6].

A cascade 1,5-hydrogen shift/ring closure for the synthesis of spirooxindole tetrahydroquinolines **23** under Sc-catalysis was developed by Feng in 2015 [7]. The chiral N,N'-dioxide **VIII**-Sc(OTf)$_3$ complex promoted the reaction with up to 97% yields and >20:1 diastereoselectivities. An interesting observation was that partial self-disproportionation of enantiomers (SDEs) occurs during silica column purification of the products (Scheme 2.6).

2.2.4. *Nickel-catalyzed methodologies*

3-Alkenyl-oxindoles **24** react with phenyliodonium ylides **25** to afford the spirooxindoles **26** under nickel catalysis [8]. Thermal decomposition of the phenyliodonium ylide malonate under mild conditions is followed by

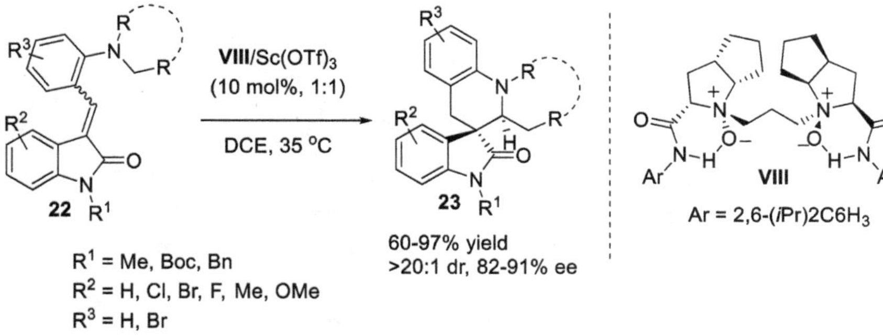

Scheme 2.6.   Spirocyclization reported by Feng.

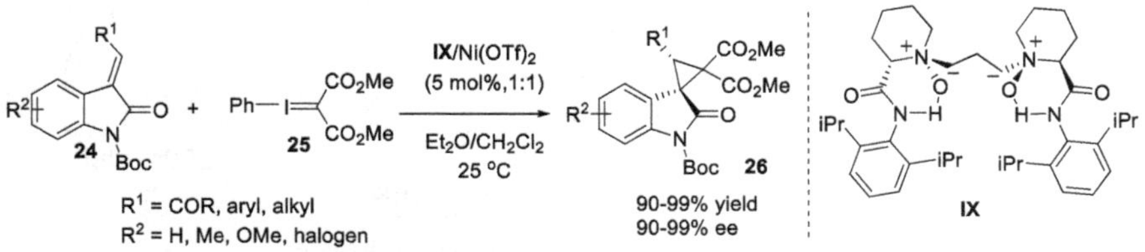

Scheme 2.7.   Spirocyclization reported by Feng and Liu.

the formation of the carbene, which reacts with the 3-alkylidene oxindole catalyzed by an *N,N′*-dioxide **IX**-nickel complex to form the spirocyclopropanes **26** in good yields (82–99%) and excellent stereoselectivities (complete diastereoselectivity and up to 99% ee; Scheme 2.7).

A [3+2] cycloaddition between 3-arylidene oxindoles **27** and iminoesters **28** catalyzed by chiral Ni complexes was reported by Arai [9]. Spirooxindoles **29** were synthesized in excellent yields (87–99%), diastereoselectivities (up to 99:1 d.r.) and enantioselectivities (up to >99% ee; Scheme 2.8). Soon after, the same research group developed a synthesis of spirooxindoles based on a thio-Michael/aldol cascade reaction catalyzed by Ni complexes [10]. 3-Alkylidene oxindoles react with thiosalicylaldehyde *via* a thio-Michael/aldol sequence to generate a thiochromanyl-spirooxindole with three contiguous stereogenic centers with excellent yields (73–99%) and stereoselectivities (up to 98:2 d.r. and up to 94% ee) using *bis*(imidazoline)pyridine derivatives as chiral ligands.

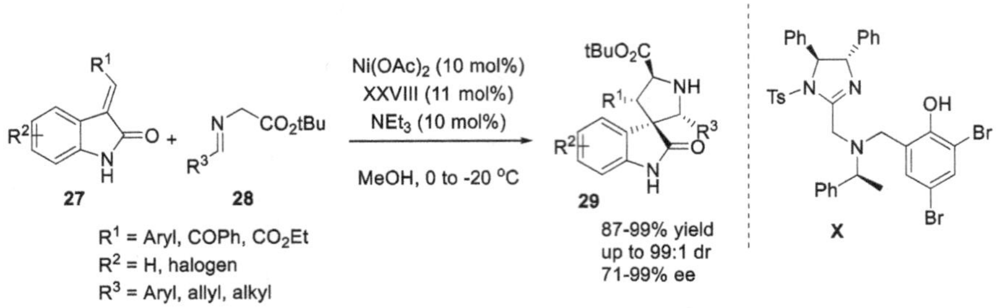

Scheme 2.8. Spirocyclization reported by Arai.

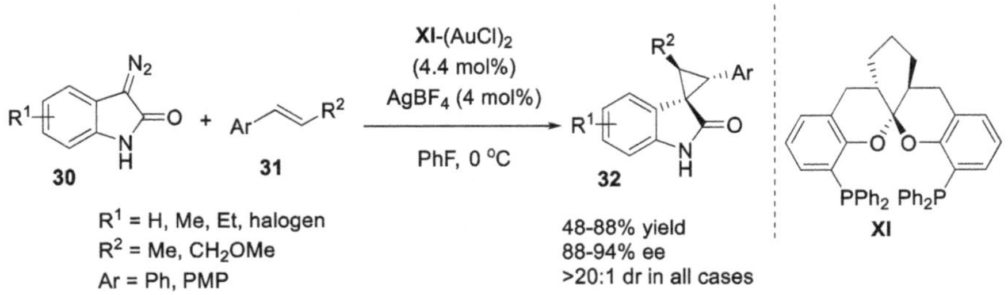

Scheme 2.9. Cyclopropanation reported by Zhou.

2.2.5. *Gold-catalyzed methodologies*

Diazooxindoles **30** and olefins **31** react under Au catalysis to form spiro-cyclopropanes **32** in moderate to good yields (48–88%), good enantioselectivities (88–94% ee) and complete diastereoselectivities when Spiroketalbiphosphine **XI** was used as a ligand [11]. The reaction presents a broad substrate scope including *cis*- and *trans*-disubstituted alkenes (Scheme 2.9).

2.2.6. *Rhodium-catalyzed methodologies*

In 2016, Chi *et al.* developed a chiral cyclopropanation to synthesize spirocyclopropaneoxindoles under Rh catalysis [12]. *N*-Boc-diazooxindoles **33** and alkenes **34** react to produce spirooxindolic cyclopropane derivatives **35** in good to high yields (71–97%) with good to excellent diastereoselectivities and enantioselectivities (up to 20:1 d.r. and up to 93% ee),

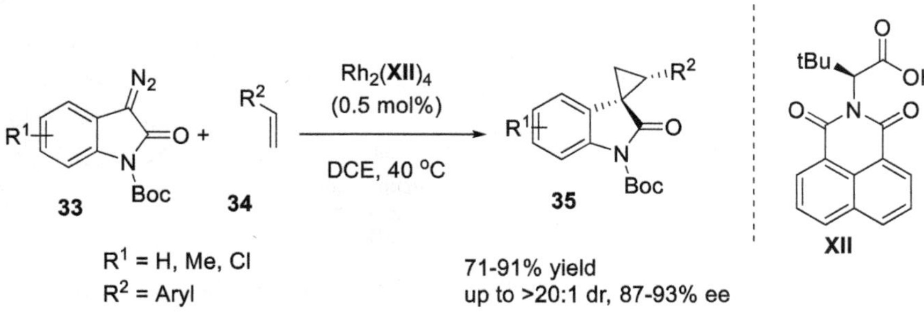

Scheme 2.10.   Spirocyclization reported by Chi *et al.*

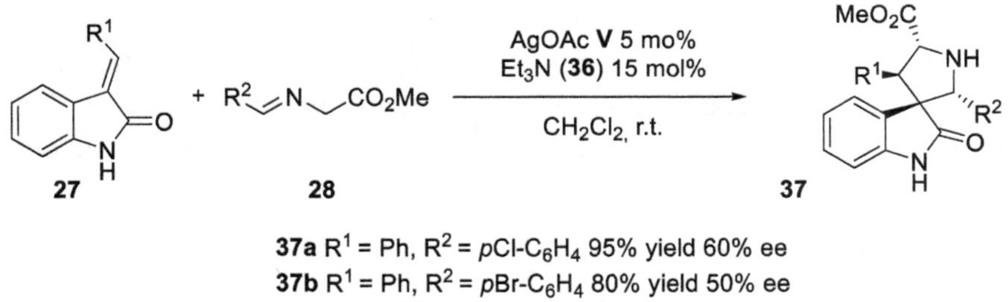

Scheme 2.11.   Ag catalyzed [3+2] cycloaddition reported by Wang.

when chiral dirhodium carboxylate complexes are used as catalysts. The reaction scope is broad allowing the use of aryl, alkyl, cyclic, and disubstituted olefines (Scheme 2.10).

2.2.7. *Other metals*

An elegant enantioselective synthesis of spirooxindole pyrrolidines was reported by Wang [13].

A [3+2] cycloaddition between 2-oxoindolin-3-ylidene **27** and azomethine ylides **28** was catalyzed by Ag(I)/(S)-TF-BiphamPhos derivatives (**V**) yielding the spirooxindole pyrrolidines 37 in excellent yields, albeit with moderate enantioselectivities (Scheme 2.11).

A singular reaction was reported by Zhou *et al.* using Hg complexes for the synthesis of spirooxindoles [14]. In this reaction, alkenes **38** and diazooxindoles **30** react together to form spirocyclopropane **39**. The Hg

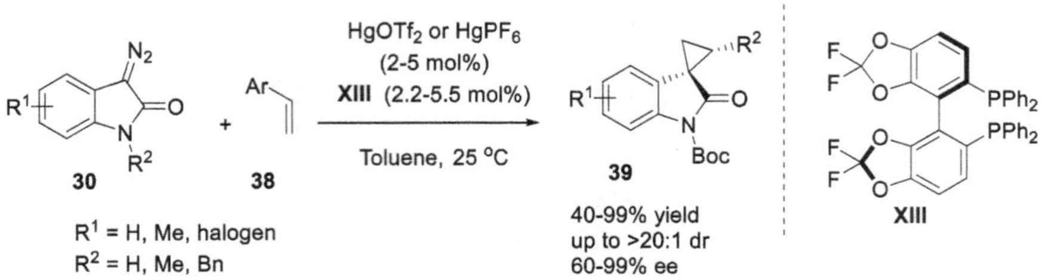

Scheme 2.12.   Spirocyclopropanation reported by Zhou.

complex acts as a soft Lewis acid, decomposing the diazooxindole to generate the active species for the cyclopropanation reaction with a series of different alkenes. Chiral diphosphine **XIII** is crucial to obtain high stereoselectivities. The reaction presents a good substrate scope affording the final cyclopropanes in good to excellent yields (40–99%) and enantioselectivities (60–99% ee; Scheme 2.12).

2.3. Organocatalytic Approaches

In this section, we present a detailed overview of the organocatalytic syntheses of spirooxindoles.

In 2000, after the pioneering works of List [15] and MacMillan [16] on enamine and iminium catalysis, a renaissance of organocatalysis occurred. Since then, organocatalysis has become the third pillar in asymmetric catalysis, complementing the organometallic and enzymatic catalysis for the assembly of chiral compounds. The easy stereoprediction of the formed stereocenters, green approach that avoids the use of metals and an often great functional group tolerance, makes organocatalytic methodologies an excellent approach for the synthesis of spiro compounds. The enantioselective synthesis of spirooxindoles has become an important goal in organic chemistry. Spirocyclic oxindoles that contain varied spiro rings fused at the C-3 atom represent privileged structures that are part of several pharmaceutical and natural products. Their prevalence in privileged 3D scaffolds has attracted a notable synthetic methodology interest. These efforts contribute to the creation of libraries that registered a notable success in the pursuit of new candidates for drug discovery.

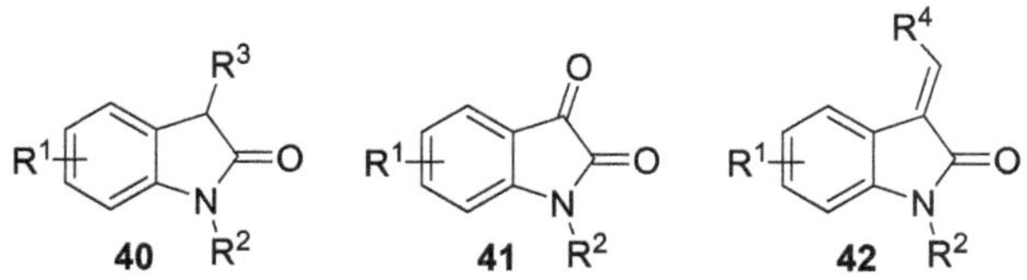

Figure 2.2.   Starting oxindoles.

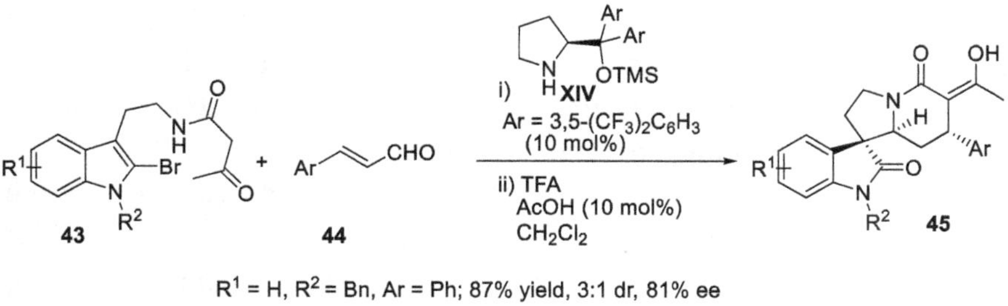

Scheme 2.13.   Spirocyclization reported by Zhou *et al.*

Organocatalysis has emerged as a prominent tool to design new spirooxindole motifs.

Herein, we have divided the synthesis of spirooxindoles based on the nature of the starting oxindole: 3-mono-substituted oxindoles **40**, conjugated oxindoles **42**, and isatin **41** (Figure 2.2).

2.3.1. *Methodologies with 3-monosubstituted oxindoles*

One of the first examples of the synthesis of oxindoles based on this approach was reported by Wu *et al.* [17]. In this example, 2-bromo indolines **43** react with enals **44** under secondary amine catalysis *via* a Michael/hemiaminal formation/dehydration/intramolecular Mannich sequence to furnish the spirooxindoles **156**, after hydrolysis, in excellent yields (81–95%), moderate diastereoslectivities (2:1–6:1 d.r.) and good enantioselectivities (77–94% ee; Scheme 2.13).

A multicomponent organocascade between oxindoles **47**, enals **44**, and nitrostyrenes **46** was reported by Zhou *et al.* The reaction rendered spirocyclohexanones (**48** and **49**) with good yields (90–45%) and stereoselectivities (up to 8:2:1 d.r. and up to >99% ee) [18]. It starts with a Michael addition between the oxindole and the nitrostyrene catalyzed by thiourea catalyst **XVII**. Next, a Michael reaction with the enal activated by the secondary amine catalyst (**XV** or **XVI**), followed by intramolecular aldol reaction, furnished the spirooxindole. The reaction has a broad scope tolerating aromatic and aliphatic enals as well as aromatic and aliphatic nitroalkenes; however, a decrease in diastereoselectivity and yield was observed when aliphatic starting materials were employed. A limitation of the present methodology is the need to use *N*-Boc-protected oxindoles. Remarkably, as shown in Scheme 2.14, the nature of the secondary amine catalysts determines the diastereomer obtained.

A [3+2] cycloaddition reaction based on a tandem reaction between 3-hydroxyoxindoles **50** decorated with an indolyl moiety and 3-methyl-2-vinylindoles **51** was reported by Shi in 2016 [19]. Chiral phosphoric acid derivative **XVIII** promotes the reaction, obtaining the final spiro compounds **52** in good yields (72–99%) and excellent stereoselectivities (>95:5 d.r. and 90–98% ee). The mechanism consists of an acid-catalyzed

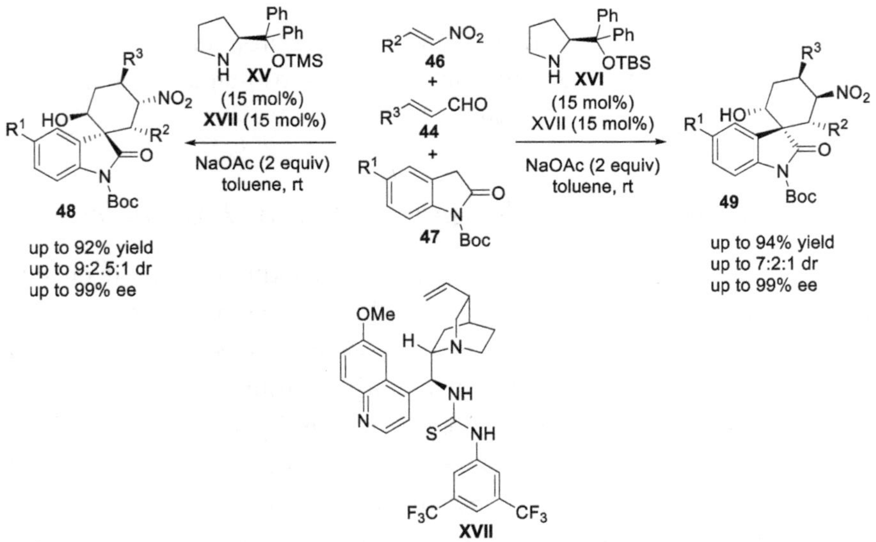

Scheme 2.14. Spirocyclization reported by Li and Zhou.

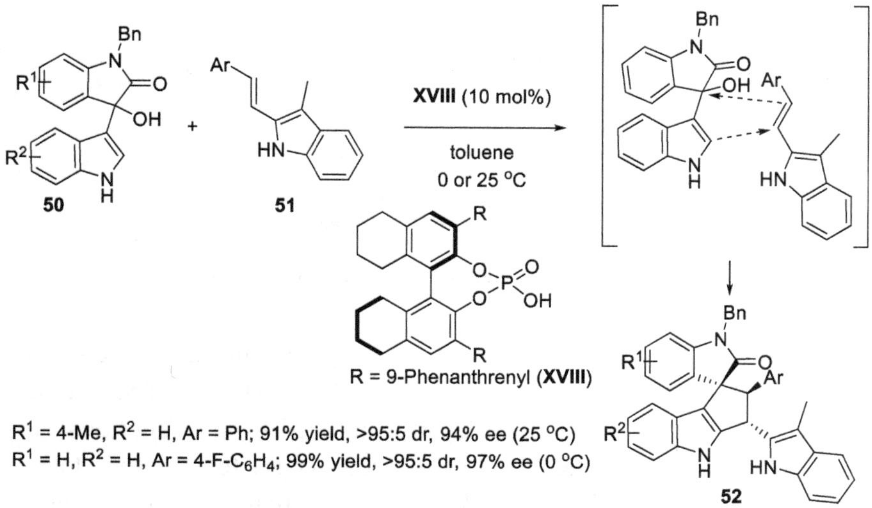

Scheme 2.15.   Spirocyclization reported by Shi.

dehydration to form the carbocation intermediate, which subsequently forms an ion pair with the anion of the phosphoric acid. The NH moiety in both indoles is crucial for the stabilization of the transition state through hydrogen bonding and is key to afford higher ee (Scheme 2.15).

The same group reported a similar organocascade reaction, starting this time with 7-vinylindoles [20]. Chiral phosphoric acids promote the generation of vinyliminium ion from isatin-derived 3-indolylmethanols. Next, a [3+2] cycloaddition takes place between the *in situ*-formed vinyliminium ion and the 7-vinylindole to furnish the spirooxindole in moderate yields (43–68%), total diastereoselectivities and excellent enantioselectivities (94–99% ee). Wang *et al.* reported a double Michael addition between oxindole and dienones promoted by Soos' catalyst, with excellent results (up to 98% yield, 20:1 d.r., and 99% ee) [21].

Soon later, Yao *et al.* reported the desymmetrization of achiral spirooxindoles by a sulfa-Michael addition [22]. Spirocyclohexadienone oxindoles **53** react with thiols **54**, catalyzed by bifunctional tertiary amine/thiourea catalyst **XIX**. The reaction affords the spirooxindoles **55** in good yields (77–95%) and stereoselectivities (>20:1 d.r. and 82–95% ee; Scheme 2.16).

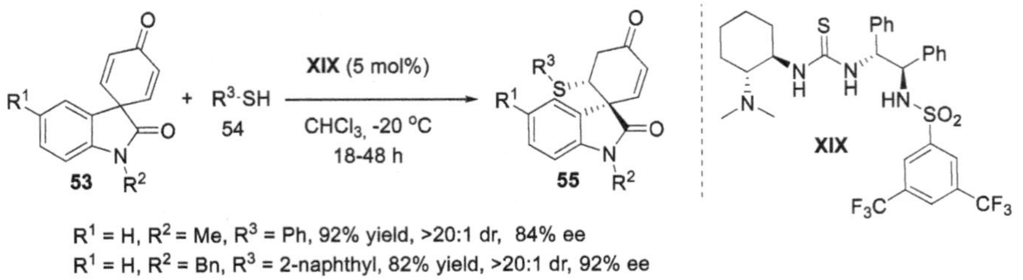

R^1 = H, R^2 = Me, R^3 = Ph, 92% yield, >20:1 dr, 84% ee
R^1 = H, R^2 = Bn, R^3 = 2-naphthyl, 82% yield, >20:1 dr, 92% ee

Scheme 2.16. Desymmetrization reported by Wang.

Xu and Liang reported a formal [5+1] cycloaddition between oxindoles and ester-linked bisenones with good results [23]. The reaction comprised a double Michael addition, catalyzed by a bifunctional tertiary amine/thiourea catalyst and provided access to spirooxindole-δ-lactones in good yields (71–94%) and excellent stereoselectivities (>25:1 d.r. and 88–97% ee).

A different approach was developed by Wang consisting an enantioselective Michael–Povarov cascade reaction [24]. 3-Allyl oxindole derivatives **56** react with enals **44** catalyzed by the Jørgensen–Hayashi catalyst **XV** to form intermediate **57**, followed by aniline addition to form the imine, which then undergoes a Povarov reaction with the allyl moiety to generate spirooxindole **59**. High stereoselectivity is achieved through a $\pi-\pi$ interaction between the aniline **58** and the Ph group located in the double bond. The spirooxindoles **59** were produced in good yields (68–86%) and excellent stereoselectivities (up to >20:1 d.r. and 98% ee; Scheme 2.17). Soon after, Ding and Wang designed a similar strategy using oxindoles with a side chain in the 3 position bearing α,β-unsaturated esters and enals [25]. A Michael/Michael cascade reaction takes place, catalyzed by the Jørgensen–Hayashi catalysts, affording the final spirooxindoles in good yields (70–86%) and excellent stereoselectivities (up to >99:1 d.r. and >99% ee).

Morita–Baylis–Hillman (MBH) carbonates derived from oxindoles have proved to be an excellent platform for the synthesis of spirooxindoles. Chen *et al.* reported a [3+2] cycloaddition with maleimides using a diphosphine catalyst with excellent results (up to 84% yields and >99% ee) [26]. The same research group presented a similar approach, consisting of the reaction of trifluoropyruvate **61** with MBH carbonate **60**.

$R^1 = H$, $R^2 = $ n-Pr, $R^3 = H$, 82% yield, >20:1 dr, 98% ee
$R^1 = H$, $R^2 = $ n-Pr, $R^3 = $ 4-CF$_3$, 69% yield, 5:1 dr, 97% ee

Scheme 2.17.   Spirocyclization reported by Wang.

$R^1 = H$, $R^2 = H$; 94% yield, 19:1 dr, 97% ee
$R^1 = CF_3O$, $R^2 = H$; 89% yield, 9:1 dr, 96% ee

Scheme 2.18.   Spirocyclization reported by Liu.

The reaction was catalyzed by an isocupreidine derivative **XXII**, obtaining the spirooxindoles **62** in good yields (76–94%) and excellent enantioselectivities (93–99% ee) [27]. The mechanism undergoes an activation of the MBH carbonate by isocupreine, which reacts with trifluoropyruvate *via* an aldol fashion, followed by an intramolecular Michael addition, followed by the release of the isocupreine derivative to generate the final compound (Scheme 2.18).

A similar approach was reported by Quyang and Chen *et al.* starting from the MBH carbonates derived from oxindoles. This time MBH carbonates **64** reacts with 2-alkylidene-1H-indene-1,3(2H)-diones **63** to afford the spirooxindoles with excellent results (Scheme 2.19) [28]. A singularity of this reaction is that using different catalysts (phosphine derivative **XXIII**, phosphine **XXIV**, or DMAP derivative **XXV**), different diastereomers were obtained. The result, determined by DFT calculations,

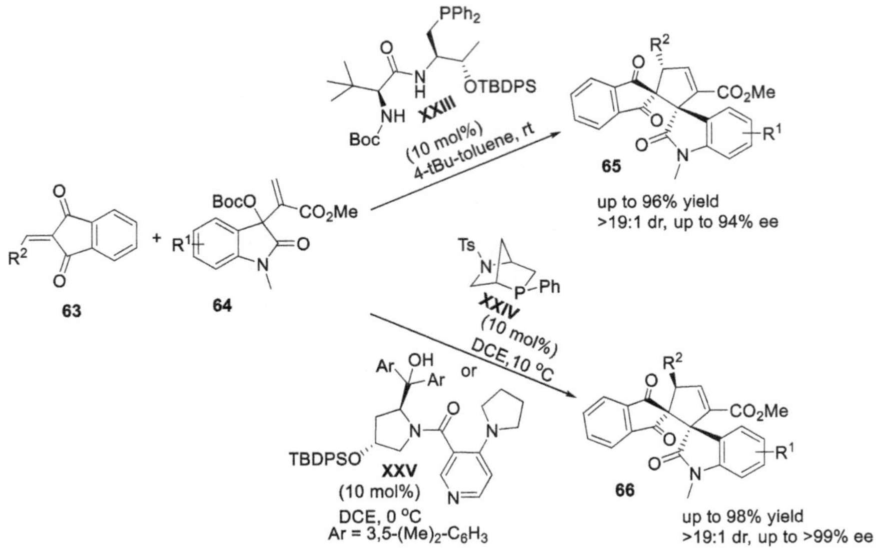

Scheme 2.19. Spirocyclization reported by Quyang and Chen.

is the catalyst-based switch in the annulation mechanism. The same research group reported a very similar approach using 1,2-benzoisothiazole 1,1-dioxide or 1,2,3-benzoxathiazine 2,2-dioxide derivatives as the activated alkene. After cycloaddition, the spirooxindoles were produced in good yields (75–97%) and excellent stereoselectivities (>19:1 d.r. and up to 99% ee) [29].

In the synthesis of spirocycloprpanoxindoles, a common approach is the use of 3-chloro oxindoles **67**. For example, Lu *et al.* reported the cycloaddition of 3-chloro oxindoles **67** to nitrostyrenes **46** catalyzed by bifunctional thiourea/tertiary amine catalyst **LXVII** [30]. The reaction undergoes a Michael–intramoleular alkylation sequence affording the cyclopropanes **68** with good yields (74–96%) and excellent steroselectivities (up to >25:1 d.r. and 99% ee; Scheme 2.20). Later, Du *et al.* reported a similar reaction using alkylidene pyrazolones [31]. The Michael–α-alkylation cascade reaction was catalyzed by squaramides and produced the cyclopropanes in excellent yields (up to 99%) and moderate stereoselctivities (up to 87:13 d.r. and 74% ee).

A similar reaction was reported by Barbas *et al.* consisting the use of oxindoles decorated with a ketone moiety (**69**), to generate spirooxindoles

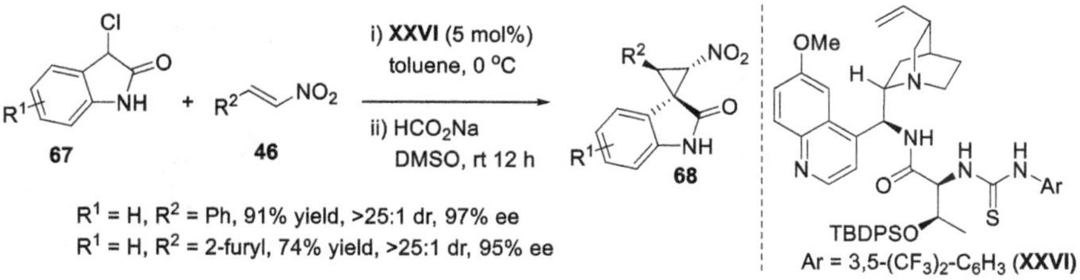

Scheme 2.20. Spirocyclization reported by Lu.

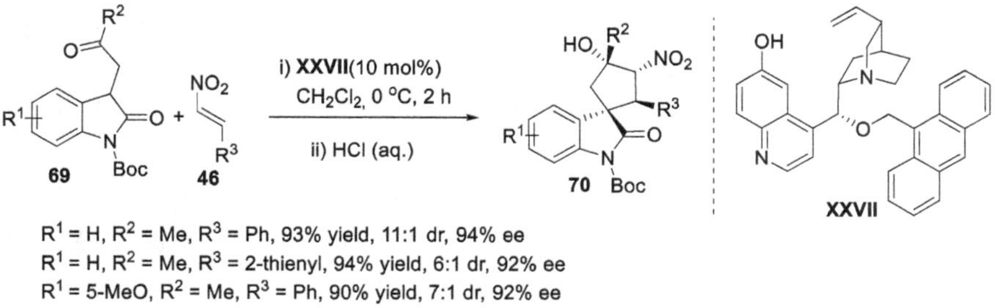

Scheme 2.21. Spirocyclization reported by Barbas.

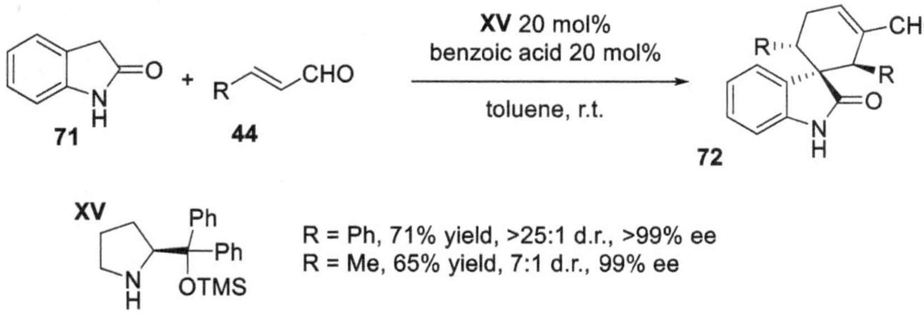

Scheme 2.22. Synthesis of spirooxindoles developed by Rios.

70 after reaction with nitro styrenes, through a Michael–Henry tandem reaction catalyzed by quinine-derived catalyst **XXVII** [32]. The reaction rendered the corresponding spirooxindoles in excellent yields (89–97%) and stereoselectivities (up to 18:1 d.r. and 97% ee; Scheme 2.21).

In 2010, Rios reported a multicomponent cascade reaction based on a double Michael–intramolecular aldol reaction for the synthesis of spirooxindoles [38]. The reaction was efficiently catalyzed by the Jørgensen–Hayashi catalysts with excellent results (Scheme 2.22).

2.3.2. *Organocatalytic methodologies starting with α,β-unsaturated oxindoles*

One of the first organocatalytic examples of synthesis of oxindoles was developed by Melchiorre in 2009 *via* a three-component cascade Michael–Michael–aldol reaction or consecutive Michael–Michael reactions with enones [33].

Enolizable aldehydes react with alkylidene oxindoles *via* an enamine intermediate. Next, the intermediate reacts with an enal *via* a Michael–intramolecular aldol reaction to form intermediate **77**, which will dehydrate to form the final spirocyclic compound. The reaction is efficiently catalyzed by diphenylprolinol derivatives such as **XV** with excellent yields and stereoselectivities (Scheme 2.23).

In a similar way, enones react with alkylidene oxindoles to furnish spirooxindoles under primary amine catalyst **XXVIII** with good yields and excellent stereoselectivities (Scheme 2.24).

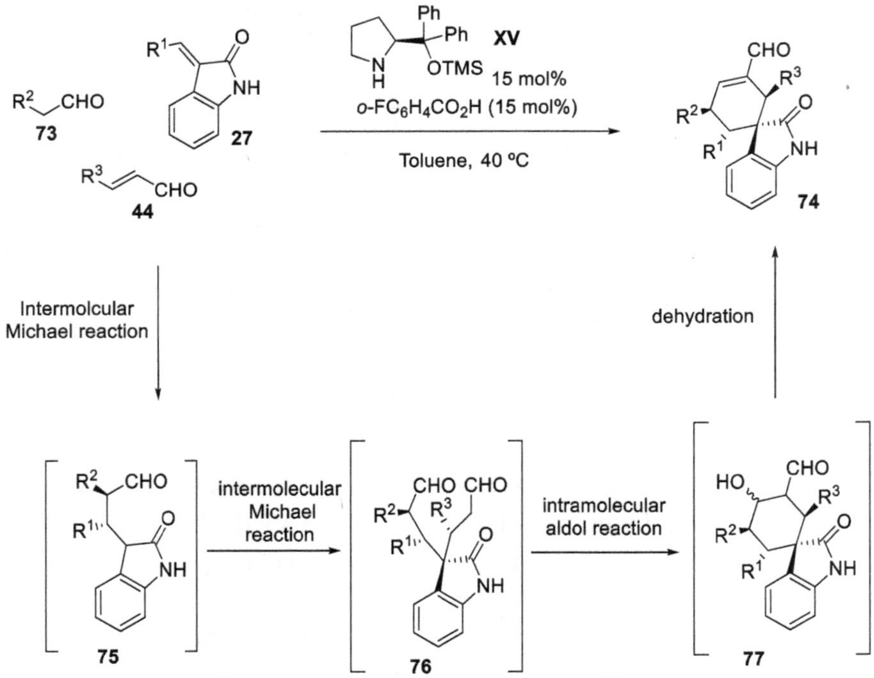

Scheme 2.23. Proposed mechanism for the spirocyclization reported by Melchiorre.

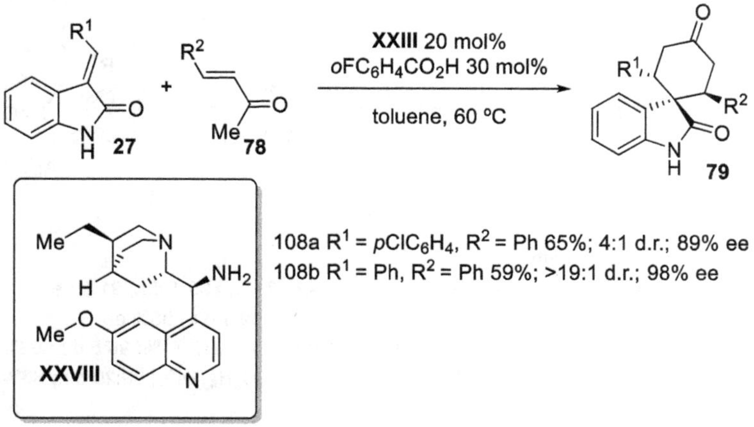

XXIII 20 mol%
oFC6H4CO2H 30 mol%
toluene, 60 °C
27
78
79
XXVIII
108a R1 = pClC6H4, R2 = Ph 65%; 4:1 d.r.; 89% ee
108b R1 = Ph, R2 = Ph 59%; >19:1 d.r.; 98% ee

Scheme 2.24. Spirocyclization of enones reported by Melchiorre.

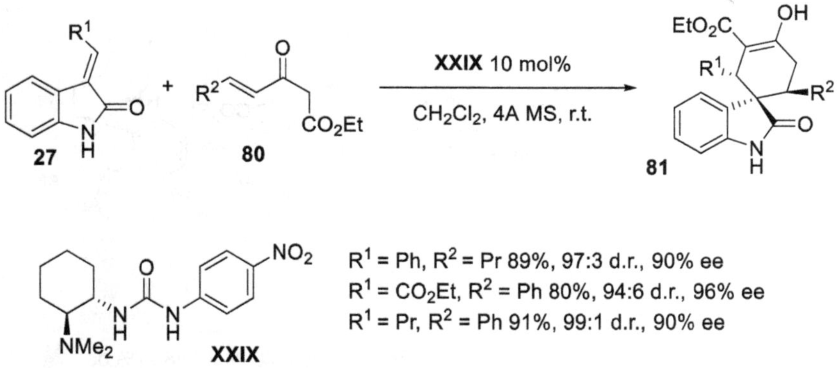

XXIX 10 mol%
CH2Cl2, 4A MS, r.t.
27
80
81
XXIX
R1 = Ph, R2 = Pr 89%, 97:3 d.r., 90% ee
R1 = CO2Et, R2 = Ph 80%, 94:6 d.r., 96% ee
R1 = Pr, R2 = Ph 91%, 99:1 d.r., 90% ee

Scheme 2.25. Spirocyclization reported by Gong.

A similar approach using Nazarov reagents was reported by Gong using a hydrogen bonding catalysis (Scheme 2.25) [34].

The organocatalytic synthesis of spiro[pyrrolidin-3,3′-oxindoles] *via* a [3+2] cycloaddition between methyleneindolines **83** and *in situ* azomethine ylide derivatives, yielding the spiro[pyrrolidin-3,3′-oxindoles] **85** in good yields and with excellent stereoselectivities (Scheme 2.26) was reported by Gong and co-workers [35].

In 2009, Marinetti and co-workers developed a nice [3+2] cycloaddition between 3-benzylideneindolin-2-ones (**84**) and allenes (**86**).

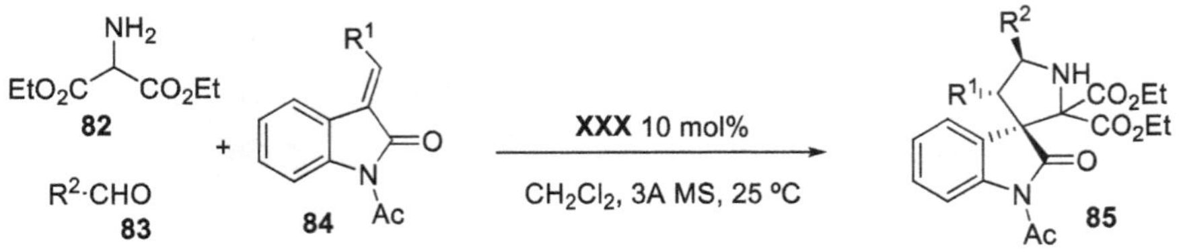

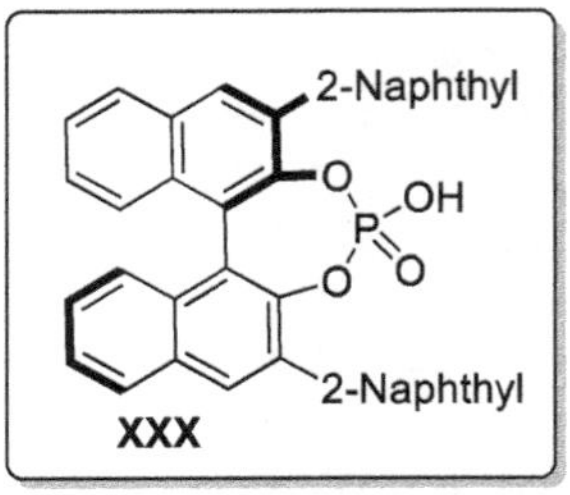

Scheme 2.26.   [3+2] Cycloaddition reported by Gong.

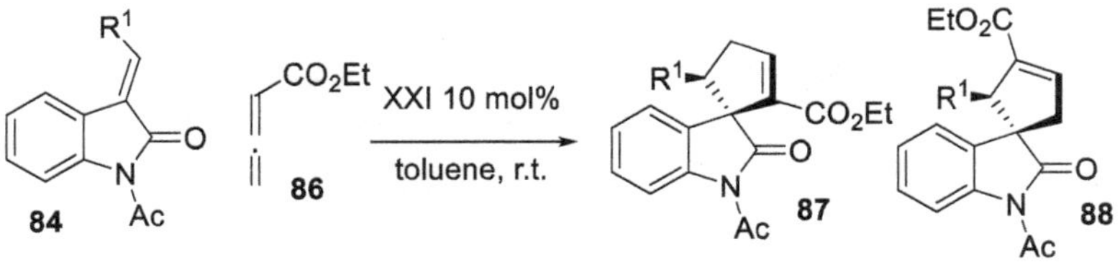

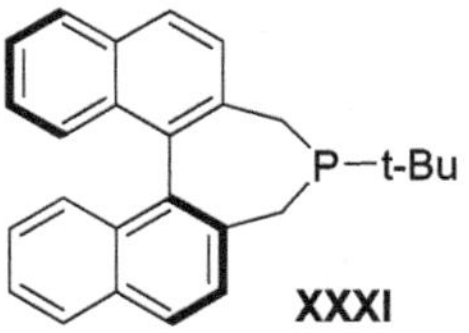

R^1 = Ph, R^2 = n-Pr, 71%; >99:1 d.r.; 91% ee
R^1 = R^2 = Ph, 87%; >99:1 d.r.; 85% ee
R^1= 2-furyl, R^2 = p-NO$_2$C$_6$H$_4$, 74%; 95:5 d.r.; 93% ee
R^1 = n-Pr, R^2 = p-NO$_2$C$_6$H$_4$, 94%; 80:20 d.r.; 83% ee

R = 2-furyl, 25% yield 76:24 87:88 97% ee
R = 2-quinolyl, 75% yield 90:10 87:88 97% ee

Scheme 2.27.   [3+2] Cycloaddition reported by Marinetti.

The reaction is efficiently catalyzed by chiral phosphine **XXXI** giving the spirooxindoles **87/88** in good yields and good stereoselectivities (Scheme 2.27) [36]. Barbas and co-workers reported a similar reaction this time using MBH carbonates with good results [37].

One of the first organocatalytic methodologies based on NHC catalysis for the synthesis of spirooxindoles was developed by Chi in 2012 [39]. Conjugated oxindole **89** bearing a conjugate imine reacts with enals **44** (in its homoenolate form after NHC addition) *via* a Michael addition followed by an intramolecular Mannich reaction to form the spirocyclic ring. The last step is the nucleophilic attack of the resulting Ts-amine to the carbonyl group, forming the β-lactam and releasing the NHC carbene. Only two chiral examples were reported and the final products were afforded in high yields (89% and 93%) and diastereoselectivities (>20:1 d.r.), albeit moderate to low enantioselectivities (39% and 51% ee; Scheme 2.28, top).

A similar approach was developed years later by Xie *et al.* [40]. Ketimines **92** and oxindole-derived enals **91** undergo a [3+3] cycloaddition catalyzed by NHC catalyst **XXXIV**. One equivalent of oxidant **94** is needed to achieve the spirooxindoles **93** in moderate yields (40–69%) and excellent enantioselectivities (96–99% ee; Scheme 2.28, bottom).

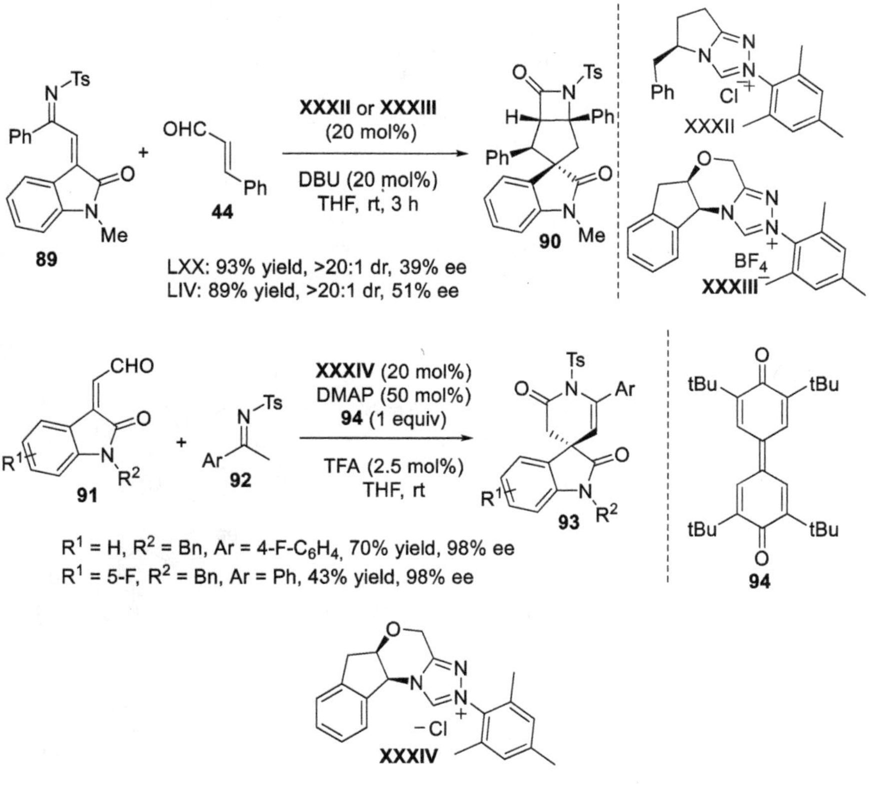

Scheme 2.28.   Spirocyclization reported by Chi (top) and Yang *et al.* (bottom).

A new tandem reaction was reported by Wang and Hong for the synthesis of oxindoles. Methyleneindolinones **95** react with oxindoles decorated at the C-3 position with an alkyl chain containing a leaving group (Br) **96** [41]. The reaction starts with a Michael addition followed by intramolecular 5-*exo-tet* alkylation to afford the *bis*-pirobycycles **97**. The reaction is efficiently catalyzed by a chiral bifunctional squaramide bearing a tertiary amine (Brønsted base) **XXXV** affording the products in moderate to excellent yields (54–95%), good diastereoselectivities (up to 13.7:2:1 d.r.), and excellent enantioselectivities (up to 96% ee; Scheme 2.29, top). Soon after, the same research group reported a similar

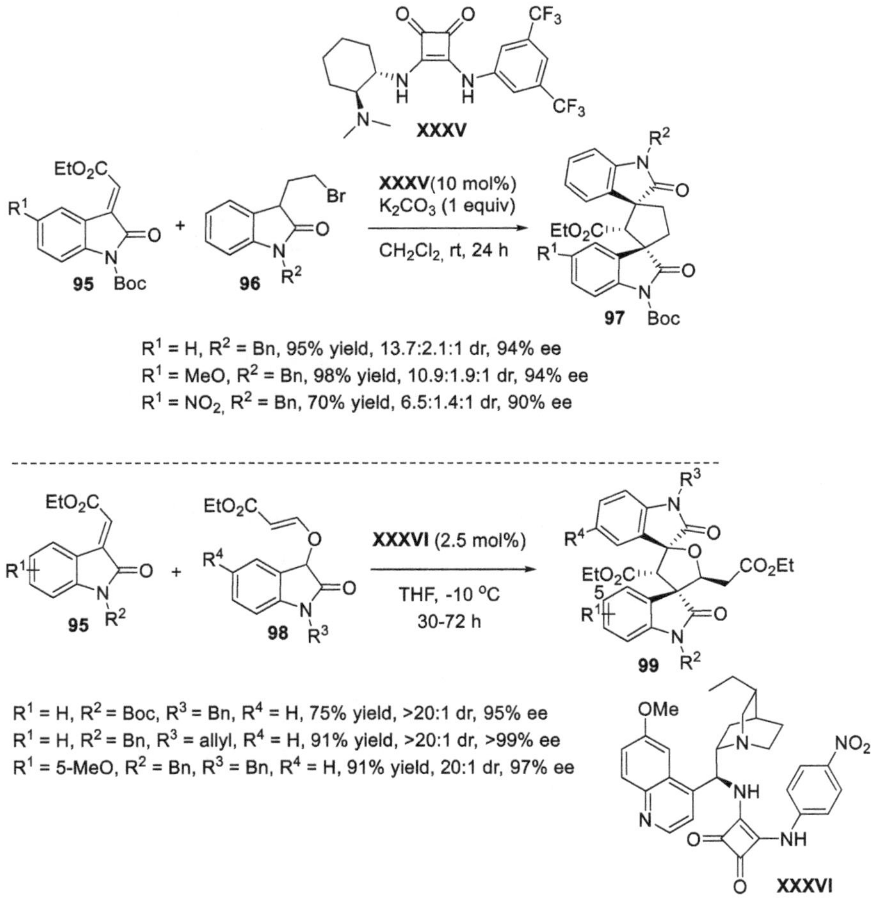

XXXV
XXXV(10 mol%)
K2CO3 (1 equiv)
CH2Cl2, rt, 24 h
95
96
97
R1 = H, R2 = Bn, 95% yield, 13.7:2.1:1 dr, 94% ee
R1 = MeO, R2 = Bn, 98% yield, 10.9:1.9:1 dr, 94% ee
R1 = NO2, R2 = Bn, 70% yield, 6.5:1.4:1 dr, 90% ee
XXXVI (2.5 mol%)
THF, -10 °C
30-72 h
95
98
99
R1 = H, R2 = Boc, R3 = Bn, R4 = H, 75% yield, >20:1 dr, 95% ee
R1 = H, R2 = Bn, R3 = allyl, R4 = H, 91% yield, >20:1 dr, >99% ee
R1 = 5-MeO, R2 = Bn, R3 = Bn, R4 = H, 91% yield, 20:1 dr, 97% ee
XXXVI

Scheme 2.29. Spirocyclization reported by Wang and Hong (top) and Du (bottom).

Michael–Michael cascade reaction using a Michael acceptor instead of the Br leaving group [42]. The reaction was again catalyzed by bifunctional squaramide/tertiary amine catalyst with good yields (up to 76%) and excellent stereoselectivities (up to 15:1 d.r. and >99% ee).

Due *et al.* reported a Michael–Michael consecutive reaction for the formation of *bis*-spirooxindoles using the bifunctional squaramide/tertiary amine catalyst **XXXVI**. α,β-unsaturated oxindoles **95** react with 3-hydroxy oxindole derivatives **98** bearing a α,β-unsaturated ester, which is directly attached to the alcohol [43]. *Bis*-spirooxindoles **99** were obtained with excellent results (up to >20:1 d.r. and >99% ee; Scheme 2.29, bottom). Years later, a very similar approach was published where α-alkylidene succinimides were reacted with α,β-unsaturated oxindoles under squaramide/tertiary amine catalysts. The spirooxindoles were obtained with good to excellent yields (up to 93%) and excellent stereoselectivities (up to 99:1 d.r. and 98% ee) [44]. The reaction was tested in a multigram scale to establish its practicality.

Acyl chlorides **103** react with 3-alkylenyloxindoles **100** under Lewis base catalysis to afford spirooxindoles **102** in very good yields (74–91%) and enantioselectivities (87–93% ee) [45]. The reaction starts with the activation of the acid chloride by the Lewis base (cinchona alkaloid derivative; **XXXVII**) to form the γ-enolate **104**, followed by a Michael addition with the 3-alkylenyloxindole **100**. Next, intramolecular cyclization and release of the catalysts lead to the formation of spirooxindole **102** (Scheme 2.30).

The amino-catalyzed reaction of enynes and alkylidene oxindoles was reported by Ramachary *et al.* [46]. Primary amine catalysts derived from cinchona alkaloids reacted with enynes to generate the enamine, which reacts with the alkylidene oxindole in a Michael fashion. Then, an intramolecular Michael addition between the formed enolate and the conjugated triple bond affords the cyclohexene-fused spirooxindole in moderate yields (40–60%) and high enantioselectivities (81–96% ee).

Alkylidene oxindoles **95** and nitrostyrenes decorated with a hydroxyl **105** or amino-protected group **106** react under cinchona alkaloid derivatives to form spirooxindoles [47]. The mechanism consists of a hetero-Michael addition between the hydroxyl or *N*-protected moiety to the alkylidene oxindole, followed by an intramolecular Michael addition

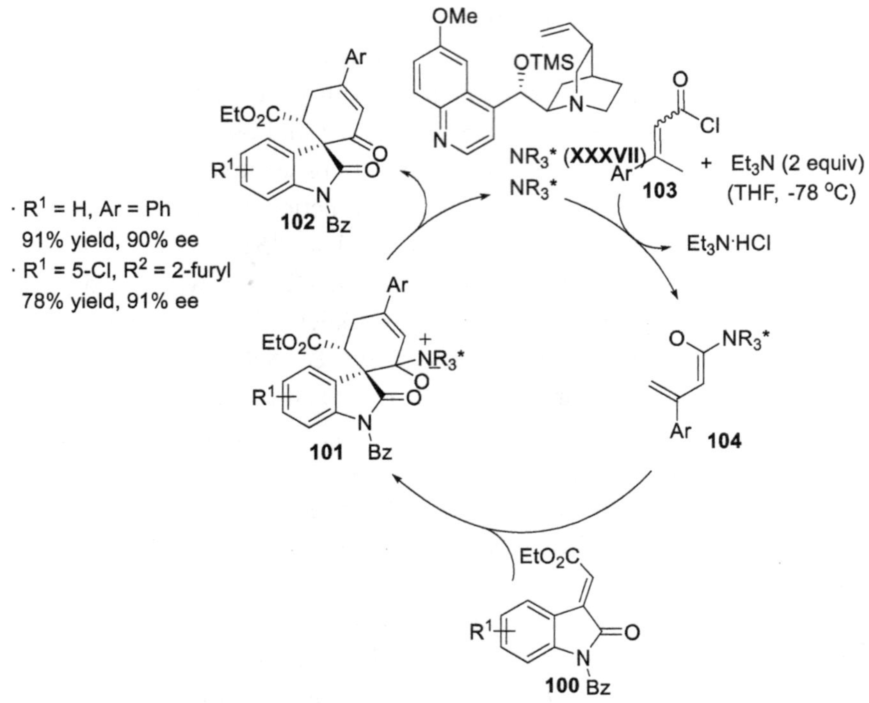

· R^1 = H, Ar = Ph
  91% yield, 90% ee
· R^1 = 5-Cl, R^2 = 2-furyl
  78% yield, 91% ee

102 Bz

101 Bz

100 Bz

103   + Et$_3$N (2 equiv)
           (THF, -78 °C)

Et$_3$N·HCl

104

Scheme 2.30.   Spirocyclization reported by Ye.

between the previously formed enolate of the oxindole and the nitrostyrene furnishing the corresponding oxindoles with excellent results. For the chromanes formation, a cinchona alkaloid decorated with squaramide **XXXVII** is necessary to achieve good yields (61–75%) and stereoselectivities (up to >99% ee). On the other hand, for the synthesis of tetrahydroquinolines, the Sharpless ligand [(DHQD)$_2$PHAL; **XXVIII**] emerged as the best organocatalyst, achieving excellent yields (87–96%) and enantioselectivities (up to 98% ee; Scheme 2.31).

Alkylidene oxindoles **19** have been the starting compound for the synthesis of spirooxindoles, taking advantage of their reactivity towards malonate derivatives. For example, they reacted with chloro-malonates **109** to render the corresponding spirooxindole cyclopropanes **110** in good yields (72–97%) and stereoselectivities (up to 93:7 d.r. and 94% ee) using the thiourea/tertiary amine bifunctional catalyst **XXXIX** [48]; or with γ-halogenated-β-ketoesters **111** *via* a domino Michael addition

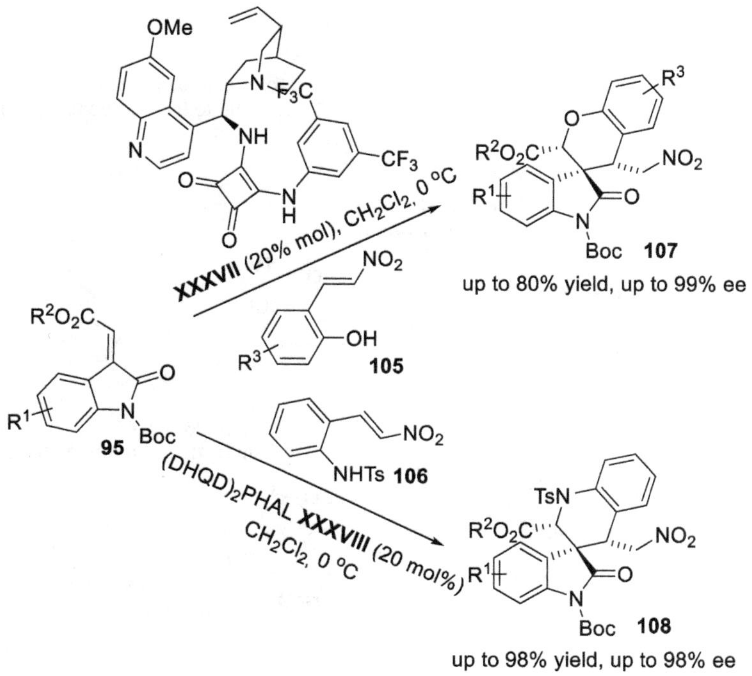

Scheme 2.31.   Spirocyclization reported by Zhu.

alkylation reaction, catalyzed by a bifunctional tertiary amine/thiourea catalyst **XXXX**, affording the final spiro compounds **112** in excellent yields (84–96%) and moderate to good stereoselectivities (up to 10:1 d.r. and 99% ee) [49]. Finally, they react with malonate derivatives bearing an enone in the side chain **113** [50]; the Michael–Michael tandem reaction was catalyzed by a multifunctional quaternary phosphonium compound **XXXXI**, affording the spirooxindoles **114** with excellent yields (80–98%) and stereoselectivities (up to >19:1 d.r. and 99% ee; Scheme 2.32).

A close cascade reaction was reported by Kanger *et al.* comprising a consecutive Michael addition/aldol reaction between nitro ketones **115** and alkylidene oxindoles **24** [51]. The reaction is catalyzed by the bifunctional thiourea/Brønsted base catalyst **XLII**, derived from cinchona alkaloids, rendering the final spiro compounds **116** in good yields (85–95%) and excellent stereoselectivities (up to d.r. 20:1 and up to 98% ee). the brackets should be: (up to d.r. 20:1 and up to 98% ee; Scheme 2.33).

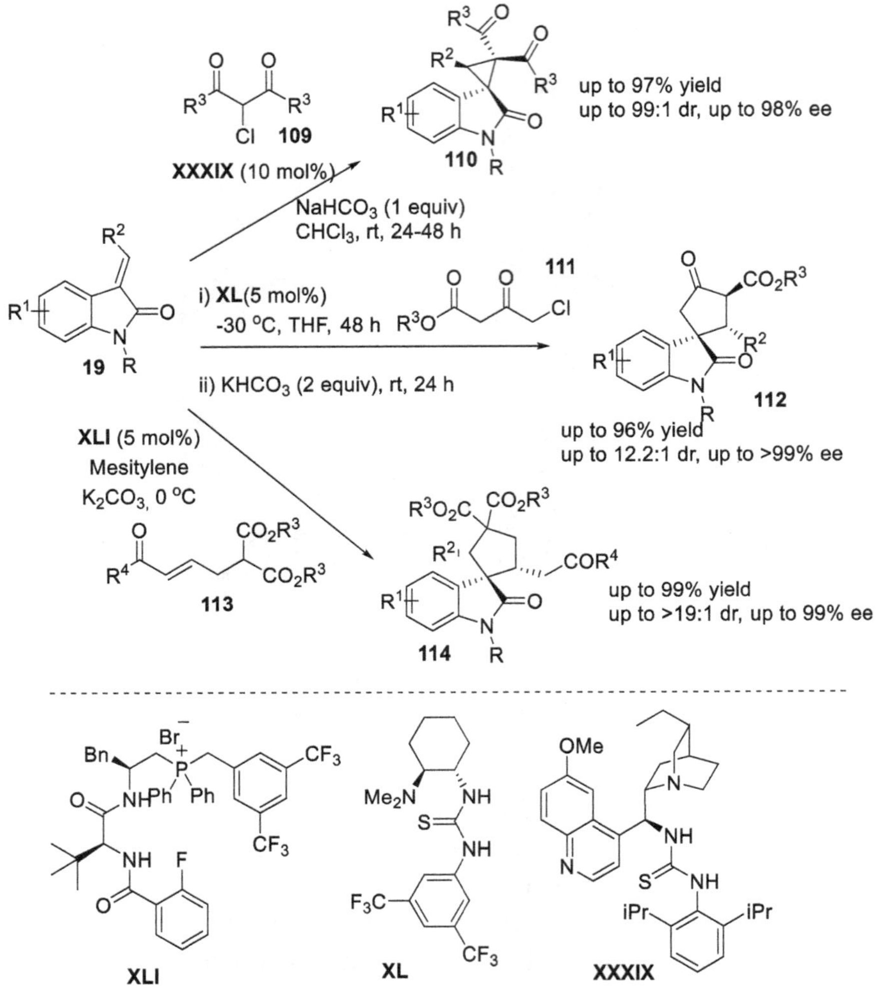

Scheme 2.32. Several methodologies leading to spirooxindoles.

Years later, a Michael–Michael tandem reaction using enones instead of ketones was reported by Zhao *et al.* [52]. The reaction, catalyzed by a bifunctional catalyst **XLIII**, furnished the corresponding spirooxindoles in good yields (76–86%) and excellent stereoselectivities (up to 95:5 d.r. and 98% ee; Scheme 2.34). Soon later, Quintivalla *et al.* reported a similar approach using unsaturated esters [53].

XLII
(10 mol%)
CH_2Cl_2, rt
24
115
116
XLII
R^1 = H, R^2 = CO_2Me, R^3 = Me, 87% yield, 10:1 dr, 95% ee
R^1 = 3-F, R^2 = CO_2Et, R^3 = Me, 79% yield, 10:1 dr, 97% ee

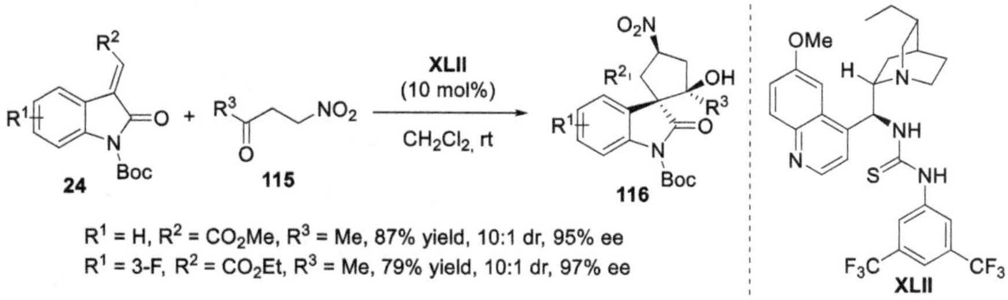

Scheme 2.33. Spirocyclization reported by Kanger.

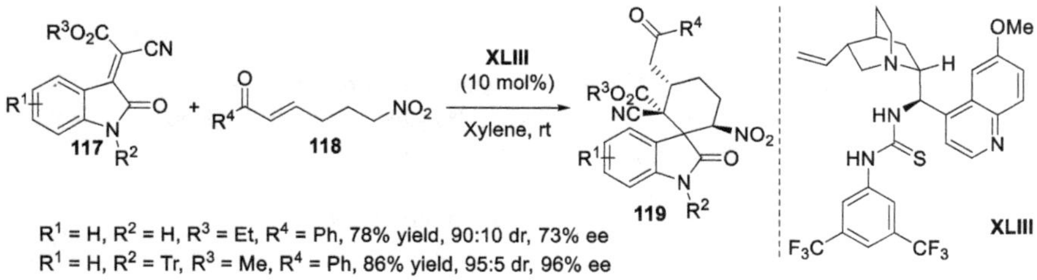

XLIII
(10 mol%)
Xylene, rt
117
118
119
XLIII
R^1 = H, R^2 = H, R^3 = Et, R^4 = Ph, 78% yield, 90:10 dr, 73% ee
R^1 = H, R^2 = Tr, R^3 = Me, R^4 = Ph, 86% yield, 95:5 dr, 96% ee

Scheme 2.34. Spirocyclization reported by Zhao.

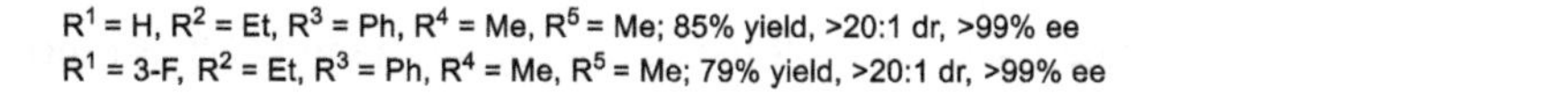

24
46
120
XXXVII (5 mol%)
CH_2Cl_2, rt, 1 h
121
XXXVII
R^1 = H, R^2 = Et, R^3 = Ph, R^4 = Me, R^5 = Me; 85% yield, >20:1 dr, >99% ee
R^1 = 3-F, R^2 = Et, R^3 = Ph, R^4 = Me, R^5 = Me; 79% yield, >20:1 dr, >99% ee

Scheme 2.35. Spirocyclization reported by Sun.

A multicomponent Michael–Michael–aldol reaction was reported by Sun *et al.* [54]. The 1,2-dicarbonyl compounds **120**, nitroalkene **46**, and methyleneindolines **24** generated the spirooxindoles **121**, under squaramide catalysis, in excellent yields (up to 85%) and with excellent stereoselectivities (up to >20:1 d.r. and 99% ee; Scheme 2.35).

A 1,3-dipolar cycloaddition between alkylidene oxindoles **95** and cyclic imino esters **122** catalyzed by a bifunctional thiourea/tertiary amine

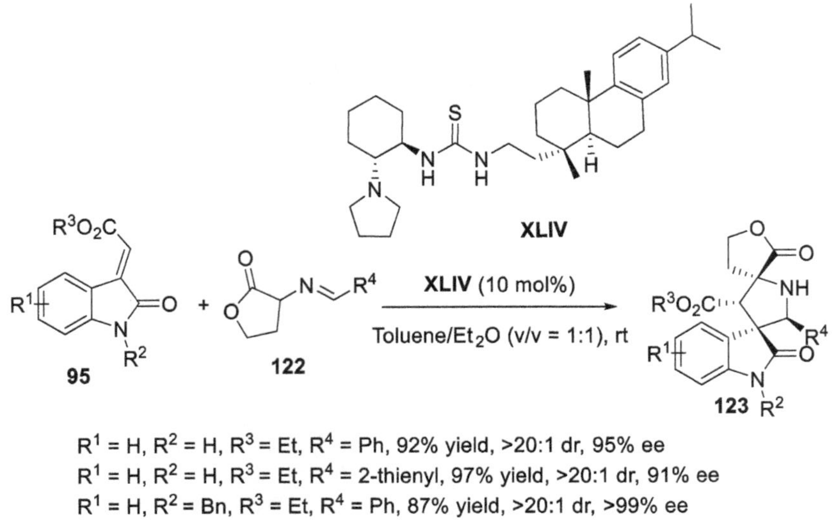

XLIV
R^3O_2C
R^1
O
N
R^2
95
N
R^4
122
XLIV (10 mol%)
Toluene/Et_2O (v/v = 1:1), rt
NH
R^3O_2C
R^4
123
R^1 = H, R^2 = H, R^3 = Et, R^4 = Ph, 92% yield, >20:1 dr, 95% ee
R^1 = H, R^2 = H, R^3 = Et, R^4 = 2-thienyl, 97% yield, >20:1 dr, 91% ee
R^1 = H, R^2 = Bn, R^3 = Et, R^4 = Ph, 87% yield, >20:1 dr, >99% ee

Scheme 2.36. Spirocyclization reported by Wang.

XLIV was reported by Wang *et al.* [55]. The reaction afforded the final spirooxindoles **125** in good yields (83–97%) and excellent stereoselectivities (up to >20:1 d.r. and >99% ee; Scheme 2.36). A similar [3+2] cycloaddition using nitrones as a 1,3-dipole was reported by Shi *et al.* [56]. This time the reaction was catalyzed by *bis*-thiourea affording the final spirooxindoles in moderate yields (45–78%) and excellent stereoselectivities (up to 99:1 d.r. and 99% ee).

Zhao reported a 1,3-dipolar cycloaddition between noncyclic imino esters **36** and alkylidene oxindoles **19** using a thiourea/quaternary ammonium salt **XLV** as the catalyst. This reaction provided the final spiro compounds **252** with excellent results (up to 98% yield, >19:1 d.r., and 99% ee; Scheme 2.37) [57].

A similar reaction using azalactones **125** was reported by Sun *et al.* [58]. The reaction is catalyzed by a bifunctional thiourea/Brønsted base catalyst **XLVI,** which deprotonates the azalactone to form an enolate species in which the C2 and C4 positions are activated. The enolate of the azalactone undergoes an enantioselective 1,3-dipolar cycloaddition with the alkylidene oxindole, furnishing the spiro-3,3′-pyrrolidonyl spirooxindoles **127** in good yields (70–95%), excellent stereoselectivities (up to

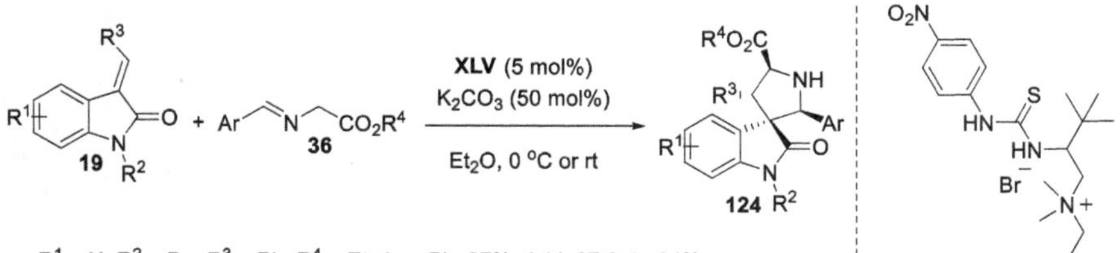

Scheme 2.37.   Spirocyclization reported by Shang and Zhao.

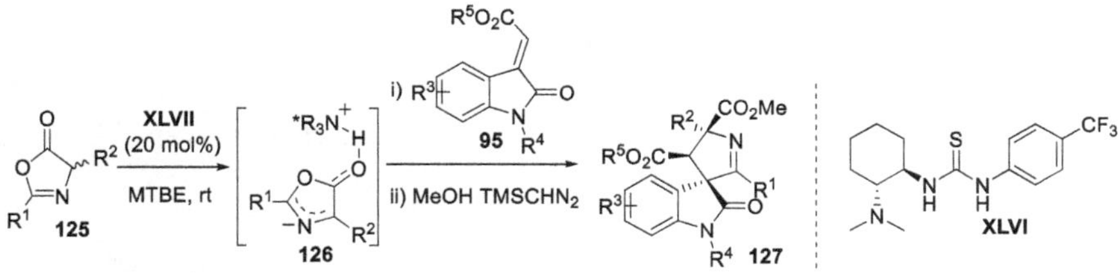

R¹ = H, R² = Bn, R³ = Ph, R⁴ = Et, Ar = Ph; 97% yield, 97:3 dr, 94% ee
R¹ = 4-F, R² = Bn, R³ = Ph, R⁴ = tBu, Ar = Ph, 90% yield, 95:5 dr, 96% ee
R¹ = H, R² = Boc, R³ = Ph, R⁴ = Me, Ar = Ph, 85% yield, 96:4 dr, 81% ee

Scheme 2.38.   Spirocyclization reported by Wang.

R¹ = Ph, R² = Bn, R³ = H, R⁴ = Boc, R⁵ = Et; 86% yield, 90:10 dr, 94% ee
R¹ = Ph, R² = iBu, R³ = H, R⁴ = Boc, R⁵ = Et, 81% yield, 86:14 dr, 94% ee
R¹ = Ph, R² = Bn, R³ = H, R⁴ = Boc, R⁵ = tBu, 81% yield, 91:9 dr, 97% ee

93:7 d.r. and up to 98% ee) and a good group compatibililty (Scheme 2.38). Later, the same research group reported the same reaction, catalyzed by chiral phosphoric acids, with similar results [59].

N,N′-Cyclic azomethine imines **128** in conjuntion with alkylideneoxindoles have been reported for the synthesis of spirooxindoles using diphosphoric acid derivatives **XLVII** as catalysts [60]. The 1,3-dipolar cycloaddition renders the final spiro compounds **257** in excellent yields (84–94%) and good to excellent stereoselectivities (up to 20:1 d.r. and 99% ee; Scheme 2.39).

A multicomponent tandem reaction was developed by Ren *et al.* using benzylidene oxindoles decorated with a hydroxyl group in the *ortho*-position of the phenyl ring **130** [61]. The reaction comprises an oxa-Michael addition followed by an intramolecular Michael addition. Next, an intermolecular Michael addition with a new molecule of enal occurs, followed by an intramolecular aldol reaction to form the cyclohexene ring

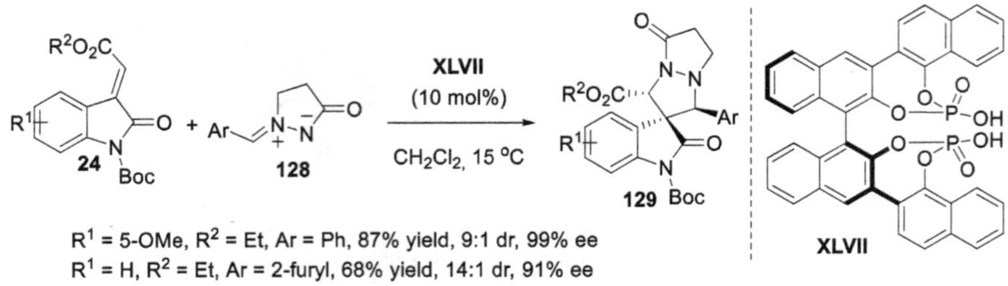

XLVII
(10 mol%)
CH2Cl2, 15 °C
24
128
129
XLVII
R1 = 5-OMe, R2 = Et, Ar = Ph, 87% yield, 9:1 dr, 99% ee
R1 = H, R2 = Et, Ar = 2-furyl, 68% yield, 14:1 dr, 91% ee

Scheme 2.39. Spirocyclization reported by Hong and Wang.

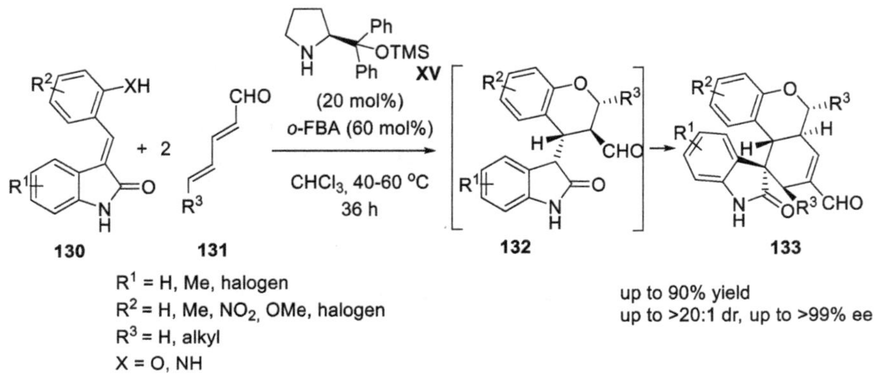

XV
(20 mol%)
o-FBA (60 mol%)
CHCl3, 40-60 °C
36 h
130
131
132
133
R1 = H, Me, halogen
R2 = H, Me, NO2, OMe, halogen
R3 = H, alkyl
X = O, NH
up to 90% yield
up to >20:1 dr, up to >99% ee

Scheme 2.40. Spirocyclization reported by Liu and Wang.

after dehydration. The reaction afforded the spirooxindoles **133** in moderate to good yields (31–90% yield) with good to excellent stereoselectivities (up to 12:1 d.r. and >99% ee). A clear limitation is the need to use aliphatic enals. Remarkably, when β,β-disubstituted enals were used, the reaction stopped after the intramolecular Michael addition, preventing the formation of the spirocenter (Scheme 2.40).

Ortho-hydroxy chalcones were used, under a bifunctional tertiaryamine/thiourea catalyst, to synthesize spirooxindoles with good yields (60–95%) and moderate to high stereoselectivities (up to >20:1 d.r. and 90% ee) by reacting with alkylidene oxindoles [62]. The mechanism comprises an oxa-Michael addition, followed by an intramolecular Michael addition between the enolate generated *in situ* and the enone to build the spirooxindole. Yang and Du [63] and Huang *et al.* [64] independently reported an almost identical reaction, using *ortho*-amino chalcones **135**, this time

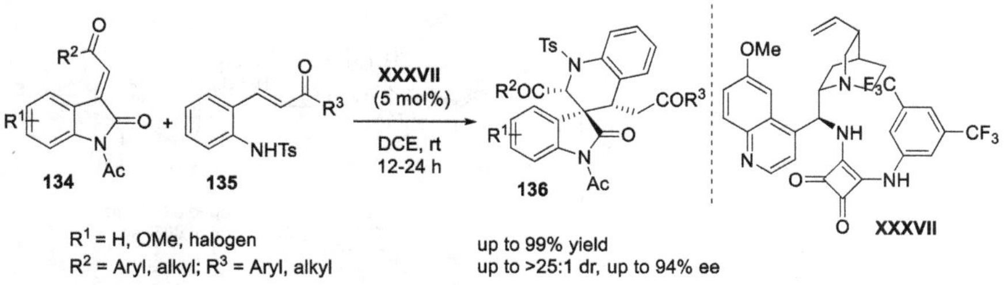

Scheme 2.41.   Spirocyclization reported by Du and Zhao.

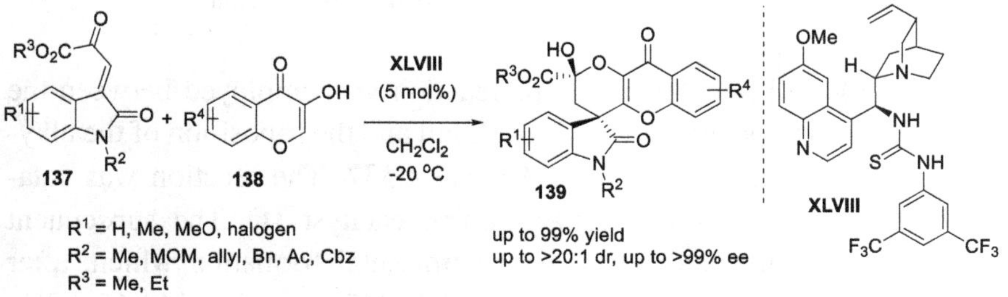

Scheme 2.42.   Spirocyclization reported by Wang.

with excellent results (85–99% yields, >25:1 d.r., and up to 90% ee; Scheme 2.41).

Wang *et al.* developed a tandem reaction comprising the addition of 2-hydroxy-1,4-naphthoquinone or 3-hydroxy-4H-chromene-4-ones **138** to benzylidene oxindoles decorated with a ketoester (**137**) [65]. The reaction started with an initial Michael reaction between the enolate of the 2-hydroxy-1,4-naphthoquinone or 3-hydroxy-4H-chromen-4-ones to the benzylidene oxindole, followed by intramolecular hemiacetal formation, to furnish the corresponding spirooxindoles. The reaction is efficiently catalyzed by a bifunctional thiourea/tertiary amine catalyst **XLVIII**, affording the corresponding spiro compounds **139** in moderate to good yields (55–92%) and excellent stereoselectivities (up to >20:1 d.r. and >99% ee). The functional group tolerance is excellent, while the only limitation is the need for *N*-substituted oxindoles (Scheme 2.42).

Kesavan *et al.* reported the synthesis of naphthopyran-spirooxindoles in 2016 [66]. The reaction consists of a tandem Friedel–Crafts/ hemiketalization reaction. Taking advantage of the electron-rich nature of

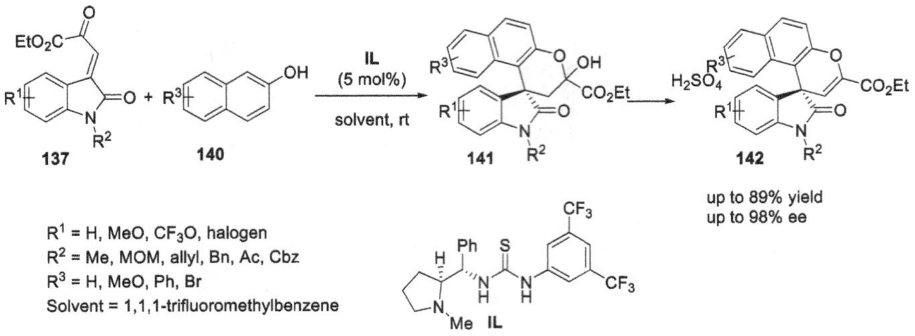

Scheme 2.43.   Spirocyclization reported by Kesavan.

naphthol **140,** a Friedel–Crafts type reaction was employed between the most active *meta*-position of the 2-naphthol and the β-position of the alkylidene oxindole decorated with a ketoester **137**. The reaction was catalyzed by a chiral thiourea/tertiary amine catalyst **IL**. The subsequent addition of sulfuric acid resulted in hemiacetal formation which, after dehydration, rendered the final spirooxindole **142** in good yields (65–89%) and good to excellent enantioselectivities (70–98% ee). The tertiary amine acts as a base to activate the nucleophile, while the thiourea moiety activates the ketoester for the Michael addition (Scheme 2.43).

N-protected isatylidenemalononitriles have become popular reactants for the synthesis of spirooxindoles. In 2012, a cycloaddition reaction between isocyanoacetates **146** and *N*-protected isatylidenemalononitriles, prepared *in situ via* the Knoevenagel reaction between isatin **143** and malononitrile **144**, was reported by Wei *et al.* [67] consisting of a formal [3+2] cycloaddition catalyzed by a bifunctional tertiary amine/thiourea catalyst **XLIX**, to afford the final spirooxindoles **147** in good yields (73–92%), moderate diastereoselectivities (up to 88:12 d.r.) and high enantioselectivities (80–97% ee; Scheme 2.44).

Soon afterwards, Shi and Wei *et al.* reported a Rauhut–Currier/ Michael/Rauhut–Currier cascade reaction between dienones **148** and *N*-protected isatylidenemalononitriles **145**. The reaction is catalyzed by chiral phosphine **L** with good yields (63–92%) and excellent stereoselectivities (up to 20:1 d.r. and 97% ee; Scheme 2.45) [68]. The mechanism starts with the nucleophilic attack of the phosphine to the dienone to form the enolate **150,** which attacks the *N*-protected isatylidenemalononitriles.

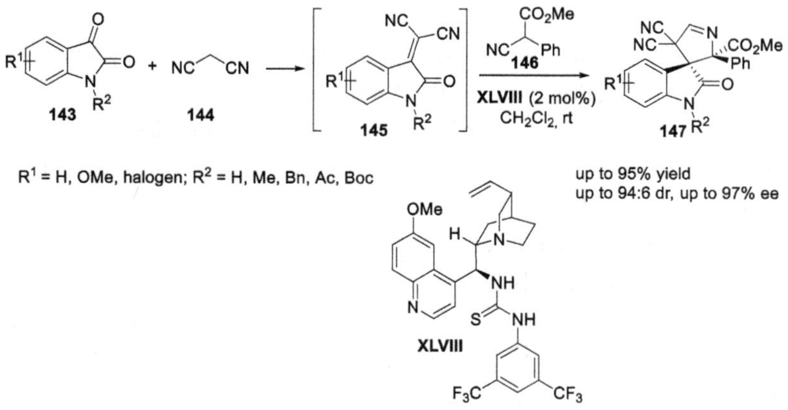

143
144
145
146
147
XLVIII (2 mol%)
CH2Cl2, rt
R1 = H, OMe, halogen; R2 = H, Me, Bn, Ac, Boc
up to 95% yield
up to 94:6 dr, up to 97% ee
XLVIII

Scheme 2.44. Spirocyclization reported by Yan.

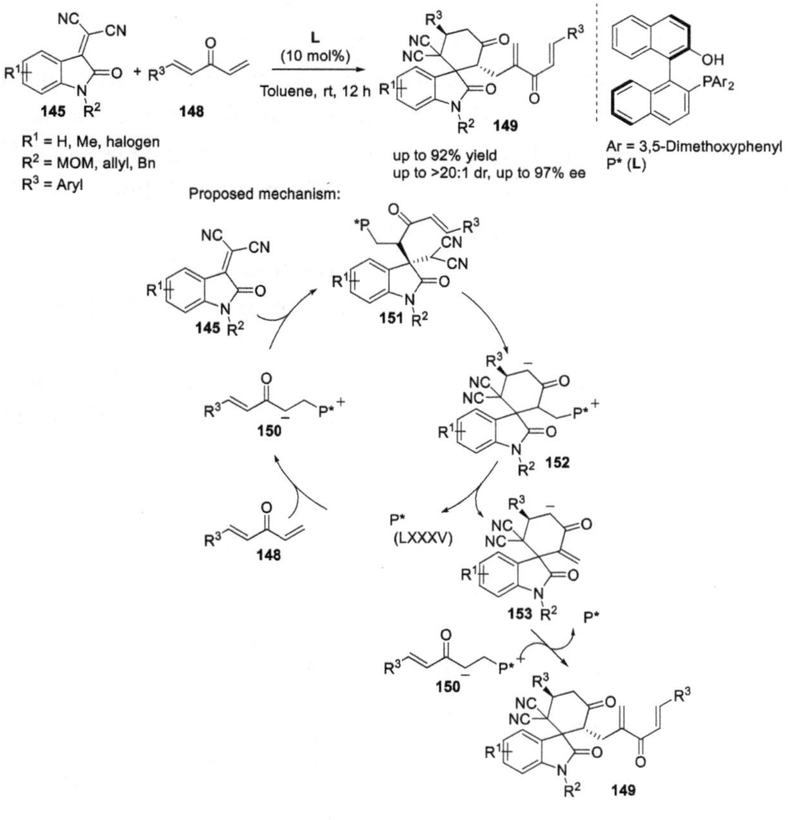

L
(10 mol%)
Toluene, rt, 12 h
145
148
149
R1 = H, Me, halogen
R2 = MOM, allyl, Bn
R3 = Aryl
up to 92% yield
up to >20:1 dr, up to 97% ee
Ar = 3,5-Dimethoxyphenyl
P* (L)
Proposed mechanism:
150
151
152
153
P*
(LXXXV)

Scheme 2.45. Spirocyclization reported by Shi and Wei.

Next, intramolecular Michael addition to the enone to form the spirooxindole **153** followed by the release of the catalyst and an intermolecular Rauhut–Currier reaction, render the final compounds **149**. Soon later, the same group reported the use of similar alkylidenoxindoles reacting with MBH carbonates in a [4+1] cycloaddition to generate spirooxindoles in good yields, moderate diastereoselectivities and high enantioselectivities [69].

N-protected isatylidenemalononitriles **154** reacted with α,α-dicyanoalkenes **155**, catalyzed by the cinchona alkaloid-derived bifunctional tertiary amine/thiourea **LI** [70]. The mechanism comprises a deprotonation of the cyclohexylidene malononitrile by the tertiary amine of the catalyst, which undergoes a Michael addition to the *N*-protected isatylidenemalononitrile, followed by intramolecular nucleophilic addition to the CN group to form the imine which, upon isomerization, generates the spirocyclic oxindole **156** with moderate to good yields (51–93%) and high enantioselectivities (74–95% ee; Scheme 2.46).

Enolizable aldehydes decorated with an enone **157** were used in combination with *N*-protected isatylidenemalononitriles **145** for the synthesis of spirooxindoles through a double Michael cascade reaction [71]. The reaction is efficiently catalyzed by TMS-protected diphenylprolinol **XV** and requires the use of Schreiner thiourea **LII** as a cocatalyst for optimum results, furnishing the final spirooxindole **158** in good yields (76–95%) and excellent stereoselectivities (up to 97:3 d.r. and 98% ee; Scheme 2.47).

A vinylogous Michael/cyclization cascade reaction of acyclic β,γ-unsaturated amides **159** with isatylidene malonitriles **145** to build

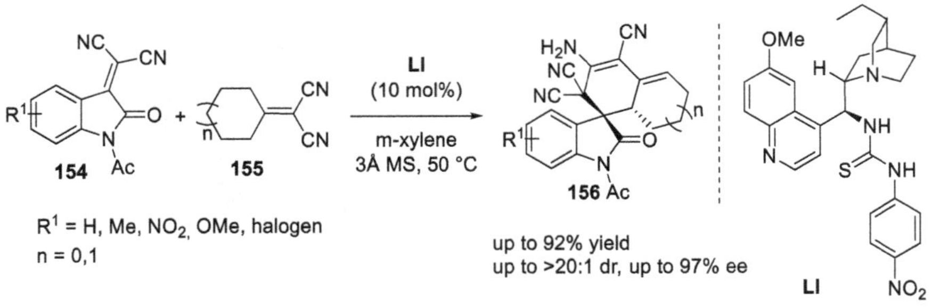

Scheme 2.46. Spirocyclization reported by Wang.

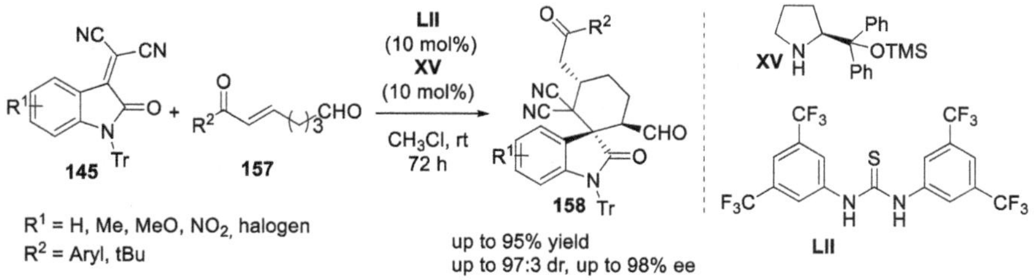

Scheme 2.47.   Spirocyclization reported by Zhao.

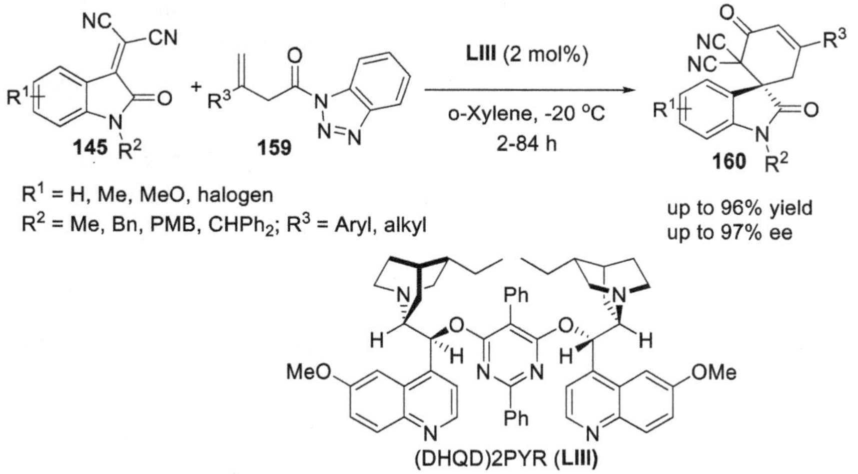

Scheme 2.48.   Spirocyclization reported by Wu and Sha.

spirooxindoles **160** was reported by Li *et al.* [72]. The reaction is catalyzed by [(DHQD)$_2$PYR, **LIII**] with good yields (84–96%) and enatioselectivities (85–97% ee; Scheme 2.48).

Seyferth–Gilbert reagent **161** has been used in 1,3-dipolar cycloadditions with isatylidene malononitriles **145** to produce spirooxindoles **162** with good yields (up to 99%) and enantioselectivities (up to 99% ee; Scheme 2.49) when catalyzed by cinchona alkaloids **LIV** [73].

Zhang *et al.* reported a formal [4+2] cycloaddition for the synthesis of spirooxindoles [74]. Vinyl ketones **164** react with oxindole-derived α,β-unsaturated imines **163** *via* a Rauhut–Currier reaction promoted by phosphine **XC**, followed by a Michael addition between the vinyl ketone and

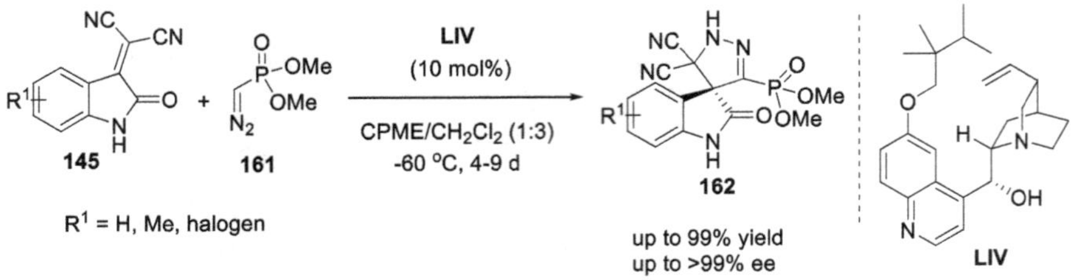

Scheme 2.49. Spirocyclization reported by Peng.

the tosyl amine, achieving the spiro compounds **165** in good yields (67–87%) and excellent stereoselectivities (up to 20:1 d.r. and 98% ee). A limitation of this methodology is that only unsubstituted vinyl ketones can be used (Scheme 2.50).

Benzylidene oxindoles **24** react with 2-vinylindoles **172** in a Diels–Alder fashion, catalyzed by bifunctional tertiary amine/squaramide catalysts **LV** [75]. The reaction affords spiro[tetrahydrocarbazole-3,3′-oxindole] **173** in moderate yields (52–84%) and good stereoselectivities (4:1 to >20:1 d.r. and 65–99% ee; Scheme 2.51). Soon after, almost the same reaction was reported by Wang *et al.* using chiral phosphoric acid derivatives [76].

A thio-Michael reaction between 1,4-dithiane-2,5-diol **175** and isoindigos **174** followed by an intramolecular aldol reaction to synthesize the spirotetrahydrothiophene ring **176** under quinidine **LIII** catalysis, was reported by Gui *et al.* [77]. The spirooxindoles were obtained with good yields (70–91%) and high stereoselectivities (up to 20:1 d.r. and 90% ee; Scheme 2.52).

A formal [2+2] cycloaddition based on dienamine catalysis was developed by Qi *et al.* for the synthesis of spirooxindoles [78]. In this method, enals possessing a δ-hydrogen **177** reacted with diphenyl prolinol **XV** to form the dienamine intermediate, which undergoes a formal [2+2] cycloaddition with alkylidene oxindoles **24** to furnish the butanocycle-bearing spirooxindoles **178** in good yields (66–83%) and excellent enantioselectivities (up to >19:1 d.r. and 97% ee). Surprisingly, the use of a silyl-protected diphenyl prolinol derivatives gave lower stereoselectivities, suggesting that H-bonding plays a crucial role in this reaction (Scheme 2.53).

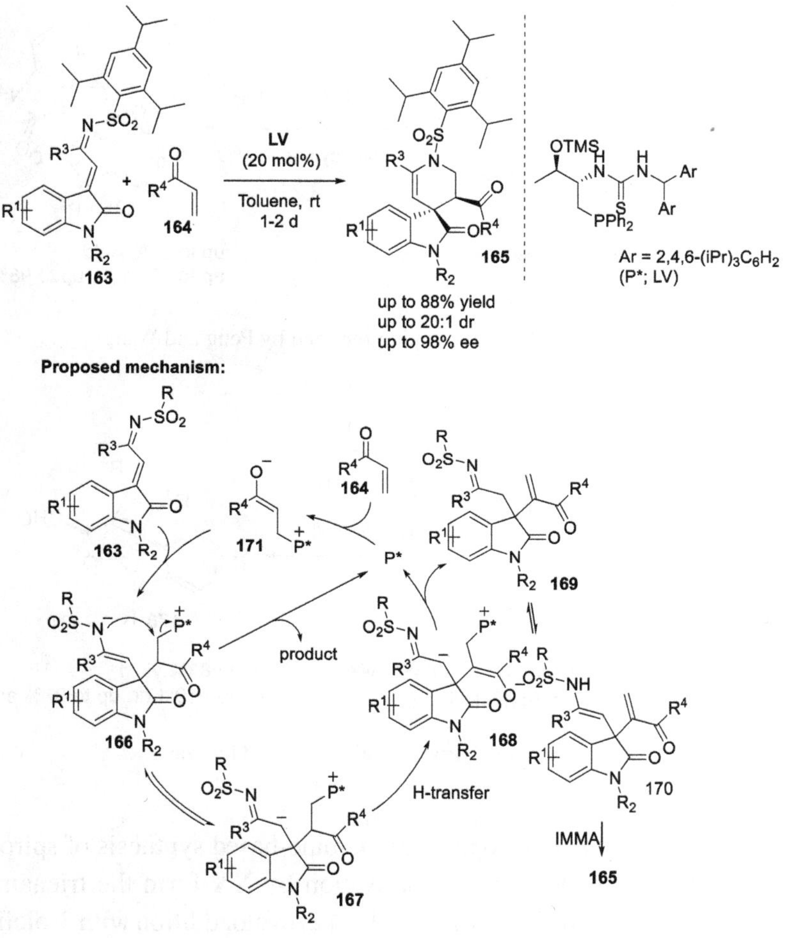

Scheme 2.50.   Spirocyclization reported by Shi and Wei.

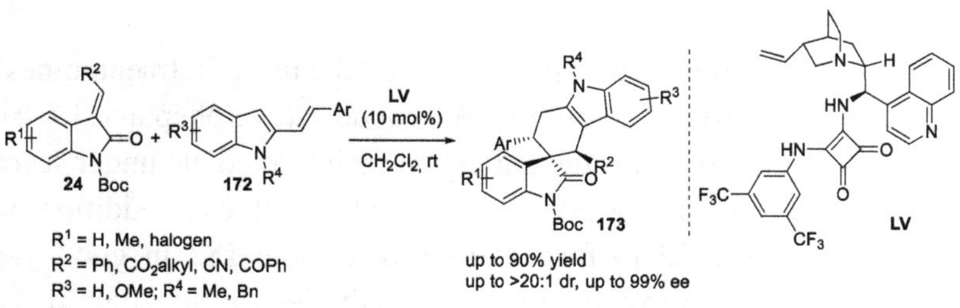

Scheme 2.51.   Spirocyclization reported by Lu.

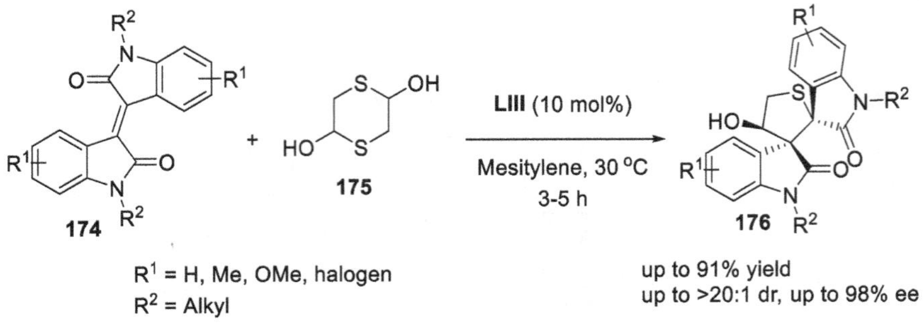

R¹ = H, Me, OMe, halogen
R² = Alkyl

up to 91% yield
up to >20:1 dr, up to 98% ee

Scheme 2.52.   Spirocyclization reported by Peng and Wang.

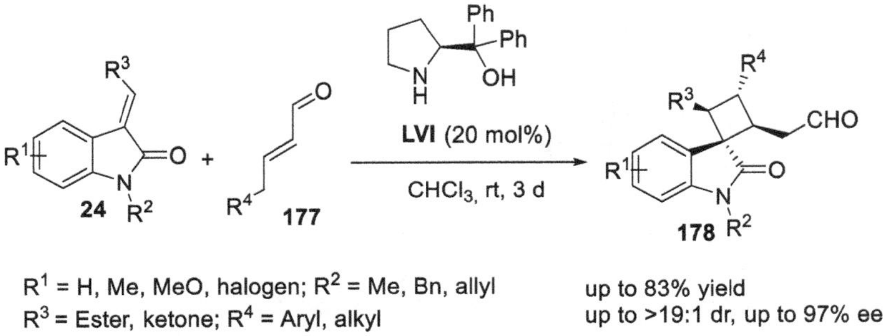

R¹ = H, Me, MeO, halogen; R² = Me, Bn, allyl
R³ = Ester, ketone; R⁴ = Aryl, alkyl

up to 83% yield
up to >19:1 dr, up to 97% ee

Scheme 2.53.   Spirocyclization reported by Wang.

Later on, Jia *et al.* developed a trienamine-based synthesis of spirooxindoles [79]. Polyenals **179** after activation by **XX** form the trienamine intermediate, which undergoes a formal [4+2] cycloaddition with 3-olefinic oxindoles **24**, affording spirooxindoles **180** in good yields (60–99%) and excellent stereoselectivities (up to >95:5 d.r. and 98% ee; Scheme 2.54). Soon after, Chen *et al.* reported a similar reaction, based on tetraenamine catalysis, with similar results [80].

In a similar fashion, Stiller *et al.* reported the use of tetraenamines for the synthesis of spirooxindoles [81]. Aldehyde 2-(cyclohepta-1,3,5-trien-1-yl)acetaldehyde **181** forms the tetraenamine intermediate under secondary amine catalysis, followed by a formal [4+2] cycloaddition with alkylidene oxindoles **25** to form the spirooxindole **182** in good yields (51–84%) and good to excellent stereoselectivities (up to >95:5 d.r. and 95% ee). Unfortunately, the scope of the reaction is quite narrow and only

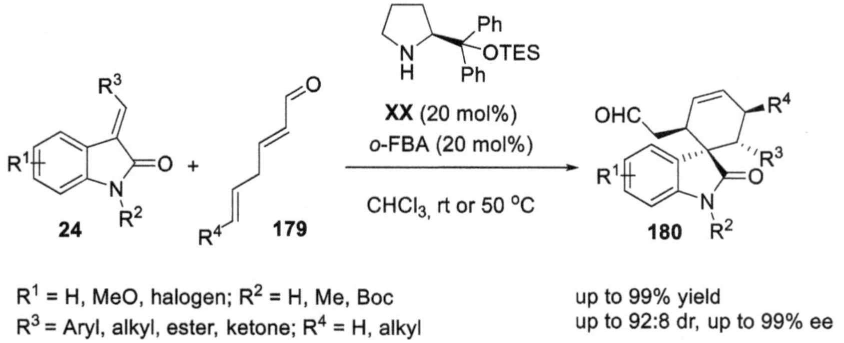

Scheme 2.54. Spirocyclization reported by Chen and Jørgensen.

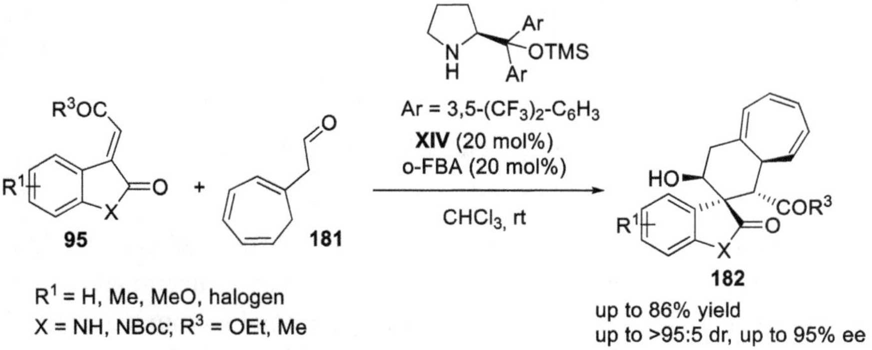

Scheme 2.55. Spirocyclization reported by Jørgensen.

ester-substituted *N*-Boc-protected oxindoles render excellent stereoselectivities. On the other hand, when ketones or non-protected oxindoles are used the enantioselectivity drops considerably (Scheme 2.55).

Wang *et al.* reported a formal [3+4] cycloaddition between alkylidene oxindoles **183** and aza-*o*-quinone methides (generated *in situ* from **184**), catalyzed by chiral NHC carbenes [82]. Isatin-derived enals **183** reacted with a NHC catalyst **LVII** to generate the Breslow intermediate. Next, the homoenolate reacts with the *in situ*-formed aza-*o*-quinone methide to build a new C–C bond. Next, the acyl azolium species undergoes lactamization to release the NHC catalyst and forms the spirobenzazepinone **185** in moderate yields (52–73%) and excellent enantioselectivities (up to 98% ee; Scheme 2.56).

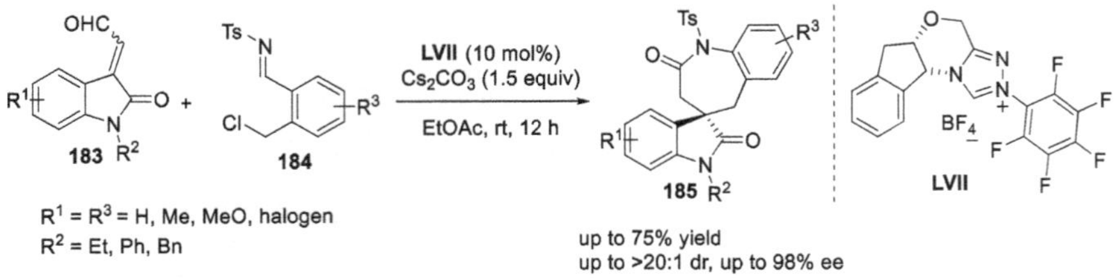

Scheme 2.56.   Spirocyclization reported by Enders.

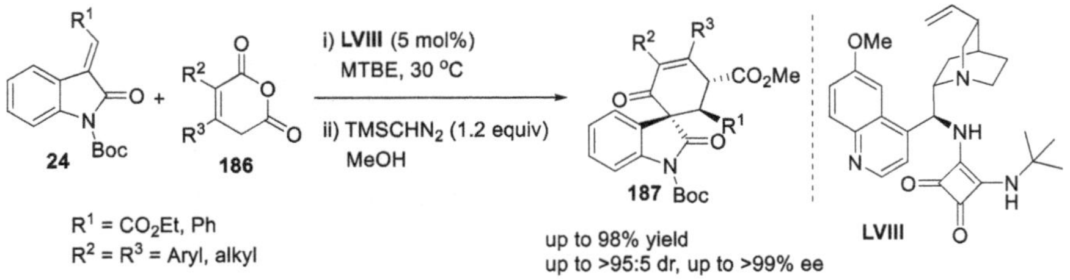

Scheme 2.57.   Spirocyclization reported by Connon.

A Tamura cycloaddition between cyclic anhydrides **186** and alkylide-noxindoles **24** was reported by Connon, catalyzed by bifunctional tertiary amine/squaramide catalyst **LVIII**, affording the final spirooxindoles **187** in excellent yields (82–98%) and stereoselectivities (only 1 diastereomer and up to 99% ee; Scheme 2.57) [83].

A multicomponent cascade reaction between alkylidene oxindoles and enolizable aldehydes was developed by Zhu *et al.* [84]. After an initial Michael addition, catalyzed by the Jørgensen–Hayashi catalyst **XV**, between the enamine of aldehyde **71** and alkylidene oxindole **95**, an aldol reaction between the *in situ*-generated enolate and another aliphatic aldehyde molecule took place, followed by intramolecular hemiacetal formation to furnish the spirooxindoles **188**. The reaction works well with small aldehydes (up to 91% yield >20:1 d.r., and 99% ee). However, stereoselectivities decrease with longer chains (Scheme 2.58). Enders reported a similar organocascade reaction using alkylidene oxindoles and aliphatic aldehydes, catalyzed by secondary amine catalysts [85]. The aliphatic aldehyde in the enamine form undergo a Michael addition with the unsaturated oxindole. Subsequently, another aliphatic aldehyde molecule undergoes oxidation to form the corresponding enal by reaction with

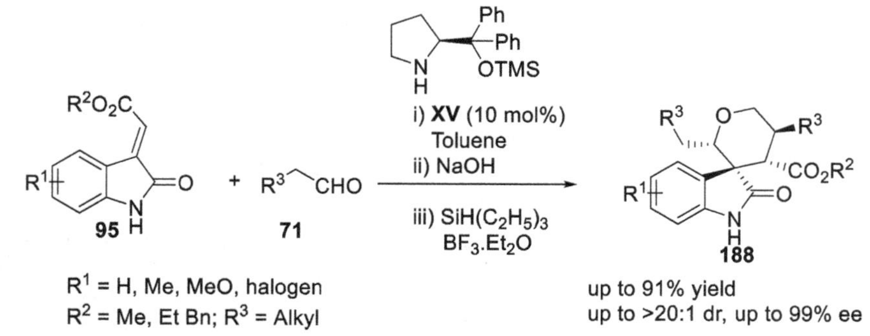

Scheme 2.58. Spirocyclization reported by Zeng and Zhong.

IBX. This enal, which is in the iminium form, reacts with the *in situ*-formed enolate of the oxindole. Finally, an intramolecular aldol reaction and dehydration furnish the spiro compounds in moderate yields (50–78%) and excellent stereoselectivities (up to 20:1 d.r. and 99% ee).

A one-pot thia-Michael/Mannich/lactamization cascade reaction for the synthesis of spirooxindoles was reported by Huang *et al.* [86]. Thiols **54** reacted with (*E*)-ethyl 2-(1-methyl-2-oxoindolin-3-ylidene) acetate derivatives **25** *via* a thio-Michael addition. Next, the *in situ*-formed enolate reacts with the formed imine (reaction of aromatic aldehydes **73** and NH$_4$OAc). The resulting amine attacks the ester moiety to form the final lactam-spirooxindole **189** in reasonable yields (59–81%) and good stereoselectivities (up to 6:1 d.r. and 95% ee; Scheme 2.59).

A new cascade cycloaddition consisting of the reaction between ketimines and alkylidene oxindoles **25**, catalyzed by bifunctional quinine-derived squaramide **XCVI**, was developed by Sun *et al.* [87]. 1,3-Diketones **190** and nitrosobenzene **191** generate ketimines, which react with alkylidene oxindoles through a Michael–Mannich cascade reaction to obtain the spirooxindoles **192** in moderate to good yields (41–94%) and excellent stereoselectivities (4:1 to >20:1 d.r. and 92–99% ee; Scheme 2.60).

2.3.3. *Organocatalytic methodologies starting with isatin*

Isatins react with malononitrile and the resultant Michael acceptor reacts with 2-hydroxynaphthalene-1,4-diones **193** catalyzed by a bifunctional thiourea/tertiary amine catalyst **LXI** [88]. The reaction renders the final

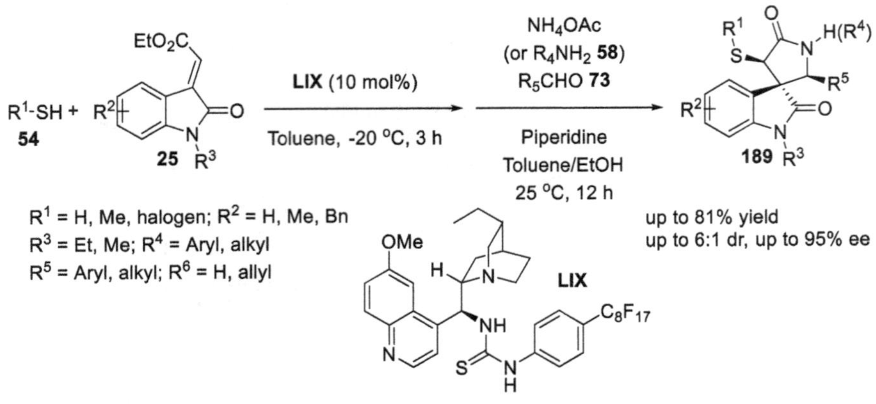

Scheme 2.59. Spirocyclization reported by Zhang.

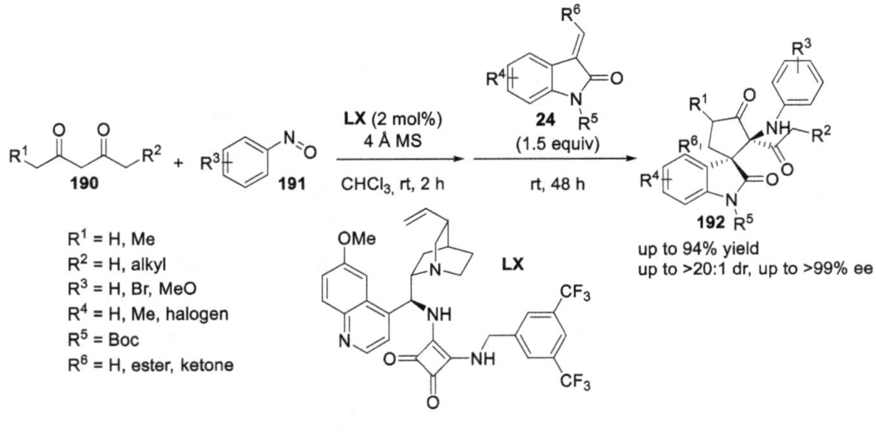

Scheme 2.60. Spirocyclization reported by Sun.

spiro compounds **194** in good to excellent yields (81–99%) and stereose-lectivities (81–97% ee; Scheme 2.61).

A multicomponent [3+3] cyclization, using cinchona alkaloid **LXII** as the catalyst, for the enantioselective synthesis of spirooxindoles bearing a tetrahydroquinolin-3-one scaffold was reported by Zhu *et al.* [89]. The reaction starts with the condensation of malononitrile **144** with isatin **143**. Next, enaminone **196** attacks the newly formed Michael acceptor **195**,

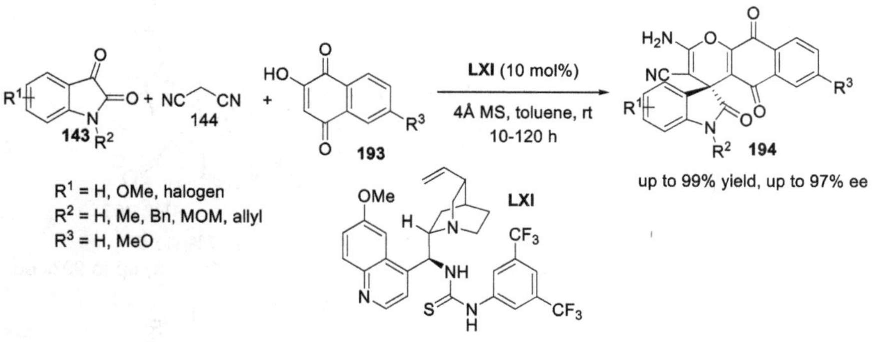

Scheme 2.61.   Spirocyclization reported by Zhao.

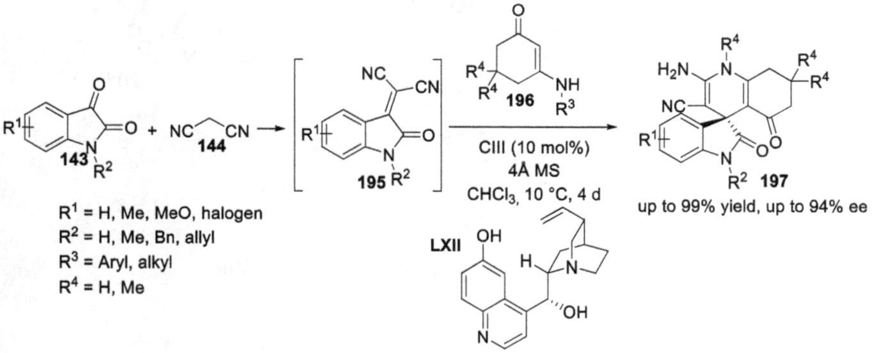

Scheme 2.62.   Spirocyclization reported by Shi.

catalyzed by **LXII**. This is followed by isomerization and intramolecular nucleophilic addition to form spiro compound **196** in moderate to excellent yields (43–99%) and high enantioselectivities (80–92% ee; Scheme 2.62).

Liu *et al.* developed an organocascade reaction starting from isatins **143** for the synthesis of spirooxindoles [90]. Crotonaldehyde reacts with isatin *via* the MBH reaction. Next, a bromination took place and finally a [3+2] cycloaddition with another molecule of isatin. The cascade reaction involves two distinct catalytic steps that use the same catalyst **LXIII**. The reaction furnishes spirooxindoles **202** in moderate to good yields (44–83%), excellent enantioselectivities (90–99% ee) and a good group tolerance (Scheme 2.63).

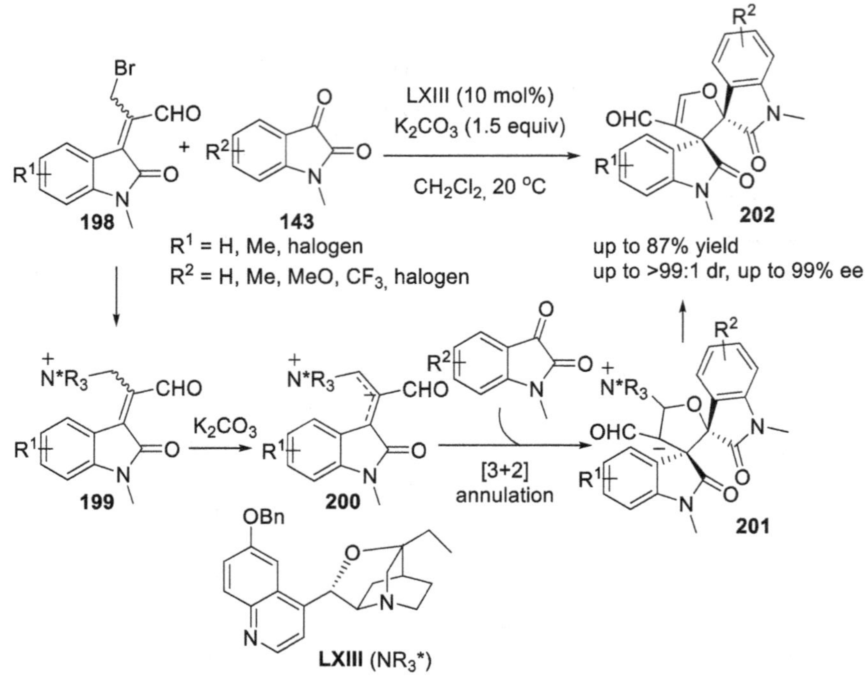

Scheme 2.63.   Spirocyclization reported by Zhou.

2.4. Synergistic Methodologies for the Synthesis of Spirooxindoles

Recently, a new catalytic approach comprising the coexistence of two catalytic cycles that work concurrently to generate one or more new bonds using a metal catalyst and an organocatalyst was analyzed. In this way, the rich chemistry of the metals is combined with the cheap and easy stereoprediction of the organocatalysts.

One of the first examples was performed by Córdova in 2015: a synergistic approach for the synthesis of spirooxindoles based on the combination of imine/enamine activation, together with Pd-allylation [91]. Enals **44** were reacted with oxindoles **203** decorated with an allylic acetate *via* a Michael/Tsuji–Trost allylation, rendering the final spirooxindoles **204** in good yields (90–76%), moderate diastereoselectivities (up to 5:1 d.r.) and excellent enenatioselectivities (up to 99% ee; Scheme 2.64).

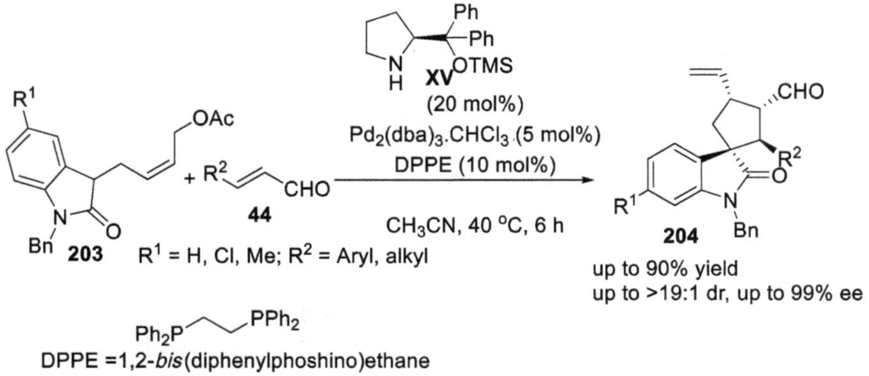

Scheme 2.64.   Spirocyclization reported by Córdova.

Lu *et al.* developed a cycloaddition reaction between isatins and isocyanoacetates to build spirooxindoles under guanidine/Ag catalysis [92]. The reaction starts with an aldol reaction between isocyanoacetate and isatin, with guanidine acting as a bifunctional catalyst, activating the isatin and the isocyanoacetate through organized multipoint hydrogen bonding. Next, intramolecular cyclization between the hydroxyl and isocyano groups, activated by the Ag salt, affords the final spirooxindole in good yields (60–99%), moderate diastereoslectivities (up to 88:12 d.r.) and good enantioselectivities (76–90% ee).

Sun *et al.* [93] reported a combination of Pd(II) salts with a secondary amine catalyst (**XV**) for the synthesis of spirooxindoles **210**. Propargylated oxindoles **205** were reacted with enals **44** *via* a Michael/Conia-ene reaction pathway. The procedure began with a Michael reaction, followed by intramolecular 5-*exo*-dig cyclization between the enamine intermediate and the triple bond activated by the Pd salt to produce intermediate **209** after reductive elimination. Subsequent isomerization of the double bond and hydrolysis provide the most stable conjugated product **210**. The reaction afforded the spirooxindole derivatives in good yields and with excellent enantioselectivities. Later, Córdova *et al.* reported a similar cascade reaction starting from allylic alcohols that were oxidized to enals. These reacted with propargylated oxindoles in the same way as the one reported by Liu and Feng *et al.* Recently, the authors confirmed the mechanism by DFT studies [94]. The same authors expanded the substrate scope of the

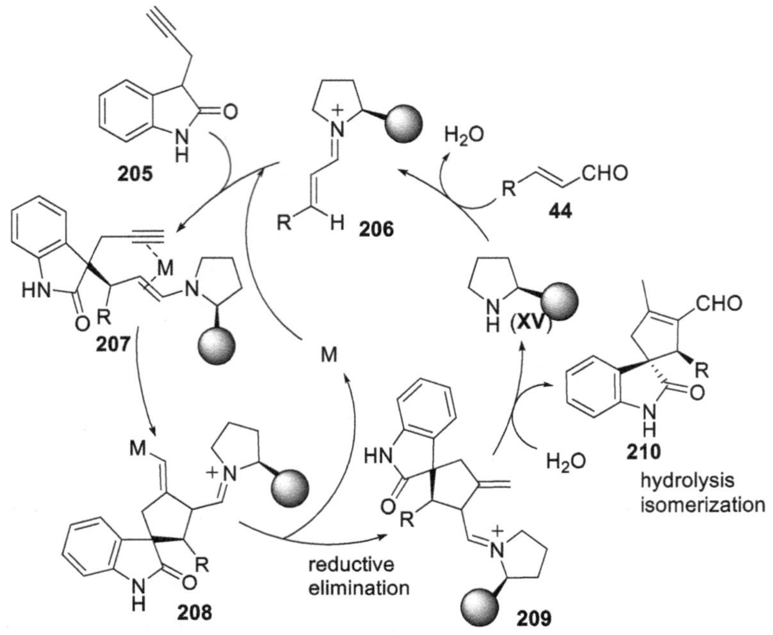

Scheme 2.65. Spirocyclization reported by Córdova and Wang.

reaction with supported secondary amines and/or supported Pd catalysts (Scheme 2.65) [95].

2.5. Conclusions

The enantioselective synthesis of spirooxindoles has been a pursued goal for organic chemists. The difficulties associated with their synthesis have made these compounds underrepresented in screening libraries, despite their unique 3D properties and presence in several biological active natural products. With the last developments in organometallic chemistry, organocatalysis and lately synergistic catalysis, a plethora of new methods have emerged to build these scaffolds derived from oxindoles with a perfect stereocontrol. As we have shown in this chapter, desymmetrization, ring-closing, cycloaddition, annulation and multicomponent reactions are methods that have been efficiently used for this purpose. Clearly, the highlighted methodologies have some drawbacks, including poor structural diversity and the need to use highly complex starting materials. Nevertheless, the achievements have been immense. In the last 5 years, the development of new methodologies for the synthesis of spirooxindoles has increased exponentially, with many improvements, such as new multicomponent reactions and the synthesis of spiro compounds using supported or flow chemistry, expected in the future.

References

1. B. Yu, D. Yu, H. Liu, *Eur. J. Med. Chem.* **2015**, 673.
2. J. Zheng, L. Lin, K. Fu, H. Zheng, X. Liu, X. Feng, *J. Org. Chem.* **2015**, *80*, 8836.
3. Y.-F. Zhang, S.-J. Yin, M. Zhao, J.-Q. Zhang, H.-Y. Li, X.-W. Wang, *RSC Adv.* **2016**, *6*, 30683.
4. B. M. Trost, N. Cramer, S. M. Silverman, *J. Am. Chem. Soc.* **2007**, *129*, 12396.
5. T.-R. Li, B.-Y. Cheng, S.-Q. Fan, Y.-N. Wang, L.-Q. Lu, W.-J. Xiao, *Chem. Eur. J.* **2016**, *22*, 6243.
6. N. R. Ball-Jones, J. J. Badillo, N. T. Tran, A. K. Franz, *Angew. Chem. Int. Ed.* **2014**, *53*, 9462.
7. W. Cao, X. Liu, J. Guo, L. Lin, X. Feng, *Chem. Eur. J.* **2015**, *21*, 1632.
8. J. Guo, Y. Liu, X. Li, X. Liu, L. Lin, X. Feng, *Chem. Sci.* **2016**, *7*, 2717.
9. A. Awata, T. Arai, *Chem. Eur. J.* **2012**, *18*, 8278.
10. T. Arai, T. Miyazaki, H. Ogawa, H. Masu, *Org. Lett.* **2016**, *18*, 5824.
11. Z.-Y. Cao, X. Wang, C. Tan, X.-L. Zhao, J. Zhou, K. Ding, *J. Am. Chem. Soc.* **2013**, *135*, 8197.
12. Y. Chi, L. Qiu, X. Xu, *Org. Biomol. Chem.* **2016**, *14*, 10357.
13. T.-L. Liu, Z.-Y. Xue, H.-Y. Tao, C.-J. Wang, *Org. Biomol. Chem.* **2011**, *9*, 1980.
14. Z.-Y. Cao, F. Zhou, Y.-H. Yu, J. Zhou, *Org. Lett.* **2013**, *15*, 42.
15. B. List, R. A. Lerner, C. F. Barbas III, *J. Am. Chem. Soc.* **2000**, *122*, 2395.
16. K. A. Ahrendt, C. J. Borths, D. W. C. MacMillan, *J. Am. Chem. Soc.* **2000**, *122*, 4243.
17. X. Wu, Q. Liu, H. Fang, J. Chen, W. Cao, G. Zhao, *Chem. Eur. J.* **2012**, *18*, 12196.
18. B. Zhou, Y. Yang, J. Shi, Z. Luo, Y. Li, *J. Org. Chem.* **2013**, *78*, 2897.
19. W. Tan, X. Li, Y.-X. Gong, M.-D. Ge, F. Shi, *Chem. Commun.* **2014**, *50*, 15901.

20. F. Shi, H.-H. Zhang, X.-X. Sun, J. Liang, T. Fan, S.-J. Tu, *Chem. Eur. J.* **2015**, *21*, 3465.

21. B. Wu, J. Chen, M.-Q. Li, J.-X. Zhang, X.-P. Xu, S.-J. Ji, X.-W. Wang, *Eur. J. Org. Chem.* **2012**, 1318.

22. L. Yao, K. Liu, H.-Y. Tao, G.-F. Qiu, X. Zhou, C.-J. Wang, *Chem. Commun.* **2013**, *49*, 6078.

23. S. Zhao, J.-B. Lin, Y.-Y. Zhao, Y.-M. Liang, P.-F. Xu, *Org. Lett.* **2014**, *16*, 1802.

24. H. Wu, Y.-M. Wang, *Chem. Eur. J.* **2014**, *20*, 5899.

25. L.-Z. Ding, T.-S. Zhong, H. Wu, Y.-M. Wang, *Eur. J. Org. Chem.* **2014**, 5139.

26. Y. Wang, L. Liu, T. Zhang, N.-J. Zhong, D. Wang, Y.-J. Chen, *J. Org. Chem.* **2012**, *77*, 4143.

27. N.-J. Zhong, F. Wei, Q.-Q. Xuan, L. Liu, D. Wang, Y.-J. Chen, *Chem. Commun.* **2013**, *49*, 11071.

28. G. Zhan, M.-L. Shi, Q. He, W.-J. Lin, Q. Ouyang, W. Du, Y.-C. Chen, *Angew. Chem. Int. Ed.* **2016**, *55*, 2147.

29. K.-K. Wang, T. Jin, X. Huang, Q. Ouyang, W. Du, Y.-C. Chen, *Org. Lett.* **2016**, *18*, 872.

30. X. Dou, W. Yao, B. Zhou, Y. Lu, *Chem. Commun.* **2013**, *49*, 9224.

31. J.-H. Li, T.-F. Feng, D.-M. Du, *J. Org. Chem.* **2015**, *80*, 11369.

32. K. Albertshofer, B. Tan, C. F. Barbas, *Org. Lett.* **2012**, *14*, 1834.

33. G. Bencivenni, L.-Y. Wu, A. Mazzanti, B. Giannichi, F. Pesciaioli, M.-P. Song, G. Bartoli, P. Melchiorre, *Angew. Chem. Int. Ed.* **2009**, *48*, 7200.

34. Q. Wei, L.-Z. Gong, *Org. Lett.* **2010**, *12*, 1008.

35. X.-H. Chen, Q. Wei, S.-W. Luo, H. Xiao, L.-Z. Gong, *J. Am. Chem. Soc.* **2009**, *131*, 13819.

36. A. Voituriez, N. Pinto, M. Neel, P. Retailleau, A. Marinetti, *Chem. Eur. J.* **2010**, *16*, 12541.

37. B. Tan, N. R. Candeias, C. F. Barbas, *J. Am. Chem. Soc.* **2011**, *133*, 4672.

38. X. Companyo, A. Zea, A.-N. R. Alba, A. Mazzanti, A. Moyano, R. Rios, *Chem. Commun.* **2010**, *46*, 6953.

39. K. Jiang, B. Tiwari, Y. R. Chi, *Org. Lett.* **2012**, *14*, 2382.

40. D. Xie, L. Yang, Y. Lin, Z. Zhang, D. Chen, X. Zeng, G. Zhong, *Org. Lett.* **2015**, *17*, 2318.

41. W. Sun, G. Zhu, C. Wu, L. Hong, R. Wang, *Chem. Eur. J.* **2012**, *18*, 6737.

42. W. Sun, L. Hong, G. Zhu, Z. Wang, X. Wei, J. Ni, R. Wang, *Org. Lett.* **2014**, *16*, 544.

43. B.-L. Zhao, D.-M. Du, *Adv. Synth. Catal.* **2016**, *358*, 3992.

44. B.-L. Zhao, D.-M. Du, *Chem. Commun.* **2016**, *52*, 6162.

45. L.-T. Shen, W.-Q. Jia, S. Ye, *Angew. Chem. Int. Ed.* **2013**, *52*, 585.

46. D. B. Ramachary, C. Venkaiah, R. Madhavachary, *Org. Lett.* **2013**, *15*, 3042.

47. H. Mao, A. Lin, Y. Tang, Y. Shi, H. Hu, Y. Cheng, C. Zhu, *Org. Lett.* **2013**, *15*, 4062.

48. A. Noole, N. S. Sucman, M. A. Kabeshov, T. Kanger, F. Z. Macaev, A. V. Malkov, *Chem. Eur. J.* **2012**, *18*, 14929.

49. J. Zhou, Q.-L. Wang, L. Peng, F. Tian, X.-Y. Xu, L.-X. Wang, *Chem. Commun.* **2014**, *50*, 14601.

50. J. Zhang, D. Cao, H. Wang, C. Zheng, G. Zhao, Y. Shang, *J. Org. Chem.* **2016**, *81*, 10558.

51. A. Noole, K. Ilmarinen, I. Jarving, M. Lopp, T. Kanger, *J. Org. Chem.* **2013**, *78*, 8117.

52. S. Abbaraju, N. Ramireddy, N. K. Rana, H. Arman, J. C. G. Zhao, *Adv. Synth. Catal.* **2015**, *357*, 2633.

53. M. Monari, E. Montroni, A. Nitti, M. Lombardo, C. Trombini, A. Quintavalla, *Chem. Eur. J.* **2015**, *21*, 11038.

54. Q.-S. Sun, X.-Y. Chen, H. Zhu, H. Lin, X.-W. Sun, G.-Q. Lin, *Org. Chem. Front.* **2015**, *2*, 110.

55. L. Wang, X.-M. Shi, W.-P. Dong, L.-P. Zhu, R. Wang, *Chem. Commun.* **2013**, *49*, 3458.

56. Y. Shi, A. Lin, H. Mao, Z. Mao, W. Li, H. Hu, C. Zhu, Y. Cheng, *Chem. Eur. J.* **2013**, *19*, 1914.

57. J.-X. Zhang, H.-Y. Wang, Q.-W. Jin, C.-W. Zheng, G. Zhao, Y.-J. Shang, *Org. Lett.* **2016**, *18*, 4774.

58. W. Sun, G. Zhu, C. Wu, G. Li, L. Hong, R. Wang, *Angew. Chem. Int. Ed.* **2013**, *52*, 8633.

59. Z. Zhang, W. Sun, G. Zhu, J. Yang, M. Zhang, L. Hong, R. Wang, *Chem. Commun.* **2016**, *52*, 1377.

60. L. Hong, M. Kai, C. Wu, W. Sun, G. Zhu, G. Li, X. Yao, R. Wang, *Chem. Commun.* **2013**, *49*, 6713.

61. W. Ren, X.-Y. Wang, J.-J. Li, M. Tian, J. Liu, L. Ouyang, J.-H. Wang, *RSC Adv.* **2017**, *7*, 1863.

62. Y. Huang, C. Zheng, Z. Chai, G. Zhao, *Adv. Synth. Catal.* **2014**, *356*, 579.

63. W. Yang, D.-M. Du, *Chem. Commun.* **2013**, *49*, 8842.

64. Y.-M. Huang, C.-W. Zheng, G. Zhao, *RSC Adv.* **2013**, *3*, 16999.

65. S.-J. Yin, S.-Y. Zhang, J.-Q. Zhang, B.-B. Sun, W.-T. Fan, B. Wu, X.-W. Wang, *RSC Adv.* **2016**, *6*, 84248.

66. S. Muthusamy, M. Prakash, C. Ramakrishnan, M. M. Gromiha, V. Kesavan, *Chem. Cat. Chem.* **2016**, *8*, 1708.

67. W.-T. Wei, C.-X. Chen, R.-J. Lu, J.-J. Wang, X.-J. Zhang, M. Yan, *Org. Biomol. Chem.* **2012**, *10*, 5245.

68. F.-L. Hu, Y. Wei, M. Shi, *Adv. Synth. Catal.* **2014**, *356*, 736.

69. F.-L. Hu, Y. Wei, M. Shi, *Chem. Commun.* **2014**, *50*, 8912.

70. X.-F. Huang, Y.-F. Zhang, Z.-H. Qi, N.-K. Li, Z.-C. Geng, K. Li, X.-W. Wang, *Org. Biomol. Chem.* **2014**, *12*, 4372.

71. H. Huang, M. Bihani, J. C. G. Zhao, *Org. Biomol. Chem.* **2016**, *14*, 1755.

72. T.-Z. Li, J. Xie, Y. Jiang, F. Sha, X.-Y. Wu, *Adv. Synth. Catal.* **2015**, *357*, 3507.

73. T. Du, F. Du, Y. Ning, Y. Peng, *Org. Lett.* **2015**, *17*, 1308.

74. X.-N. Zhang, G.-Q. Chen, X. Dong, Y. Wei, M. Shi, *Adv. Synth. Catal.* **2013**, *355*, 3351.

75. L.-J. Huang, J. Weng, S. Wang, G. Lu, *Adv. Synth. Catal.* **2015**, *357*, 993.

76. Y. Wang, M.-S. Tu, L. Yin, M. Sun, F. Shi, *J. Org. Chem.* **2015**, *80*, 3223.

77. Y.-Y. Gui, J. Yang, L.-W. Qi, X. Wang, F. Tian, X.-N. Li, L. Peng, L.-X. Wang, *Org. Biomol. Chem.* **2015**, *13*, 6371.

78. L.-W. Qi, Y. Yang, Y.-Y. Gui, Y. Zhang, F. Chen, F. Tian, L. Peng, L.-X. Wang, *Org. Lett.* **2014**, *16*, 6436.

79. Z.-J. Jia, H. Jiang, J.-L. Li, B. Gschwend, Q.-Z. Li, X. Yin, J. Grouleff, Y.-C. Chen, K. A. Jørgensen, *J. Am. Chem. Soc.* **2011**, *133*, 5053.

80. Q.-Q. Zhou, Y.-C. Xiao, X. Yuan, Y.-C. Chen, *Asian J. Org. Chem.* **2014**, *3*, 545.

81. J. Stiller, P. H. Poulsen, D. C. Cruz, J. Dourado, R. L. Davis, K. A. Jørgensen, *Chem. Sci.* **2014**, *5*, 2052.

82. L. Wang, S. Li, M. Bluemel, A. R. Philipps, A. Wang, R. Puttreddy, K. Rissanen, D. Enders, *Angew. Chem. Int. Ed.* **2016**, *55*, 11110.

83. F. Manoni, S. J. Connon, *Angew. Chem. Int. Ed.* **2014**, *53*, 2628.

84. L. Zhu, Q. Chen, D. Shen, W. Zhang, C. Shen, X. Zeng, G. Zhong, *Org. Lett.* **2016**, *18*, 2387.

85. X. Zeng, Q. Ni, G. Raabe, D. Enders, *Angew. Chem. Int. Ed.* **2013**, *52*, 2977.

86. X. Huang, M. Liu, K. Pham, X. Zhang, W.-B. Yi, J. P. Jasinski, W. Zhang, *J. Org. Chem.* **2016**, *81*, 5362.

87. Q.-S. Sun, H. Zhu, Y.-J. Chen, X.-D. Yang, X.-W. Sun, G.-Q. Lin, *Angew. Chem. Int. Ed.* **2015**, *54*, 13253.

88. H.-W. Zhao, B. Li, T. Tian, W. Meng, Z. Yang, X.-Q. Song, X.-Q. Chen, H.-L. Pang, *Eur. J. Org. Chem.* **2015**, 3320.

89. Q.-N. Zhu, Y.-C. Zhang, M.-M. Xu, X.-X. Sun, X. Yang, F. Shi, *J. Org. Chem.* **2016**, *81*, 7898.

90. Y.-L. Liu, X. Wang, Y.-L. Zhao, F. Zhu, X.-P. Zeng, L. Chen, C.-H. Wang, X.-L. Zhao, J. Zhou, *Angew. Chem. Int. Ed.* **2013**, *52*, 13735.

91. S. Afewerki, G. Ma, I. Ibrahem, L. Liu, J. Sun, A. Córdova, *ACS Catal.* **2015**, *5*, 1266.

92. Y. Lu, M. Wang, X. Zhao, X. Liu, L. Lin, X. Feng, *Synlett.* **2015**, *26*, 1545.

93. W. Sun, G. Zhu, C. Wu, L. Hong, R. Wang, *Chem. Eur. J.* **2012**, *18*, 13959.

94. S. Santoro, L. Deiana, G.-L. Zhao, S. Lin, F. Himo, A. Córdova, *ACS Catal.* **2014**, *4*, 4474.

95. L. Deiana, L. Ghisu, S. Afewerki, O. Verho, E. V. Johnston, N. Hedin, Z. Bacsik, A. Córdova, *Adv. Synth. Catal.* **2014**, *356*, 2485.

Chapter 3

Open Chain 3,3-Dialkyloxindoles

3.1. Introduction

Enantioenriched open chain 3,3-dialkyl-substituted indoles can be obtained by cyclization of an acyclic precursor (*de novo construction*) or by modification of a pre-existing ring system mainly through nucleophilic substitution, aldol condensations and Mannich or Michael reactions, and all these reactions will be reviewed in this chapter. The catalysts employed and the structure of the prepared natural products will be collected in Appendices A and C.

3.2. *de novo* Construction

The intramolecular asymmetric carbopalladation of *N*-aryl acrylamides (Heck cyclization) is one of the most used and surely the most ancient synthetic strategy to build indole nucleus from acyclic precursors. Since 1993, Overman proposed an asymmetric Heck reaction to prepare both enantiomers of the natural products esermethole and physostigmine (Scheme 3.1) [1]. The (*Z*)-stereochemistry of the 2-butenanilide was found critical to obtain high asymmetric induction, although a partial isomerization occurred during the cyclization process, and both enantiomers can be obtained by using (*R*)- or (*S*)-BINAP, respectively. Thus, either enantiomer of esermethole and physostigmine were obtained in 35–55% and 15–20% overall yield, respectively from commercially available but-2-yn-1-ol and

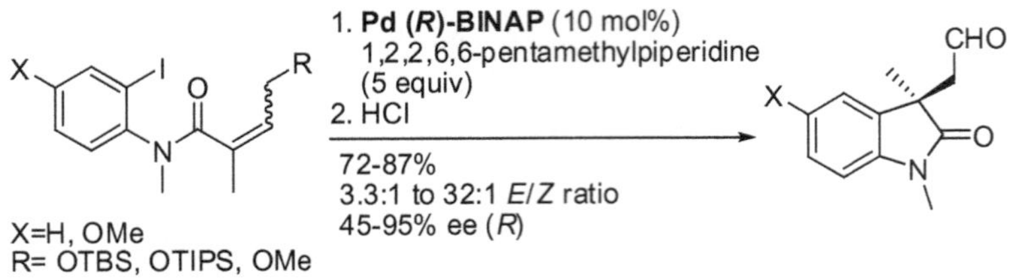

Scheme 3.1.   The most ancient asymmetric synthesis of 3,3-dialkyloxindoles.

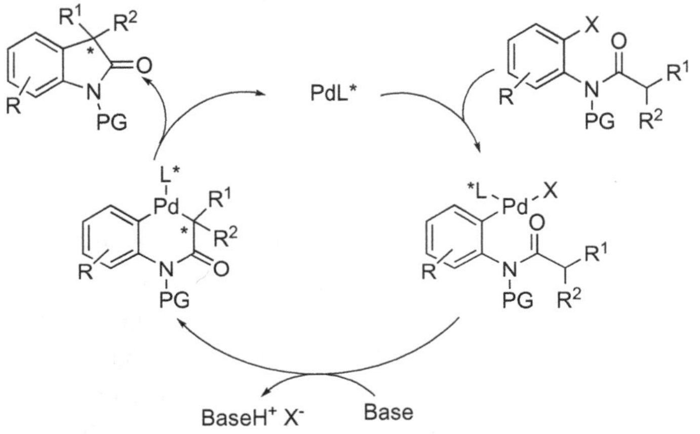

Scheme 3.2.   Envisaged mechanism for Heck cyclizations.

N-methyl-4-methoxybenzenamine. Other natural products were obtained by this method, such as quadrigemine C and psycholeine [2].

Further studies on this reaction appeared later [3], and results are collected in Table 3.1 together with many other arylation reactions. These studies allowed relating the degree of enantioselection with the electron density of the aromatic ring: electron-rich substrates gave the lowest ee. A shortening of the Pd–C bond lengths in electron-deficient substituents accounted for this observation, because a more tightened complex allowed to differentiate the two transition states in energy leading to the two enantiomers. Most of the reactions reported in Table 3.1 used palladium–NHC complexes, and the mechanism of these cyclization reactions very likely involved the steps depicted in Scheme 3.2. In these reactions, the base plays two crucial roles: generation of both the carbene ligand and the

Table 3.1.   Asymmetric arylation reactions.

X, R^a, R^1, R^2, PG	Catalyst	Yield (%)	ee (%)	Ref.
X=Br, R=H, R^1=Me, R^2=Ph, 2-MeC$_6$H$_4$, 4-i-BuC$_6$H$_4$, 1-Npt, PG=Me, Bn	**NHC-I** (10 mol%), Pd(dba)$_2$ (10 mol%), t-BuONa (1 equiv)	73–97	33–67 $(S)^b$	[4]
	NHC-II (10 mol%), Pd(dba)$_2$ (10 mol%), t-BuONa (1 equiv)	27–95	42–76 $(R)^b$	
X=Br, R=H, R^1=Me, R^2=Ph, PG=Me	**NHC-III** (10 mol%), Pd$_2$(dba)$_3$ (10 mol%), t-BuONa (1 equiv)^c	95	43 $(R)^b$	[5]
X=OTf, R=H, R^1=Ph, 4-MeOC$_6$H$_4$, 4-AcNHC$_6$H$_4$, 2-BocNHC$_6$H$_4$, 2-NO$_2$C$_6$H$_4$, 2-NO$_2$-3-MeC$_6$H$_4$, 3-pyrydyl, 1-Npt, N-Bn-indol-7-yl, N-Boc-indol-3-yl, 1,3,3-trimethylindol-7-yl, R^2=MeOCH$_2$CH=, PG=Bn	**(R)-BINAP** (10 mol%), Pd(OAc)$_2$ (5 mol%), 1,2,2,6,6-Me$_5$piperidine (4 equiv)	48–95^d	71–98 (S) (3–20:1 E/Z)	[3]
X=Br, R=H, R^1=Me, R^2=Ph, 4-i-BuC$_6$H$_4$, 3-MeC$_6$H$_4$, 2-MeC$_6$H$_4$, 1-Npt, PG=Me	**(S)-H$_8$-BINAP** (10 mol%), Pd(OAc)$_2$ (10 mol%), t-BuONa (2 equiv)	81–98	51–68 (S)	[6]
X=Br, R=H, R^1=Me, Et, R^2=Ph, 4-MeOC$_6$H$_4$, 4-FC$_6$H$_4$, 4-i-BuC$_6$H$_4$, 3-MeC$_6$H$_4$, 2-Npt, PG=Me, Et	**NHC-IV** (20 mol%), Pd(OAc)$_2$ (10 mol%), t-BuOLi (2 equiv)	49–83^e	46–66 (S)	[7]
X=Br, R=H, R^1=Me, R^2=Ph, 4-MeC$_6$H$_4$, 3-MeC$_6$H$_4$, 2-MeC$_6$H$_4$, 1-Npt, PG=Me, Bn	**NHC_V** (R=R^1=i-Pr) (5 mol%), t-BuONa (1.5 equiv)	93–99^c	80–88 (R)	[8]
X=Cl, Br, R=H, R^1=Me, Et, Bn, R^2=Ph, 2-MeC$_6$H$_4$, 2-MeOC$_6$H$_4$, 1-Npt, PG=Me, Bn	**NHC-VI** (5 mol %), [Pd(allyl)Cl]$_2$ (2.5 mol %), t-BuONa (1.5 equiv)	80–99	81–97 (S)	[9]

(*Continued*)

Table 3.1.   (*Continued*)

X, R^a, R^1, R^2, PG	Catalyst	Yield (%)	ee (%)	Ref.
X=Br, Cl, R=H, 5-Me, 5-*i*-Pr, 6-CF$_3$, 6-F, 6-MeO. 7-F, 5,7-Me$_2$, R^1=Me, Et, *i*-Pr, Bn, R^2=Ph, 4-MeC$_6$H$_4$, 4-MeOC$_6$H$_4$, 4-PhOC$_6$H$_4$, 3-MeC$_6$H$_4$, 2-MeC$_6$H$_4$, 2-MeOC$_6$H$_4$, 2-PhOC$_6$H$_4$, 1-Npt, 2-Npt, (*E*)-PhCH=CH, PG=Me, Bn^f	**NHC-VII** (R=Me, MeO) (5 mol%) Pd(dba)$_2$ (5 mol%), *t*-BuONa (1.5 equiv)	21–99	26–95 (*S*)	[10]
X=Br, R=H, R^1=Me, R^2=Ph, 4-MeC$_6$H$_4$, 2-MeC$_6$H$_4$, 2-MeOC$_6$H$_4$, 2-ClC$_6$H$_4$, 1-Npt, PG=Me, Bn	**NHC-VIII** (5 mol %), TMEDA·PdMe$_2$ (5mol%), *t*-BuONa (1.5 equiv)	75–99	71–98 (*R*)	[11]
X=Br, Cl, R=H, 4-Me, 5-MeO, R^1=allyl, CH$_2$=C(Me)CH$_2$, CMe$_2$=CHCH$_2$, R^2=Ph, 4-MeC$_6$H$_4$, 4-PhC$_6$H$_4$, 4-CF$_3$C$_6$H$_4$, 4-FC$_6$H$_4$, 3-MeC$_6$H$_4$, 3-FC$_6$H$_4$, 3-MeOC$_6$H$_4$, 2-MeC$_6$H$_4$, 2-FC$_6$H$_4$, 2-MeOC$_6$H$_4$, 3,5-Me$_2$C$_6$H$_3$, 1-Npt, 2-Npt, *N*-methylindol-3-yl, Me, PG=Me	**NHC-V** (R=3-pentyl, R^1=H) (5 mol%), *t*-BuONa (1.5 equiv)	55–98	38–94 (*R*)	[12]
X=Br, R=6-Cl, R^1=3-ClC$_6$H$_4$CH$_2$, R^2=3-MeOC$_6$H$_4$, PG=CH$_2$OMe	**NHC-IX** (5 mol%), Pd(dba)$_2$ (5 mol%), *t*-BuONa (1.5 equiv)	86^g	90 (*R*)	[13]
X=OTf, R=H, 5-Me, 5-F, 6-MeO, 6-Me, 6-Ph, 6-Cl, 7-Me, benzo[e], benzo[g], R^1=Me, R^2=Ph, 4-FC$_6$H$_4$, 4-MeOC$_6$H$_4$, 4-PhC$_6$H$_4$, 3-ClC$_6$H$_4$, 2-MeC$_6$H$_4$, 3,4-(OCH$_2$O)C$_6$H$_3$, 1-Npt, MeOCH$_2$, Bn, NTs(Bn)CH$_2$, R^1-R^2=CH$_2$CMe$_2$(CH$_2$)$_2$, PG=Me, Bn	**(*S*)-PHOS-I** (20 mol%), PdCl$_2$(MeCN)$_2$ (10 mol%) B$_2$(OH)$_4$, H$_2$O (2 equiv each), DABCO (4 equiv)	72–90	70–94 (*S*)	[14]

(*Continued*)

Table 3.1.   (Continued)

X, R^a, R^1, R^2, PG	Catalyst	Yield (%)	ee (%)	Ref.
X=Br, Cl, R=H, R^1=Me, cyclobutyl, R^2=Ph, 1-Npt, PG=Me	**Pd** (2.5 mol%), *t*-BuONa (1.5 equiv)	95–98	55–63 (*R*)	[15]
X=Br, R=H, 6-MeO, 6-F, R^1=Ph, R^2=Me, Et, PG=Me	**NHC-X** (5 mol%), Pd(dba)$_2$ (5 mol%), *t*-BuONa (1.5 equiv)	85–99	93–97 (*S*)	[16]

Notes: ^{a}The number refers to the position in the final indole. ^{b}Later other authors demonstrated that (3*S*)-1,3-dimethyl-3-(3-tolyl)oxindole have a positive α_D, while (3*S*)-1,3-dimethyl-3-phenyloxindole have a negative α_D. Thus, these configurations were attributed by comparing the reported α_D with these observations. ^{c}Other catalysts were tested but with worse results. ^{d}Aryl substituents containing basic amine functionalities did not give enantioselectivity. ^{e}Conversion at 12 h, what happens after prolonged reaction times is not reported. ^{f}1-(8-bromo-6-methyl-3,4-dihydroquinolin-1(2H)-yl)-2-phenylpropan-1-one gave (*S*)-1,8-dimethyl-1-phenyl-5,6-dihydro-1H-pyrrolo[3,2,1-*ij*]quinolin-2(4H)-one in 87% yield and 94% ee. Unprotected amides did not react. ^{g}6-Chloro-3-(3-chlorobenzyl)-1,3-dihydro-3-(3-methoxyphenyl)-2*H*-indol-2-one, a MDM2 antagonists with application in cancer therapy, was prepared in 45% overall yield from 2-(3-methoxyphenyl)acetic acid and 96% ee after recrystallization.

enolate. DFT computational analysis demonstrated that the oxidative insertion of palladium–NHC to the C–Br bond is the rate-determining step and the conversion of the *O*-enolate into the *C*-enolate accounted for the asymmetric induction [16]. Many authors invoked high "buried volume" in the catalyst as a necessary requirement for obtaining high enantiomeric excesses. However, studies demonstrated that less bulky ligands could anyway give high asymmetric induction if they are placed judiciously [8]. Some data collected in Table 3.1 deserve to be commented on.

Kündig obtained always (*S*)-configured products from two catalysts with similar structure, but opposite stereochemistry [10, 16]. Chiral NHCs having fused rings that hamper rotation around the N–C bond gave the highest ee's for more hindered substrates [9, 11], while NHCs with free rotation gave the lowest ee's. Only four procedures worked with the less reactive chlorides [9, 10b, 12, 15]. Dorta and co-workers were able to separate the three different atropisomers of their three NHC catalysts and they were tested separately [8]. In Table 3.1, the best results with the (R_a,R_a)-di-*iso*-propylnaphthyl derivative are reported, but all the nine combinations are described in the paper. A careful analysis of the catalytic data demonstrated that the chiral groups on the *N*-heterocycle determined the absolute configuration of the carbon center. However, the amount of steric pressure exerted by phenyl groups onto the different oriented naphthyl side chains increased or lowered the proportions of the intermediate, which led to the (*R*)-product.

These catalysts were unable to give good results when R^1=allyl, thus Dorta introduced another catalyst [12]. If the (R_a,R_a)-isomer was once more the most efficient in most cases, it should be noted that, enantioselectivities with *meta*-substituted aromatic moieties are slightly higher with the (R_a,S_a)-isomer. With the highly congested 3-allyl-3-(2-fluorophenyl)-4-methyloxindole both atropisomers gave very low enantioselectivity (4–8%), but the (R_a,S_a)-isomer allowed recovering the product in 95% yield, with respect to 40% of the (R_a,R_a)-isomer.

Mizoroki–Heck reaction is a modification, which involves alkylidenedihydropyridines as *N*-arylacrylamides. Thus, enantioenriched oxindole derivatives can be obtained if an Ag(I) salt allows obtaining the cationic intermediate necessary for the cyclization process (Scheme 3.3) [17]. The intermolecular trapping of the resulting Pd complex by nucleophiles allowed obtaining domino processes. For instance, in a study on the domino Heck/cyanation reaction, an asymmetric version has also been reported using (*S*)-DIFLUORPHOS, and the procedure was applied to a synthesis of (–)-physostigmine (Scheme 3.4) [18]. Oxadiazoles were other tested nucleophiles (Scheme 3.5) [19]. It should be noted that the recovered products underwent an unexpected reductive cyclization, which was exploited for a total synthesis of (+)-esermethole.

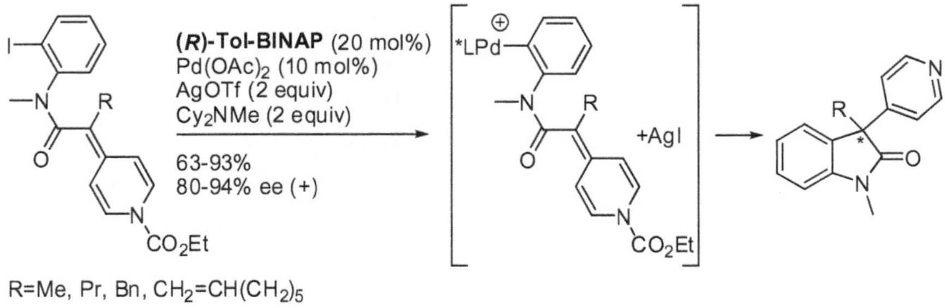

Scheme 3.3.   Asymmetric Mizoroki–Heck cyclization of alkylidenedihydropyridines.

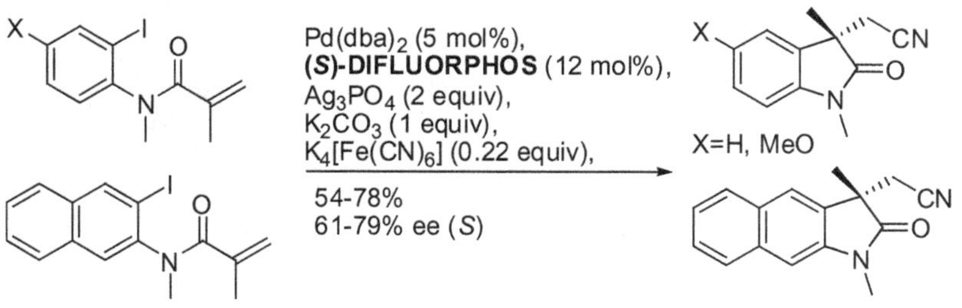

Pd(dba)2 (5 mol%),
(S)-DIFLUORPHOS (12 mol%),
Ag3PO4 (2 equiv),
K2CO3 (1 equiv),
K4[Fe(CN)6] (0.22 equiv),
54-78%
61-79% ee (S)
X=H, MeO
CN

Scheme 3.4. Insertion of cyanide to the transient α-alkylpalladium complex of the Heck reaction.

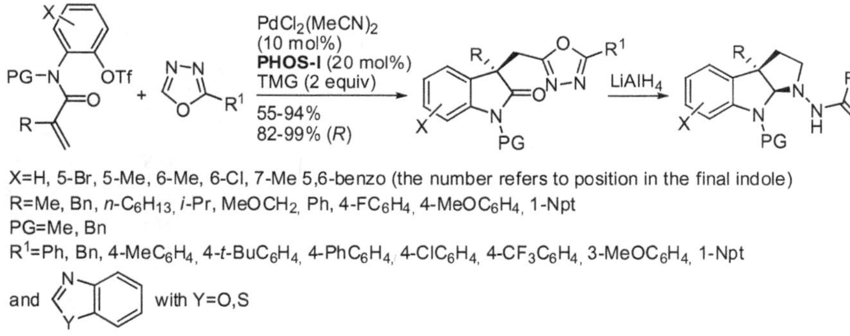

PdCl2(MeCN)2
(10 mol%)
PHOS-I (20 mol%)
TMG (2 equiv)
55-94%
82-99% (R)
LiAlH4
X=H, 5-Br, 5-Me, 6-Me, 6-Cl, 7-Me 5,6-benzo (the number refers to position in the final indole)
R=Me, Bn, n-C6H13, i-Pr, MeOCH2, Ph, 4-FC6H4, 4-MeOC6H4, 1-Npt
PG=Me, Bn
R1=Ph, Bn, 4-MeC6H4, 4-t-BuC6H4, 4-PhC6H4, 4-ClC6H4, 4-CF3C6H4, 3-MeOC6H4, 1-Npt
and
with Y=O,S

Scheme 3.5. Insertion of oxadiazoles to the transient σ-alkylpalladium complex of the Heck reaction.

Also, sterically hindered 2,6-dialkylphenylisocyanides could be asymmetrically inserted into the transient σ-alkylpalladium complex of the Heck reaction with the DUANPHOS catalyst (Scheme 3.6) [20].

Other routes to the formation of enantioenriched 3,3-disubstituted indoles by *de novo* construction of the indole nucleus are known. For instance, the same cyanomethyl products, already described in Scheme 3.4, were obtained by enantioselective intramolecular cyanoamidation of alkenyl cyanoformamides (Scheme 3.7) [21]. This approach has the merit of avoiding the use of both potentially poisonous halogens and strong bases.

3-Oxa-4-aza-[3,3]sigmatropic rearrangement of phenylnitrone with ketenes is another route to 3,3-disubstituted oxindoles. If an asymmetric nitrone is employed, enantioenriched products could be recovered. Actually, Garner's aldehyde reacted with phenylhydroxylamine to give

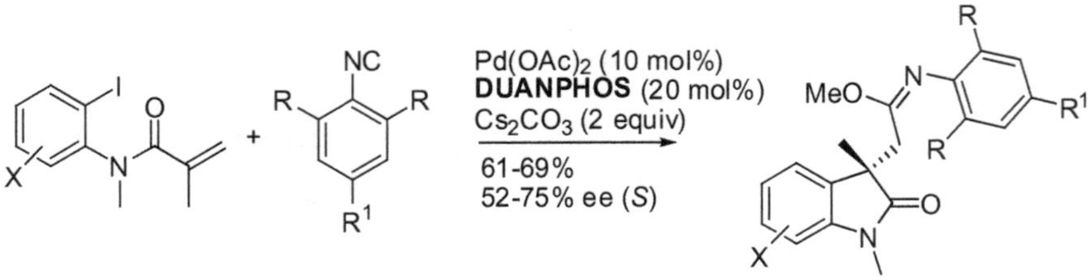

X=H, 5-MeO, 5,6-benzo (the number refers to position in the final indole)
R=i-Pr, Me, R^1=H; R=R^1=t-Bu; R=3,5-Me$_2$C$_6$H$_3$, R^1=Me; R=3,5-(CF$_3$)$_2$C$_6$H$_3$, R^1=F

Scheme 3.6.   Insertion of isocyanides to the transient α-alkylpalladium complex of the Heck reaction.

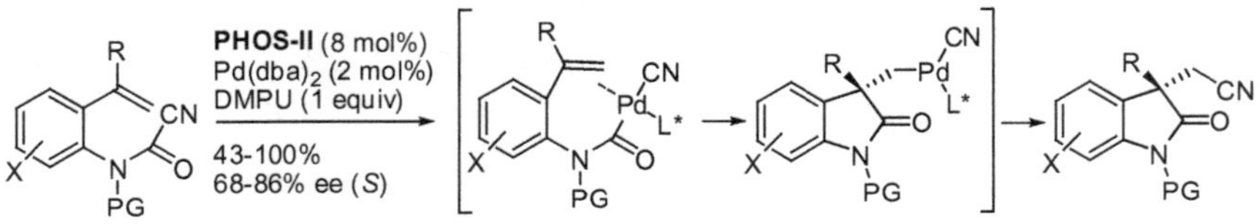

X=H, 4-Me, 5-Me, 5-MeO, 6-Cl, 5,6-(MeO)$_2$ (the number refers to position in the final indole)
PG=Bn, Me;   R=Me, Pr, TBSOCH$_2$

Scheme 3.7.   Enantioselective intramolecular cyanoamidation of alkenyl cyanoformamides.

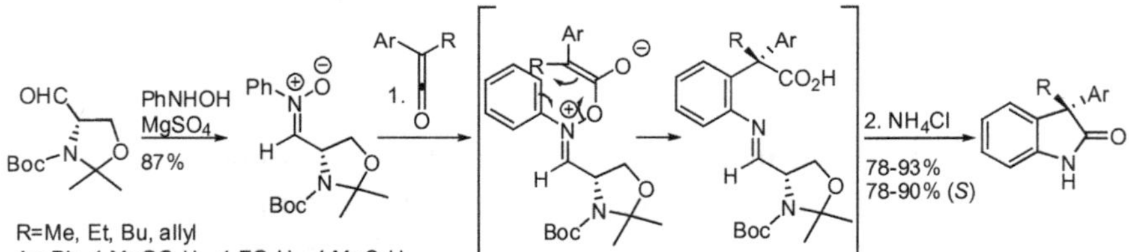

R=Me, Et, Bu, allyl
Ar=Ph, 4-MeOC$_6$H$_4$, 4-FC$_6$H$_4$, 4-MeC$_6$H$_4$

Scheme 3.8.   Asymmetric hetero-Claisen rearrangement.

(R,Z)-N-phenylnitrone, which was submitted to rearrangement (Scheme 3.8) [22]. After hydrolysis, the imino acid intermediate easily cyclized to enantioenriched oxindoles releasing the chiral auxiliary in 78–87% yield. The asymmetric induction was envisaged to take place from the preferential attack of the nitrone *anti* to the aryl substituent of the ketene. The utility of this reaction was exploited by the synthesis of 3-phenyl-hexahydropyrroloindole in 54% yield over five steps from 2-phenylpent-4-enoyl chloride.

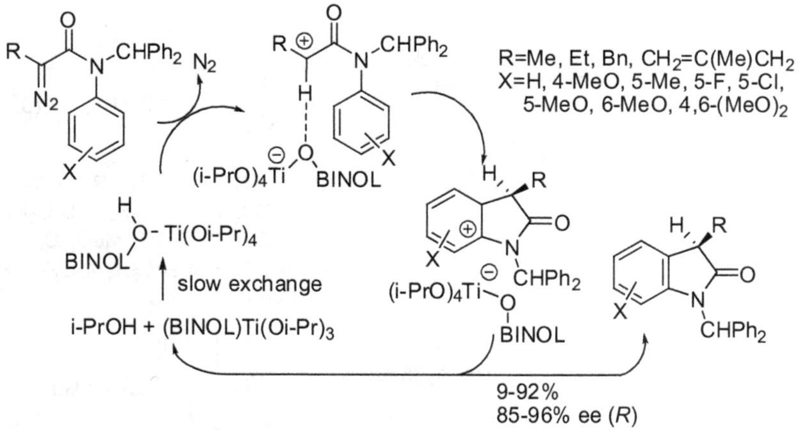

Scheme 3.9. Enantioselective intramolecular cyclization of *N*-aryldiazoamides.

A third route to the formation of enantioenriched oxindoles could be the asymmetric intramolecular cyclization of *N*-aryldiazoamides.

This route travelled in a rare synthesis of chiral 3-monosubstituted oxindoles, under (*S*)-BINOL (40 mol%) and Ti(O-*i*-Pr)$_4$ (20 mol%) catalysis (Scheme 3.9) [23]. The amount of BINOL had a drastic effect on the conversion, demonstrating the role of the proton source on the reaction rate. Moreover, the borane reduction of oxindoles gave a mixture of the corresponding indole and indolenine, the last without affecting the enantioselectivity.

The synthesis of enantioenriched 3,3-disubstituted oxindoles was instead performed *via* a Rh(II)-, Pd(II)-, or Ru(II)-catalyzed reaction in the presence of Brønsted acid-activated imines (Scheme 3.10, Eq. (3.1)) [24], a base-activated MBH carbonate (Scheme 3.10, Eq. (3.2)) [25], or allyl carbonates (Scheme 3.10, Eq. (3.3)), respectively [26]. The first reaction is an alternative to the Mannich reaction that will be described in Section 3.4, the other two reactions are alternative to the asymmetric allylic alkylation introduced by Trost and expounded in Section 3.3. Yields of the present reactions are generally lower, whereas diasteroselectivity is generally higher, and enantioselectivity comparable, with respect to those in which the indole nucleus was already built. The hypothesis that the two catalysts work synergistically is under debate. In fact, Hu and co-workers found influence of the palladium complex in the alkylation

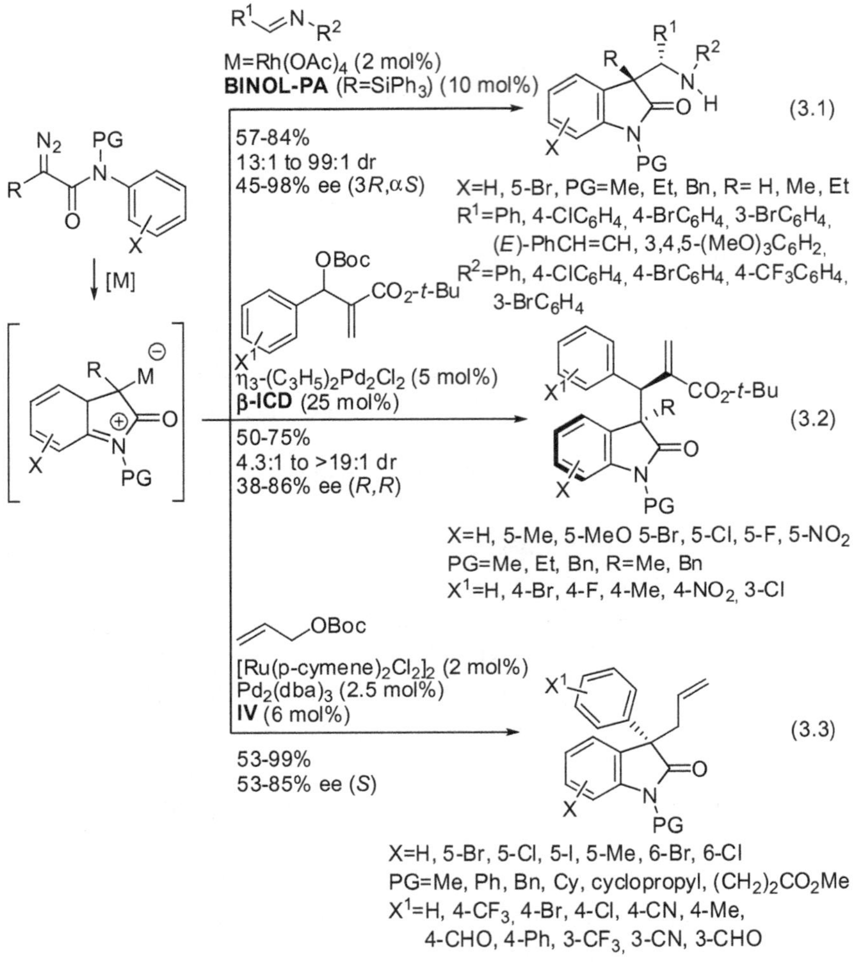

Scheme 3.10.   Combining metal-catalyzed C–H functionalization of diazoacetamides with asymmetric nucleophilic substitutions to oxindoles.

step [25], while the ruthenium catalyst did not modify the stereoselectivity of the addition step [26].

Moreover, Lautens and co-workers reported that only electron-withdrawing groups (EWGs) gave good results, and that the steric hindrance blocked up the ruthenium step, thus allowing recovering of the starting material.

3.3. Nucleophilic Substitution Reactions

If asymmetric *de novo* construction of the indole nucleus is a fascinating way, the manipulation of oxindole derivatives is more frequently used to prepare enantioenriched 3,3-disubstituted derivatives. As introduced in Chapter 1, the nucleophilic character of C-3 position in 3-prostereogenic oxindoles, allowed to setup asymmetric synthesis of all carbon quaternary 3,3-disubstituted oxindoles by nucleophilic substitution, aldol reactions or Mannich and Michael additions. Alternatively, 3-prostereogenic oxindoles with appropriate leaving groups can undergo the same reactions by nucleophiles. These reactions will be described in the present and in the following sections.

3.3.1. *Indole as the nucleophile*

Before discussing the asymmetric nucleophilic substitutions from achiral substrates, an asymmetric reaction with a chiral auxiliary should be mentioned. In 2016, Nakazaki *et al.* reported the alkylation of atropisomeric *N*-aryloxindoles with axial chirality on the C–N bond, which produced a single diastereomer in 63–85% yield. Subsequent oxidative removal of the *N*-aryl group occurred in about 50% yield, thus making this procedure interesting for the synthesis of enantioenriched 3,3-disubstituted oxindoles [27]. Unfortunately, they did not perform the separation of the atropisomeric racemate, thus this synthesis did not allow the preparation of enantioenriched products yet.

The total synthesis of (–)-esermethole, reported by Wong and co-workers in 1991, was the first example of a catalytic asymmetric alkylation of a 3-prostereogenic oxindole (Scheme 3.11, Eq. (3.4)) [28]. Many years after, Deng and Muruoka generalized this reaction, but recovered products with the opposite configuration at C-3 (Scheme 3.11, Eq. (3.5)) [29].[a]

Moreover, these compounds could be easily converted into tetrahydropyrrolo[2,3-b]indoles, the core structure of flustramines and flustramides, alkaloids isolated from *Flustra foliacea* (*Bryozoa*).

---

[a]It should be noted that the different priority of Cahn–Ingold–Prelog rules also makes (*S*)-configured these products.

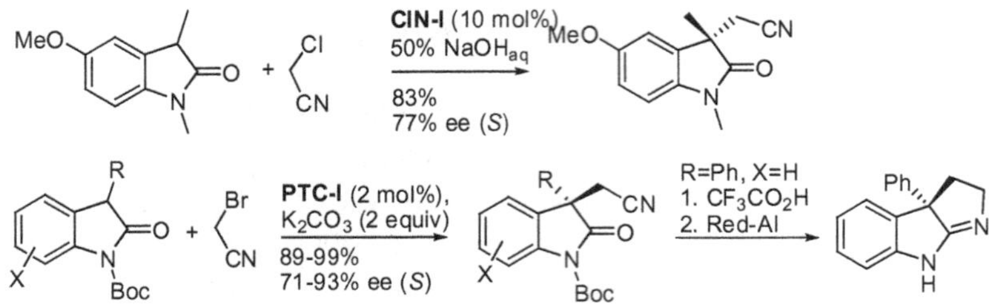

$$(3.4)$$

$$(3.5)$$

R=Ph, 3-MeC$_6$H$_4$, 3-ClC$_6$H$_4$, 4-FC$_6$H$_4$. 4-ClC$_6$H$_4$, 2-Npt
X=H, 5-Me, 5-MeO, 5-F, 6-Cl, 7-F

Scheme 3.11.   The first asymmetric alkylation of a 3-prostereogenic oxindole.

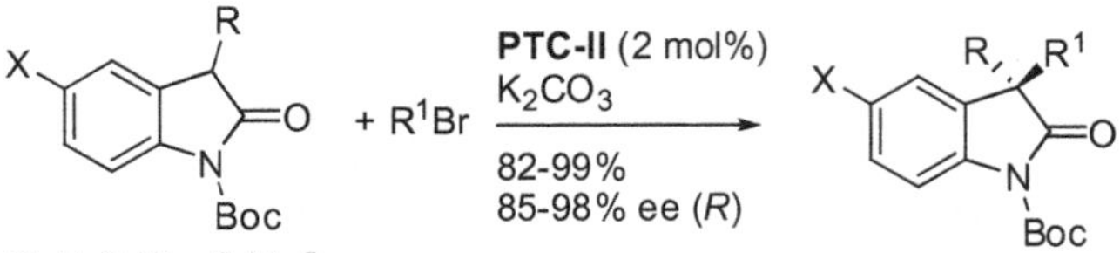

X=H, 5-Me, 5-MeO
R=Me, Et, *i*-Pr, CH$_2$=CHCH$_2$
R^1=Bn, 4-BrC$_6$H$_4$CH$_2$, 4-MeC$_6$H$_4$CH$_2$, 3-MeOC$_6$H$_4$CH$_2$, CH$_2$=CHCH$_2$, CH$_2$=C(Me)CH$_2$,
   MeO$_2$CCH$_2$, propargyl

Scheme 3.12.   Asymmetric alkylation of oxindoles with bromides.

Another phase-transfer organocatalyst was introduced by Ooi's group (Scheme 3.12) [30]. Although a mechanism was not given, studies on the substituents of the triazolium catalyst allowed establishing that the substituents at C-4 and the amide carbonyl of the catalyst exert a crucial influence to fix the stereochemistry, as well as two hydrogen bonds between C(5)–H and amide N–H.

Although halides are the classical leaving group in nucleophilic substitution reactions, other leaving groups were employed in the asymmetric synthesis of 3,3-disubstituted oxindoles. For instance, 4-acetylphenyl cyanate was used as the cyano source in the cyanation of 3-substituted oxindoles (Scheme 3.13) [31]. It should be noted that lutidine surely favored the enolate formation, but also acted as a cooperative catalyst with the zinc complex. Moreover, deprotection of the Boc group did not affect the enantioselectivity.

Tosylate leaving group was used in the reaction of 3-(1-tosylalkyl) indoles with oxindoles, under chiral bifunctional organocatalysis

Scheme 3.13. Asymmetric cyanation of 3-substituted oxindoles.

X=H, 5-F, 5-Cl, 5-Br, 5-Me, 5-MeO, 6-Cl, 6-Br
R=Ph, 4-FC$_6$H$_4$, 4-ClC$_6$H$_4$, 4-MeC$_6$H$_4$, 4-MeOC$_6$H$_4$, 3-FC$_6$H$_4$, 3-MeC$_6$H$_4$, 2-MeO-5-ClC$_6$H$_3$, 1-Npt

Scheme 3.14. Nucleophilic substitution of 3-(1-tosylalkyl)indoles with oxindoles.

(Scheme 3.14) [32]. The poor diasteroselection was explained by a low energy barrier, which did not allow discrimination between *Z*- and *E*-isomers after TsOH elimination. The addition to oxindole was instead β-selective. A further reported example was 3-cyanomethyloxindole, which afforded the expected product in 91% yield and with 2.5 (96% ee):1 (90% ee) d.r. One instance was then converted into optically pure (+)-trigolute B in 18% overall yield.

Among nucleophilic substitution reactions, Trost's asymmetric allylic alkylation played a key role (Table 3.2). Decarboxylative allylation (Table 3.2, *c,d*) as well as benzyl and MBH carbonates (Table 3.2, *f,g*) represent particular substrates for this reaction.

Table 3.2. Trost's asymmetric allylic alkylation.

X, R, R^1, LG, PG	Catalyst	Yield (%)	ee (%)	Ref.
(a)				
X=H, 5-MeO, R=Ph, 4-MeC$_6$H$_4$, 4-FC$_6$H$_4$, 4-ClC$_6$H$_4$, 4-PhC$_6$H$_4$, 3-MeC$_6$H$_4$, 2-MeC$_6$H$_4$, 2-MeOC$_6$H$_4$, 3,4-(OCH$_2$O)C$_6$H$_3$, LG=Ac, PG=Me, TMSCH$_2$, Bn, SEM, MOM	**TR-I** (5 mol%), (η^3-C$_3$H$_5$PdCl)$_2$ (2.5 mol%), *t*-BuOH (4 equiv)	96–68	66–97 (*R*)	[33]
X=H, 5-MeO, R=Me, Bn, *i*-Pr, NCCH$_2$, 2-(BocNH)C$_6$H$_4$CH$_2$, Me$_2$C=CHCH$_2$, TBSO(CH$_2$)$_2$, MeC≡CCH$_2$, TMSC≡CCH$_2$, LG=Boc, PG=Me, Bn, MOM, allyl	**TR-II** (15 mol%), Mo(C$_7$H$_8$)(CO)$_3$ (10 mol%), *t*-BuOLi (2 equiv)	92–99	74–95 (*S*)^b	[34]
X=5-MeO, R=CO$_2$Et, LG=Ac, PG=2,4-(MeO)$_2$C$_6$H$_3$CH$_2$	**TR-III** (1 mol%), [Pd(Et)Cl]$_2$ (0.25 mol%), TBAT (15 mol%)^a	100	84 (*R*)	[35]
(b)				
X=H, R=TBSO(CH$_2$)$_2$, R^1= 2-NHBoc-4-BrC$_6$H$_3$, LG=P(O)(OEt)$_2$, PG=CO$_2$Me	*ent*-**TR-II** (30 mol%), Mo(C$_7$H$_8$)(CO)$_3$ (20 mol%), NaHMDS	89^f	97 (*S,S*)	[36]
X=H, 5,7-Cl$_2$, R=Ph, 4-FC$_6$H$_4$, 4-ClC$_6$H$_4$, 4-CNC$_6$H$_4$, 4-MeOC$_6$H$_4$, 4-NMe$_2$C$_6$H$_4$, 2-MeC$_6$H$_4$, 1-Npt, 2-Npt, 3-Me-thien-2-yl, *N*-Me-2-Ph-indol-3-yl, *N*-Ts-indol-3-yl,	**TR-II** (15 mol%), Mo(C$_7$H$_8$)(CO)$_3$ (10 mol%), *t*-BuONa (1 equiv)	63–95^c	89–97 (3*S*,αR)	[37,38]

(*Continued*)

Table 3.2. (*Continued*)

X, R, R^1, LG, PG	Catalyst	Yield (%)	ee (%)	Ref.
2,4-Me$_2$-thiazol-5-yl, 2,4-Ph$_2$-oxaazol-5-yl, R^1=Ph, 4-TBSOC$_6$H$_4$, 2-NHBocC$_6$H$_4$, 3,4-Cl$_2$C$_6$H$_3$, 2-furyl, 2-thienyl, 2-(*E*)-buten-1-yl, PG= Me, MOM, Boc, Bn				
X=H, 6-Cl, 6-MeO, R= Bn, Me, Et, Allyl, *i*-Pr, TBSO(CH$_2$)$_2$, *t*-BuOCOCH$_2$, R^1=Ph, LG=Boc, PG=H	**TR-II** (15 mol%), Mo(C$_7$H$_8$)(CO)$_3$ (10 mol%), *t*-BuONa (1 equiv)	63–95^d	89–99 (3*S*,αR)	[38]
X=H, R=thiazol-5-yl, *N*-Me-indol-3-yl, 2-thienyl		84–90^e	not reported	
X=H, 6-Br, 6-Cl, 7-F, 5-MeO, 5-Me, 5-F, 5-Br R=Ph, 3-MeOC$_6$H$_4$, 4-CF$_3$C$_6$H$_4$, 4-MeC$_6$H$_4$, 4-ClC$_6$H$_4$, 4-FC$_6$H$_4$, 3,5-Me$_2$C$_6$H$_3$, 2-Npt, R^1=H, Pr, 4-BrC$_6$H$_4$, LG=Ac, PG=Boc	**TR-IV** (10 mol%), Pd(PPh$_3$)$_4$ (2.5 mol%), 4Å MS	71–92	70–92 (*S*)	[39]

(*c*)

| X=H, MeO, R=Ph, 4-MeOC$_6$H$_4$, 4-FC$_6$H$_4$, 3-MeC$_6$H$_4$, 2-MeC$_6$H$_4$, 3-Me-thien-2-yl, R^1=1-Npt, 4-MeO-naphth-1-yl, *N*-CO$_2$Me-indol-3-yl, 3-benzofuranyl, 3-furyl, 5-TMS-3-furyl, 5-C$_6$H$_{13}$-3-furyl | **TR-I** (6 mol%), η^3-C$_3$H$_5$PdCp (5 mol%), *t*-BuOH (5 equiv) | 44–98 | 76–96 (*S*) | [40] |

(*d*)

(*Continued*)

Table 3.2.   (*Continued*)

X, R, R^1, LG, PG	Catalyst	Yield (%)	ee (%)	Ref.
R=Me, CNCH$_2$, *i*-Pr, Cy, Ph, 2-MeOC$_6$H$_4$, 2-NO$_2$C$_6$H$_4$, 3-MeOC$_6$H$_4$	**TR-I** (6.5 mol%) Pd$_2$(dba)$_3$ (2.5 mol%)	65 to >99	72–95 (*R* or *S*)^g	[41]
R=2,4,6-(MeO)$_3$C$_6$H$_2$, 2,3,6-(MeO)$_3$C$_6$H$_2$, 2,3,4-(MeO)$_3$C$_6$H$_2$, 2,6-(MeO)$_2$C$_6$H$_3$, 2,4-(MeO)$_2$C$_6$H$_3$, 4-MeOC$_6$H$_4$, 4-CF$_3$C$_6$H$_4$, Ph, 2-MeO-4,6-Me$_2$C$_6$H$_2$, 2-BnO-1-Npt, 2MeO-1-Npt, piperonyl	**TR-I** (12.5 mol%) Pd$_2$(dba)$_3$ (5 mol%)	94–98	52–98 (*R*)	[42]

(*e*)

X=H, 6-Br, 5-MeO, R=Ph, 4-MeOC$_6$H$_4$, 4-CF$_3$C$_6$H$_4$, 4-BrC$_6$H$_4$, 4-ClC$_6$H$_4$, 4-FC$_6$H$_4$, 3-MeC$_6$H$_4$, 2-Npt 3-Me-thien-2-yl, 2,4-Me$_2$-thiazol-5-yl, *N*-Ts-indol-3-yl, PG=H, Me	**TR-V** (7.5 mol%), Pd$_2$(dba)$_3$•CHCl$_3$ (2.5 mol%), *t*-BuOH (5 equiv)	75–97^h	84–96 (*R*)	[43]

(*f*)

X=H, Me, F, R=Ph, Bn, 4-MeC$_6$H$_4$, 4-FC$_6$H$_4$, R^1=Ph, 4-FC$_6$H$_4$, 4-MeOC$_6$H$_4$, 3-ClC$_6$H$_4$, 3-CNC$_6$H$_4$, 3-MeC$_6$H$_4$, 2-NO$_2$C$_6$H$_4$, 3,4-Cl$_2$C$_6$H$_3$, 2-thienyl	**(DHQD)$_2$AQN** (10 mol%)	50–98^i	86–97 (*S,S*)	[44]

(*Continued*)

Table 3.2.   (*Continued*)

X, R, R^1, LG, PG	Catalyst	Yield (%)	ee (%)	Ref.
X=H, Me, F, R=Ph, 4-MeC$_6$H$_4$, 4-FC$_6$H$_4$, R^1=Ph, 4-NO$_2$C$_6$H$_4$, 4-ClC$_6$H$_4$, 4-FC$_6$H$_4$, 4-MeC$_6$H$_4$, 3-CNC$_6$H$_4$, 3-ClC$_6$H$_4$	**PHOS-III** (20 mol%)	76–93	91–98 major 80–89 minor[j]	[45]

(g)

γ-product   δ-product

| X=H, 4-Me, 4-F, 4-Cl, 4-Br, R=Ph, 2-MeC$_6$H$_4$, 3-MeC$_6$H$_4$, 3-MeOC$_6$H$_4$ | **CIN-IIIa** (20 mol%), TFA (40 mol%) | 53–74[k] | 84–98 (3*S*, α*R*) | [46] |

Notes: [a]This reaction is the key step of a total synthesis of (+)-horsfiline and oxindole is generated *in situ* from its TIPS enol. [b]Total synthesis of (–)-esermethole, (–)-physostigmine, (–)-debromoflustramine B and (–)-phenserine was performed with this method [47]. [c]branched/linear ratio (5:1 to 18:1), dr=4.5:1 to 19:1. *N,N'*-((1*R*,2*R*)-cyclohexane-1,2-diyl)*bis*(4-methoxypicolinamide) gave higher amounts of branched products. [d]Branched/linear ratio (11:1 to >15:1), dr = 4:1 to >19:1. [e]Only linear products were recovered. [f]Branched/linear ratio (19:1), dr = 8:1. Reported as the key step of ascidian natural product (–)-perophoramidine. [g]The configuration depends from the bulkiness of R: if R = linear chains, it is out-of-plain, if R = branched chains or aryl, it is behind-the-plane (Scheme 3.15). Authors explained the different configuration in the surmised mechanism. [h]Linear/branched ratio (9:1 to >19:1). *N*-unprotected oxindoles and electron-donating substituents increased branched product amounts. The products can be converted either to the aldehyde or to a protected alcohol. [i]dr = 1.7:1–11.5:1. *o*-Phenyl substituted MBH carbonates gave the highest dr ratio, but the lowest yield and the lowest dr ratio was found with R = Bn. [j]*anti/syn* ratio (1:1 to 2:1). Authors did not report the absolute configuration of their products, but, comparing the α$_D$ values, the (*S,S*)-isomer should be the most abundant. [k]δ:γ Ratio (5:1 to 10:1). The reaction was scaled up to 3 mmol. The envisaged transition state is reported in Scheme 3.15.

Trost envisaged a mechanism in which the chiral ligand replaces one of the Pd ligand creating a π-allyl palladium chiral complex and the nucleophile attack occurs from the less hindered face. In contrast with the molybdenum catalyst, nucleophiles had to coordinate to the palladium and then a reductive elimination should occur [34]. Moreover, when branched derivatives could be obtained, the nature of the nucleophile might

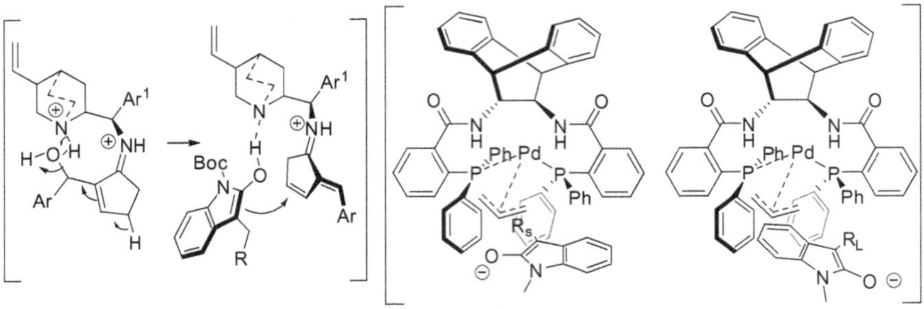

Scheme 3.15. Transition states for MBH alcohols (left) and for decarboxylative asymmetric allylic alkylation (R_s = small alkyl group, R_L = large alkyl or aryl group (right).

influence the equilibrium ratios between *O*-bound and *C*-bound molybdenum enolates, the less crowded *O*-bound enolate favoring the branched product. In particular, larger aryl groups should disfavor the crowded *C*-bound enolate, while more compact five-membered heterocycle substituted oxindoles should accommodate the *C*-bound enolate more readily. Also, the more electron-rich molybdenum complex should favor reductive elimination *via* the less hindered *O*-bound enolate [37]. Among the reactions described in Table 3.2, it should be noted that:

(i) *ter*-Butanol favored the oxindole–hydroxindole tautomerization;
(ii) Strong bases favored the formation of oxindole enolate;
(iii) In contrast with Trost, Liu and Xiao surmised that thiourea moiety created H-bonds with the oxindole shielding a face to the attack of the achiral π-allylic palladium-PPh$_3$ complex [39];
(iv) Benzylation was performed on unprotected oxindoles, while the other allylic alkylation required nitrogen protection [40];
(v) The S$_N$2′-type asymmetric allylic alkylation can efficiently overcome the asymmetric Michael addition of oxindole to acrolein, which is often made complicated by 1,2-addition and polymerization reactions (see also Section 3.5) [43].

The enantioselective α-arylation, α-vinylation, and α-alkynylation of oxindoles (SNAr-like reactions) were also reported. For instance, 3-acyloxindole enolate reacted with allyl 3-chloropropiolate under chiral phase-transfer catalysis by a salt of dihydrocinchonine (**CIN-IV**, Appendix A) [48] (–)-*tert*-Butyl 3-[3-(allyloxy)-3-oxoprop-1-ynyl]-1-methyl-2-oxoindoline-3-carboxylate was produced in 95% yield and 85% ee. Mechanistic investigations allowed supposing an attack from the *Si*-face of the enolate. It should be noted that 3-acyl oxindoles are unstable, thus the 3-acyloxindole was converted before reaction into the corresponding 2-silyl ether and then *in situ* desialylated by 33% KF solution. The reaction was then largely extended under chiral N,N'-dioxide-Sc(III) complex **NNO-I** (0.5–5 mol%, $n = 2$) and 4 Å MS, through an addition–elimination mechanism [49]. β-Halopropiolates and propynones reacted leading to eleven different 3-alkynyl-3-alkyl(aryl)oxindoles in 79–98% yields and 95–99% ee (R).

α-Arylation of *N*-unprotected 3-substituted oxindoles with diaryliodonium salts was performed with Sc(III) salts (Scheme 3.16, Eq. (3.6)) [50]. The reaction was scaled up to a gram scale, but vinylation with vinyliodonium salts failed. Moreover, *m*-substituted 3-benzyloxindoles gave slightly higher enantioselectivities than *o*- or *p*-substituted and the electron-poorer aryl moiety is preferably transferred, with salts having two different aryl groups. Authors also envisaged that the amide moiety of the ligand shielded the *Si*-face of the indole enolate, in order to explain the observed stereochemistry.

Aryl fluoride also reacted in the presence of a phase-transfer catalyst (Scheme 3.16, Eq. (3.7)) [51]. Noteworthy, a high regioselectivity was observed in poly-fluoro electron-deficient arenes, which were always recovered as a single isomer. The incoming aryl group attacked, leading to the configuration depicted in Scheme 3.16, and was demonstrated by X-ray diffraction analysis. Modification of the products, for example reducing the nitro groups to amino groups, did not affect enantiomeric excesses.

An older paper reported instead both the arylation and the vinylation of 3-prochiral oxindoles (Scheme 3.16, Eq. (3.8)) [52]. In particular, *cis*-vinyl bromides gave significantly higher enantioselectivity than *trans*-isomers. The obtained products were also manipulated without affecting enantiomeric excesses.

Only many years later, another vinylation of 3-substituted oxindoles appeared in the literature (Scheme 3.17) [53]. The *Z*-configuration

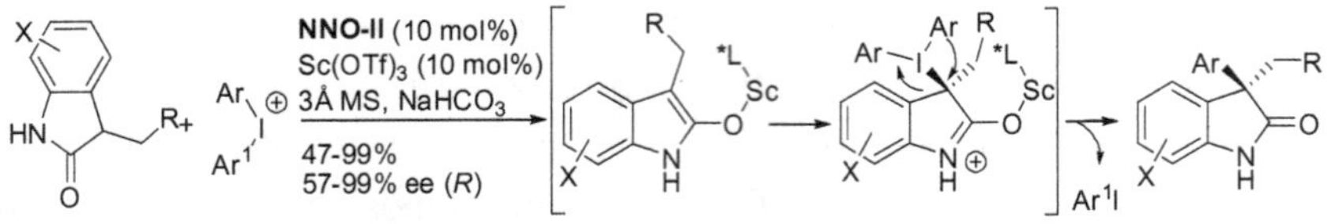

(3.6)

R=Ph, 2-MeC$_6$H$_4$, 2-ClC$_6$H$_4$, 3-MeC$_6$H$_4$, 3-ClC$_6$H$_4$, 3-FC$_6$H$_4$, 3-BrC$_6$H$_4$, 3-NO$_2$C$_6$H$_4$,
  4-MeC$_6$H$_4$, 4-ClC$_6$H$_4$, 4-FC$_6$H$_4$, 4-BrC$_6$H$_4$, 4-NO$_2$C$_6$H$_4$, 4-CNC$_6$H$_4$, 4-MeOC$_6$H$_4$,
  2,4-Cl$_2$C$_6$H$_3$, 2,6-Cl$_2$C$_6$H$_3$, 3,4-(OCH$_2$O)C$_6$H$_3$, 2-Npt, 2-thienyl, *t*-Bu, CH$_2$=CH
X=H, 6-Cl
Ar=Ar^1=4-FC$_6$H$_4$, 4-ClC$_6$H$_4$, 4-*t*-BuC$_6$H$_4$, 4-MeC$_6$H$_4$
Ar=Ph, 4-ClC$_6$H$_4$, 4-BrC$_6$H$_4$, 4-PhC$_6$H$_4$, Ar^1= Mes

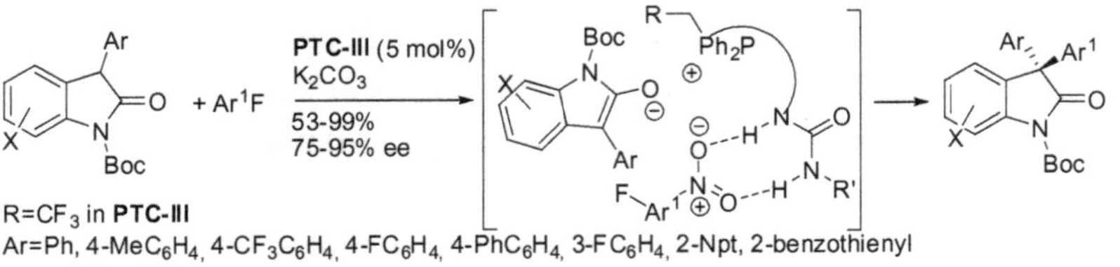

(3.7)

R=CF$_3$ in **PTC-III**
Ar=Ph, 4-MeC$_6$H$_4$, 4-CF$_3$C$_6$H$_4$, 4-FC$_6$H$_4$, 4-PhC$_6$H$_4$, 3-FC$_6$H$_4$, 2-Npt, 2-benzothienyl
X=H, 6-Me, 6-MeO
Ar^1=2,4-(NO$_2$)$_2$C$_6$H$_3$, 2,4-(NO$_2$)$_2$-5-BrC$_6$H$_2$, 2,4-(NO$_2$)$_2$-5-FC$_6$H$_2$, 4,6-difluoropyrimidin-2-yl
R=OMe in **PTC-III**
Ar=Ph, X=5-F, 6-Cl, Ar^1=2,4-(NO$_2$)$_2$C$_6$H$_3$, 4-NO$_2$C$_6$F$_4$

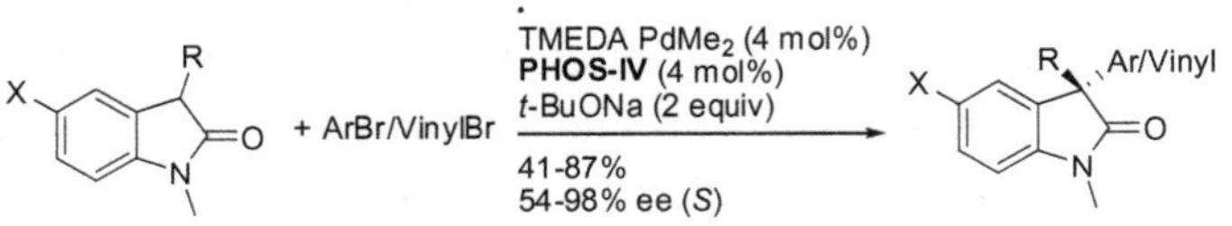

(3.8)

R=Me, Bn
X=H, MeO
Ar=4-MeOC$_6$H$_4$, 4-CF$_3$C$_6$H$_4$, 3-MeOC$_6$H$_4$, 3-ClC$_6$H$_4$, 3-(1,3-dioxolan-2-yl)C$_6$H$_4$,
  3,5-Me$_2$C$_6$H$_3$, 2-Npt, 3 thienyl
Vinyl= (*Z*)-MeCH=CH, (*E*)-MeCH=CH, (*E*)-PhCH=CH, CH$_2$=C(Me)

Scheme 3.16. Asymmetric arylation of 3-prostereogenic oxindoles.

of the starting material was completely maintained in the product, an example with a vinyl chloride gave similar results, but ethyl 3-bromoacrylate afforded the product only in 46% yield with 44% ee. Moreover, from an *E*-configured starting material, the product was recovered only in 59% ee.

Finally, authors gave a tentative mechanistic explanation of the reaction based on an addition–elimination mechanism rather than a simple α-vinylation.

An S$_N$1-type pathway is also possible with stabilized carbocations, as for instance the carbocation generated from Michler's hydrol {*bis* [4-(dimethylamino)phenyl]methanol} (Scheme 3.18, Eq. (3.9)) [54]. Only

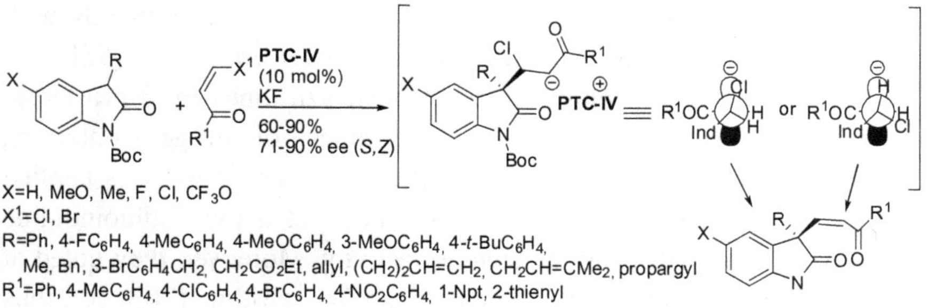
PTC-IV
(10 mol%)
KF
60-90%
71-90% ee (S,Z)
PTC-IV
R¹OC
Ind
or
X=H, MeO, Me, F, Cl, CF3O
X¹=Cl, Br
R=Ph, 4-FC6H4, 4-MeC6H4, 4-MeOC6H4, 3-MeOC6H4, 4-t-BuC6H4,
Me, Bn, 3-BrC6H4CH2, CH2CO2Et, allyl, (CH2)2CH=CH2, CH2CH=CMe2, propargyl
R¹=Ph, 4-MeC6H4, 4-ClC6H4, 4-BrC6H4, 4-NO2C6H4, 1-Npt, 2-thienyl

Scheme 3.17. Asymmetric addition of oxindoles to electron-deficient β-haloalkenes.

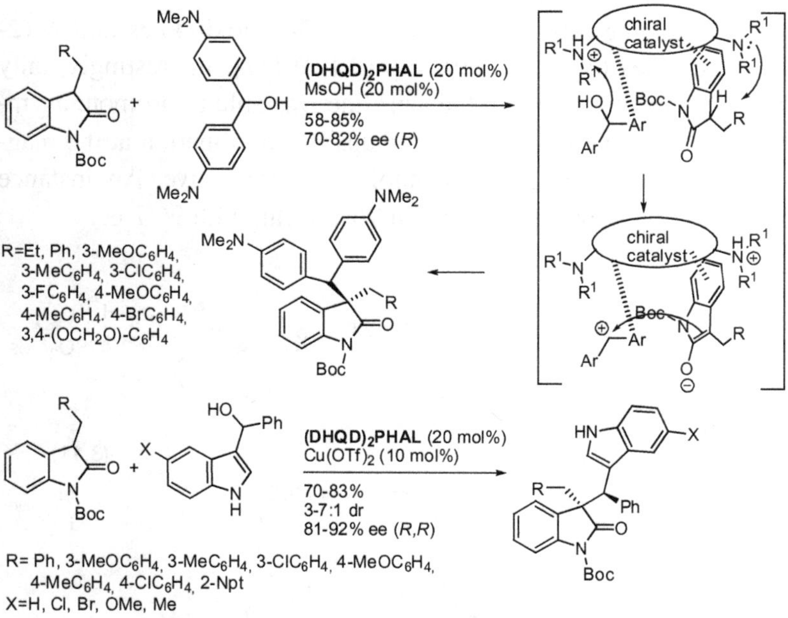
(DHQD)2PHAL (20 mol%)
MsOH (20 mol%)
58-85%
70-82% ee (R)
chiral
catalyst
R=Et, Ph, 3-MeOC6H4,
3-MeC6H4, 3-ClC6H4,
3-FC6H4, 4-MeOC6H4,
4-MeC6H4, 4-BrC6H4,
3,4-(OCH2O)-C6H4
(DHQD)2PHAL (20 mol%)
Cu(OTf)2 (10 mol%)
70-83%
3-7:1 dr
81-92% ee (R,R)
R= Ph, 3-MeOC6H4, 3-MeC6H4, 3-ClC6H4, 4-MeOC6H4,
4-MeC6H4, 4-ClC6H4, 2-Npt
X=H, Cl, Br, OMe, Me

(3.9)

(3.10)

Scheme 3.18. Alkylation of 3-substituted-2-oxindoles with an S_N1-type pathway.

bis-cinchona alkaloids can promote high enantioselectivity. In fact, one tertiary amine moiety formed the oxindole enolate, while simultaneously the other tertiary amine moiety, combined with the acid, gave the carbocation [55]. The acid anion did not influence the stereochemistry, thus a cheaper achiral acid can be used. On the other hand, π-π stacking interactions can further stabilize the transition state. Then, this procedure was applied to the synthesis of *bis*-indolylmethanes, compounds with anti-cancer and inhibition properties (Scheme 3.18, Eq. (3.10)) [55].

Asymmetric ring opening of *N*-activated aziridines by 3-prostereogenic oxindoles is a useful reaction for the synthesis of nitrogen-containing compounds (Scheme 3.19, Eq. (3.11)) [56]. Ooi and co-workers applied then this reaction to the asymmetric synthesis of a pyrrolidinoindoline derivative without affecting the stereoselectivity. Moreover, they noted at the end of the reaction an enrichment of the (*R*)-aziridine (76% ee), thus they deduced the stereochemistry of the product envisaging a (*S*)-aziridine involvement. This is an example of alkylative kinetic resolution; other examples will be reported in Section 3.8.

Another aziridine ring opening involved 3-aryloxindoles and *N*-(2-picolinoyl)aziridines (Scheme 3.19, Eq. (3.12)) [57]. Interestingly, only the pyridine-protecting group in the aziridine was able to promote asymmetrically the ring-opening process, perhaps by coordination at the magnesium center. Curiously, 3-allyloxindole was unreactive. An instance with the (*S*)-BINOL gave the product in 76% yield, with 96% ee.

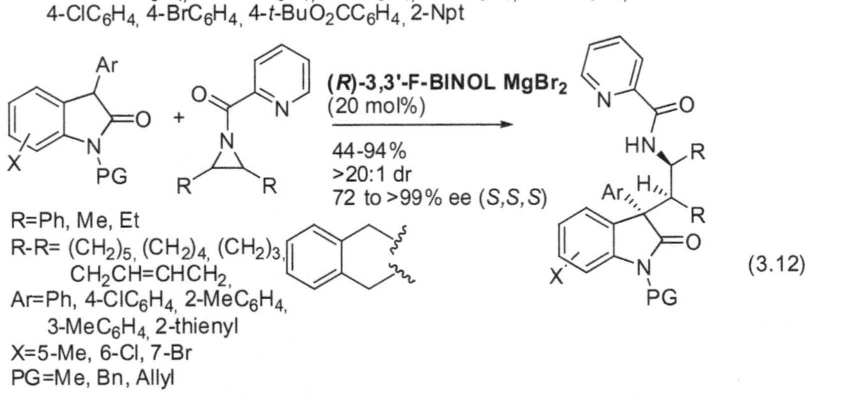

PTC-V (5 mol%), K2CO3
71-98%
11:1 to >20:1 dr
99% ee
2 equiv
(3.11)
R=Me, Et, Bn X=H, Me, OMe, F
Ar=Ph, 3-MeC6H4, 3-MeOC6H4, 3-ClC6H4, 3-BrC6H4, 4-MeC6H4, 4-ClC6H4, 4-BrC6H4, 4-t-BuO2CC6H4, 2-Npt
(R)-3,3'-F-BINOL MgBr2 (20 mol%)
44-94%
>20:1 dr
72 to >99% ee (S,S,S)
(3.12)
R=Ph, Me, Et
R-R= (CH2)5, (CH2)4, (CH2)3, CH2CH=CHCH2,
Ar=Ph, 4-ClC6H4, 2-MeC6H4, 3-MeC6H4, 2-thienyl
X=5-Me, 6-Cl, 7-Br
PG=Me, Bn, Allyl

Scheme 3.19. Asymmetric ringopening of aziridines.

3.3.2. *Indole as the Substrate*

If the 3-prostereogenic indoles have a leaving group, such as in 3-halooxindoles, they can be attacked by a nucleophile, by inversion of the polarity with respect to the syntheses reported in Section 3.3.1. An instance was the enantioselective decarboxylative reaction between β-keto acids and 3-halooxindoles (Scheme 3.20, Eq. (3.13)) [58]. The surmised mechanism supposed a HX elimination by K_3PO_4 and a decarboxylation of β-keto acid promoted by the tertiary amine of the catalyst. Then the hydroxy group of the catalyst coordinated the o-azaxylylene intermediate, arranging a transition state in which the Re-face of the o-azaxylylene is hindered. A similar reaction allowed the introduction of a monofluoroalkyl group (Scheme 3.20, Eq. (3.14)) [59]. This reaction was then applied to the preparation of 3,4′-piperidyl spirooxindoles and hexahydropyrrolo[2,3-b] indolines. In these examples, the chiral catalyst forms a complex with the nucleophile, which attacks the activated o-azaxylylene intermediate bound with secondary interactions.

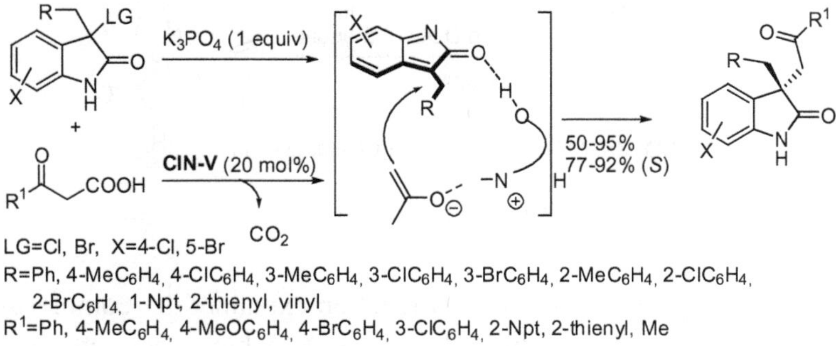

R, LG
K_3PO_4 (1 equiv)
X
N
O
H
CIN-V (20 mol%)
CO2
50-95%
77-92% (S)
(3.13)
R
O, R1
X
N
H
O
R1, COOH

LG=Cl, Br, X=4-Cl, 5-Br
R=Ph, 4-MeC$_6$H$_4$, 4-ClC$_6$H$_4$, 3-MeC$_6$H$_4$, 3-ClC$_6$H$_4$, 3-BrC$_6$H$_4$, 2-MeC$_6$H$_4$, 2-ClC$_6$H$_4$, 2-BrC$_6$H$_4$, 1-Npt, 2-thienyl, vinyl
R^1=Ph, 4-MeC$_6$H$_4$, 4-MeOC$_6$H$_4$, 4-BrC$_6$H$_4$, 3-ClC$_6$H$_4$, 2-Npt, 2-thienyl, Me

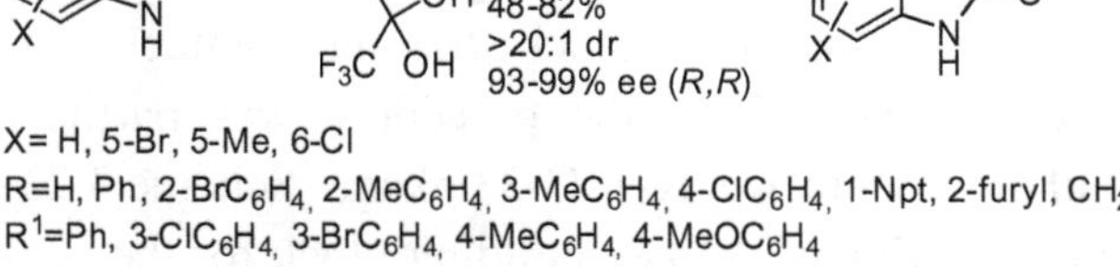

R, Br
X
N
H
O
R1
F
OH
F$_3$C OH
TAK-I (20 mol%),
DIPEA (2.5 equiv)
48-82%
>20:1 dr
93-99% ee (R,R)
(3.14)

X= H, 5-Br, 5-Me, 6-Cl
R=H, Ph, 2-BrC$_6$H$_4$, 2-MeC$_6$H$_4$, 3-MeC$_6$H$_4$, 4-ClC$_6$H$_4$, 1-Npt, 2-furyl, CH$_2$=CH, N$_3$CH$_2$
R^1=Ph, 3-ClC$_6$H$_4$, 3-BrC$_6$H$_4$, 4-MeC$_6$H$_4$, 4-MeOC$_6$H$_4$

Scheme 3.20. Enantioselective elimination reactions between β-keto derivatives and 3-halooxindoles.

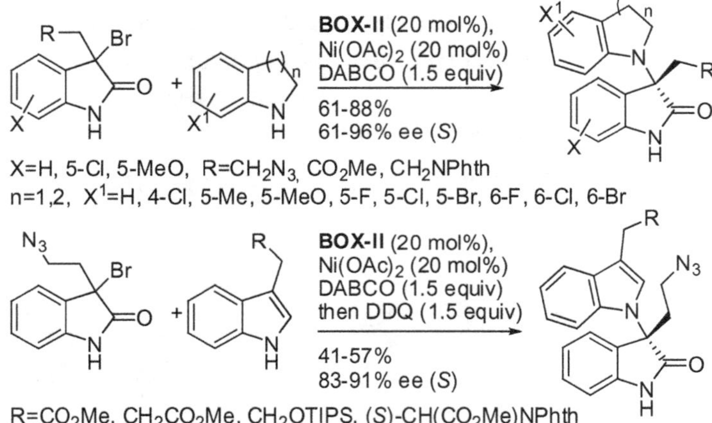

Scheme 3.21.   Asymmetric alkylation reaction of 3-bromooxindoles with 3-substituted indoles.

Scheme 3.22.   Asymmetric construction of the N-1–C-3 linkage.

However, Wang's group reported a reaction explained by complexation of the oxindole by a chiral Lewis acid, which then attacked 3-substituted indoles (Scheme 3.21) [60]. In fact, mass spectrometry gave a base peak compatible with oxindole/catalyst complex.

From this reaction, a key building block for the synthesis of (+)-perophoramidine was obtained. Then the procedure was partially modified in order to build the asymmetric N-1–C-3 linkage (Scheme 3.22) and then applied to the preparation of the key building block for the synthesis of (+)-psychotrimine [61].

Finally, the asymmetric alkylation of 3-halooxindoles with malonates will be discussed. Various 3-alkyl- and 3-arylhalooxindoles were tested

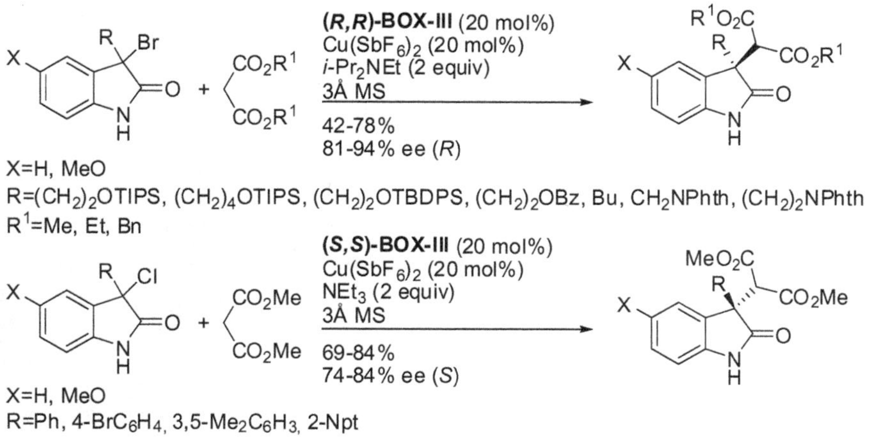

Scheme 3.23.   Asymmetric alkylation of 3-halooxindoles with malonates.

but, curiously, alkyl and aryl derivatives had bromo and chloro as the leaving group, and (*R,R*) or (*S,S*)-catalyst, respectively (Scheme 3.23) [62]. The recovered products were also converted into spirocyclic and fused pyrrolidinoindolines. Then the same research group studied the reaction with dimethyl 2-nitrophenylmalonate with different Cu(II) chiral *bis*(phosphine) complexes, under many different reaction conditions, and the best conditions [ferrocenyl phosphine **FE-I** (Appendix A, 40 mol%), Cu(isobutyrate)$_2$ (20 mol %), *t*-BuOLi (20 mol %), *i*-Pr$_2$NEt (3 equiv.)] afforded the product in 70% ee [63]. The yield of the reaction was not reported, but the racemic reaction was performed in 80% yield.

MBH carbonates derived from isatins are other substrates, which can undergo nucleophilic substitution, and the reported results have been collected in Table 3.3. The products are rich in functionality, thus they can be manipulated mainly into spirooxindoles and pyrroloindolenines.

In the previous section (Scheme 3.18), the S$_N$1-type pathway of 3-substituted-2-oxindoles with stabilized carbocations has been described, but also 3-hydroxy-3-indolyoxindole can generate stable carbocations, which can be attacked by nucleophiles. Shi and co-workers studied this reaction in racemic form, but also tested some asymmetric catalysts with scarce success [64]. Other authors engaged this challenge and their results are collected in Table 3.4. Under acidic conditions, the accepted mechanism contemplates the formation of two isomeric stable chiral ion pairs

Table 3.3.   Asymmetric nucleophilic substitution of MBH carbonates derived from isatins.

X, PG	NuH	Catalyst	Yield (%)	d.r.	ee (%)	Ref.
X=H, F, Cl, MeO PG=Boc	R=2-thienyl. Et, R^1=Et R-R^1=(CH$_2$)$_3$, (CH$_2$)$_4$, (CH$_2$)$_5$, (CH$_2$)$_2$SCH$_2$,	**β-ICD** (10 mol%)	38–98	>99:1	77–88 $(3S,\alpha R)^a$	[65]
X=H, 5-Br, 5-Cl, 5-F, 5-MeO, 7-Cl, 7-F, 5,7-Me$_2$ PG=Me	RCH$_2$NO$_2$ R=H, Me, Et	**β-ICD** (10 mol%)	40–92	2:1 to 13:1	84–98 (R,R) 81–91 $(3R,\alpha S)$	[66]
H, 5-Me, 5-Br, 5-Cl, 5-F, 5-MeO, 5-I, 5-CF$_3$O, 7-F, 5,7-Me$_2$ PG=Me	R=Me, Bu, 4-ClC$_6$H$_4$	**β-IQD/(R)-BINOL** (R=H) (10 mol% each)	70–91	>19:1	80–92 $(3R,\alpha S)$	[67]
H, 5-Me, 5-MeO, 5-CF$_3$O, 5-Br, 5-Cl, 5-F, 5-I, 7-F, 5,7-Me$_2$ PG=Me	R=Me, Et, Pr, Bu, *i*-Bu	**CIN-VI** (10 mol%), 4Å MS	46–96	>19:1	72–86 $(3R,\alpha S)$	[68]

Notes: [a]Drawbacks were found with R–R^1=(CH$_2$)$_3$ (double alkylation). R=*t*-Bu, R^1=H (steric hindrance) and when two non-equivalent enolizable positions are present (scarce regiochemistry).

Table 3.4. S_N1-type pathway reaction of *bis*-indole carbocations.

X, X^1, PG, LG	NuH	Catalyst	Yield (%)	d.r.	ee (%)	Ref.
X=H, 5-Me, 5-Cl, 5-Br, 5-Cl, 5-F, 6-Cl, 7-Cl, X^1=H, 5-MeO, 5-Me, 5-Cl, 5-Br, 6-Me, 6-F, 7-Me, PG=Me, LG=OH	Y=CH$_2$, NCO$_2$Et, CH[O(CH$_2$)$_2$O], O, S, n=1,2,3 PhthN—R R=CO$_2$Et, CO$_2$-*t*-Bu, CN	(**R**)**-BINOL-PA** (Ar=4-MeOC$_6$H$_4$) (10 mol%)	27–95	3.5:1–99:1	74–97 (*R*,*R*)	[69]
X=H, 5-F, 5-Me, X^1=H, 5-Br, 5-MeO, 5-Cl, 5-F, 6-Me, 6-F, PG=H, LG=OH	BocO$_2$CHN R=H, Me, Cl	(**S**)**-BINOL-PA** (Ar=2-Npt) (10 mol%)	70–92[a]	—	90–96 (*R*)	[70]
X=H, 5-Me, 5-F, 5-Cl, X^1=H, 5-F, 5-MeO, 5-Me, 6-Me, PG=H, LG=OH	OPMB CHO	**Pseudo-CIN-IIb** (10 mol%) (**S**)**-BINOL-PA** (Ar=9-anthryl) (30 mol%)	53–89	6:1–10:1	94–98[c] (3*S*,α*R*)	[71]
X=H, 5-Br, 5-Cl, X^1=H, 5-Me, 5-MeO, 5-Br, 7-Me, PG=H, Me, LG=OH	R CHO R=Me, Et, Bn, *n*-C$_6$H$_{13}$	/MeSO$_3$H (10 mol%)	62–98	1:2[b]–2.7:1	56 to >99 (3*R*,α*R*) 67 to >99 (3*S*,α*R*)	[72]

(Continued)

Table 3.4.   (*Continued*)

X, X^1, PG, LG	NuH	Catalyst	Yield (%)	d.r.	ee (%)	Ref.
X=H, 5-Me, 6-Cl, 6-Br, 7-Me, 7-F, X^1=H, 5-MeO, 5-Me, 5-F, 6-Me, 6-Br, 7-Me, PG=Me, Bn, 4-*t*-BuC$_6$H$_4$CH$_2$, 3-MeC$_6$H$_4$CH$_2$, 3-ClC$_6$H$_4$CH$_2$, 3,4-Cl$_2$C$_6$H$_3$CH$_2$, LG=OH	Ar^1=Ph, 4-MeOC$_6$H$_4$, 4-EtOC$_6$H$_4$, 4-*i*-PrOC$_6$H$_4$, 4-FC$_6$H$_4$, 4-ClC$_6$H$_4$, 3,4-(MeO)$_2$C$_6$H$_3$	(***R***)-**BINOL-PA** (Ar=9-anthryl) (10 mol%), 5Å MS	50–99	—	72–88 (*R*)	[73]
X=H, 5-F, 5-Me, 6-Br, 7-F, 7-Me, X^1=H, 5-Me, 5-Cl, 6-Me, 6-F, 6-Br, 7-Me, 7-Br, PG=Bn, Me, Ph, 4-BrC$_6$H$_4$CH$_2$, 4-ClC$_6$H$_4$CH$_2$, 3-ClC$_6$H$_4$CH$_2$, 3,4-Cl$_2$C$_6$H$_3$CH$_2$, LG=OH	R=H, Me, Et, Bn, R^1=H, Me, MeO^d	(***R***)-**H$_8$-BINOL-PA** (Ar=1-Npt) (10 mol%)	44–87	5:1 to >20:1	86–97 (*S,Z*)^e	[74]
X=H, 5-Me, 6-Me, 7-Me, 7-F, X^1=H, 5-Me, 5-F, 5-MeO, 6-Me, 6-F, 6-Br, 7-Br, PG=Me, *i*-Pr, Bn, Ph^f, LG=OH	Y=H, 4-Me, 4-F, 5-MeO, 5-F, 5-Br, 6-MeO, 5-F, 5-Br, 6-Me, 6-F, 7-Br	(***R***)-**BINOL-PA** (Ar=4-ClC$_6$H$_4$) (10 mol%)	80–99	—	64–82 (*R*)^g	[75]

(*Continued*)

Table 3.4. (*Continued*)

X, X^1, PG, LG	NuH	Catalyst	Yield (%)	d.r.	ee (%)	Ref.
X=H, 5-F, X^1=H, 5-F, 5-Cl, 5-Br PG=Bn^h, LG=OH	R=Me, Ph, *c*-C$_6$H$_{11}$, 2-MeC$_6$H$_4$, 3-MeC$_6$H$_4$, 3-ClC$_6$H$_4$, 4-MeC$_6$H$_4$, 4-MeOC$_6$H$_4$, 4-FC$_6$H$_4$, 4-ClC$_6$H$_4$, 4-BrC$_6$H$_4$, 2-Npt, 2-furyl	(*S*)- BINOL-PA {Ar=4-[3,5-(CF$_3$)$_2$-C$_6$H$_3$]C$_6$H$_4$} (10 mol%), Na$_2$SO$_4$	71–94	—	84–99 (*R*)	[76]
X=H, 5-MeO, 5-Me, 6-Br, X^1=H, 5-MeO, 5-Br, 5-Me, 6-Me, 6-Br, 7-Me, 7-Br, PG=Me, LG=OH	Y=H, 5-MeO, 5-Me, 5-Br, 5-NO$_2$, 6-Br, 7-Br, 7-MeO	PHOS-V (0.5 mol%), 5Å MS	75 to >99^i	—	83–97 (*R*)	[77]
X=H, 5-MeO, 5-Me, 5-Br, 5-Cl, 7-Br, X^1=H, 5-MeO, 5-Br, 5-Me, 6-Br, 7-Br, PG=Me, LG=OH		H$_{16}$-PHOS-V (2.5 mol%), hydroxypropyl-β-cyclodextrin (1.5 mol%)	80 to >99	—	85–98 (*R*)^j	

(Continued)

Table 3.4. (*Continued*)

X, X^1, PG, LG	NuH	Catalyst	Yield (%)	d.r.	ee (%)	Ref.
X=H, 5-F, 5-Cl, 5-Br, 5-Me, 5-MeO, 6-Br, 7-Br, 7-CF$_3$, X^1=H, 5-Cl, 5-Br, 5-Me, 5-MeO,6-F, 6-Cl, 7-Me, 2-Me, PG=Bn, LG=Ts	R^1, R^2 with NO$_2$, CO$_2$Bn, CO$_2$Bn; R^1=R^2=H, (CH$_2$)$_4$, R^1=Me, R^2=H	**CIN-IIa** (10 mol%), K$_3$PO$_4$ (1 equiv.)	60–98	—	78–99 (*S*)^j	[78]
X=H, 4-Br, 5-MeO, 5-Me, 5-Cl, X^1=H, 5-Br, 5-Me, 6-Me, 7-Me, PG=Me, Bn, LG= PhSO$_2$	naphthalen-2-ol with OH; Y=H, 6-Br, 7-Br	**CIN-VIIa** (10 mol%), K$_3$PO$_4$ (2 equiv.)	39–99	—	18–98 (*R*)	[79]

Notes: ^{a}Isolated as *bis*-indolylketones. DFT calculation demonstrated that the *cis* was more stable than the *trans* isomer of the dehydrated *bis*-indole. The synthesis of (+)-folicanthine was also performed. ^{b}The stereochemistry was determined because one of the products was converted into (+)-gliocladin C (19% overall yield, 12 steps). It should be noted that (*R*)-phosphoric acid completely inhibited the catalytic activity of the amine. ^{c}(3*R*,α*R*):(3*S*,α*R*) ratio, X=7-Me favors (3*S*,α*R*). ^{d}The presence of the OH group was not decisive for the reaction occurrence, since the desired product was also generated with 2-MeO-substituted styrene. ^{e}See Scheme 3.24. ^{f}When X=4-Br, X^1=H, PG=Bn, the yield was only 57% and ee 2%. ^{g}Control experiments demonstrated that the N–H group in the indole moiety of 3-indolylmethanol was important in obtaining good enantioselectivity, two N–H groups on the isatin moiety disturb the hydrogen-bond formation between 3-methylindole and the catalyst and the N–H group of 3-methylindole is crucial for hydrogen-bonding interaction with the catalyst. ^{h}The reaction released CO$_2$. A key intermediate for the synthesis of (+)-folicanthine was also obtained. ^{i}Performed also in a gram-scale in >99% yield and 94% ee. ^{j}This reaction was also applied to the enantioselective total synthesis of (+)-gliocladin C.

between the catalyst and the two isomeric cationic vinyliminium intermediates (see Table 3.4 head scheme and Scheme 3.24). Then, the nucleophile attacks the most stable ion pair at the less hindered face. *N*,*N*′-Protected 3-hydroxy-3-indolyloxindoles did not undergo any reaction, thus some authors envisaged a likely deprotonation of the vinyliminium intermediate [69, 73]. Other authors reported instead that the N–H group of the indole moiety was essential for the reaction occurrence.

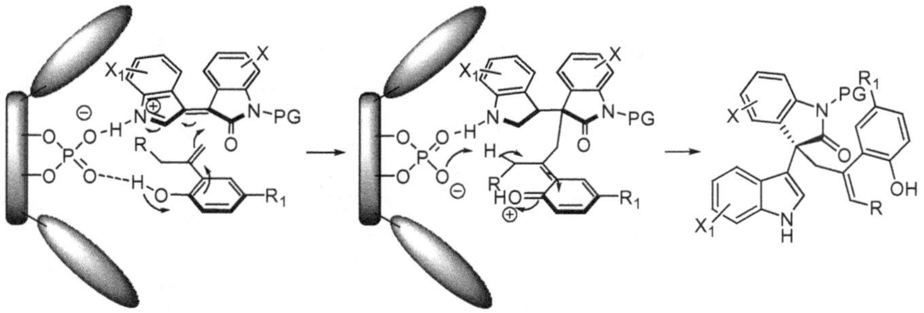

Scheme 3.24. Proposed mechanisms for the reaction of 3-hydroxy-3-indolyloxindole with nucleophiles.

3.4. Aldol Reactions

Conversely, from nucleophilic substitution and Michael addition reactions, in the aldol and Mannich reactions, 3,3-dialkyloxindoles can be obtained only from 3-alkyl/aryl prostereogenic oxindoles. In fact, reaction with isatin derivative should lead to 3-hydroxy- or 3-aminooxindoles and these reactions will be described in the following chapters.

The first example of asymmetric aldol reaction leading to 3,3-disubstituted oxindoles involved Garner's aldehyde or (R)-glyceraldehyde acetonide and 3-substituted-2-siloxyindoles, an example of chiral pool reaction (Scheme 3.25) [80].

In 2007, Shibata and Toru reported the first organocatalyzed synthesis of both enantiomers of 3,3-dialkyloxindoles starting from oxindole by the reaction with ethyl trifluoropyruvate (Scheme 3.26, Eq. (3.15)) [81].

The merit of this reaction was to prove that a chiral Brønsted base can activate oxindole derivatives having pK_a values above 18, thus opening new scenarios largely exploited later, as widely described in this book. Only 3-ethyl-2-oxindole gave low yield (37%). The CF_3 group was essential, because low conversion and enantioselectivity were observed with ethyl pyruvate. Later, Bencivenni and Bartoli found that products arising from the reaction of N-Boc-oxindoles with ethyl glyoxylate under very similar reaction conditions rapidly epimerize in the presence of protic sources [82]. Very likely, the CF_3 group prevented this by-reaction.

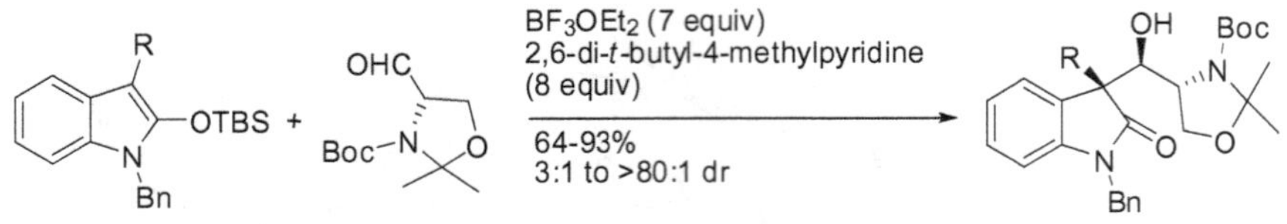

R=Bn, 4-MeOC$_6$H$_4$, 3,4-(MeO)$_2$C$_6$H$_3$, 3,4-(OCH$_2$O)C$_6$H$_3$, CH$_2$=CMe, 1-benzylindol-3yl

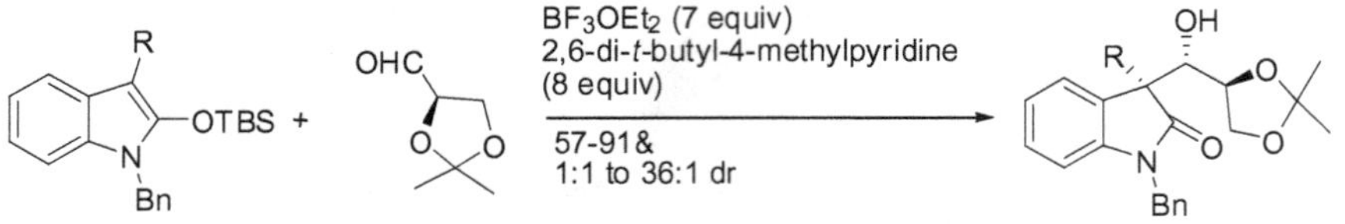

R=Bn, *i*-Pr, 4-MeOC$_6$H$_4$, 3,4-(MeO)$_2$C$_6$H$_3$, 3,4-(OCH$_2$O)C$_6$H$_3$, 1-benzylindol-3yl

Scheme 3.25. Mukaiyama aldol reactions of 2-siloxyindoles with chiral aldehydes.

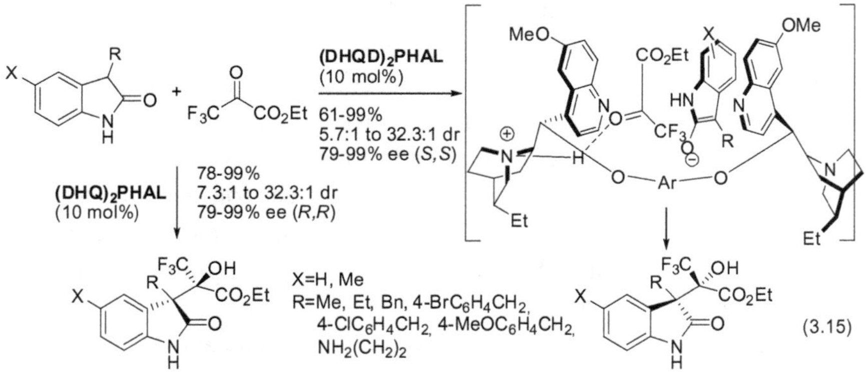

(3.15)

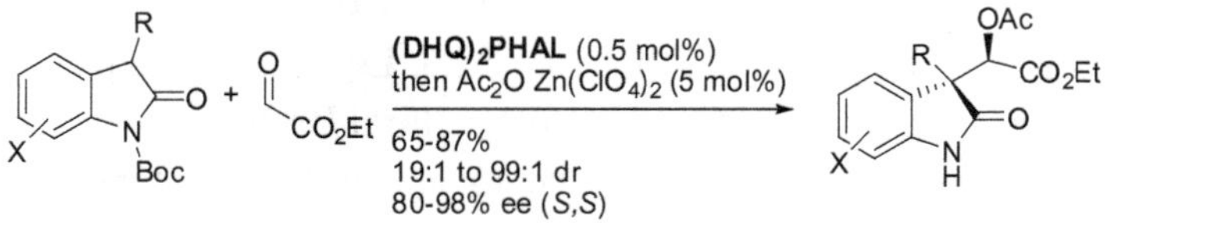

(3.16)

X=H, 5-Me, 5-F, 6-Cl, 7-F, 5,7-Me$_2$
R=Me, Bu, *i*-Bu, Bn, 4-ClC$_6$H$_4$CH$_2$, 4-CF$_3$C$_6$H$_4$CH$_2$, 4-MeOC$_6$H$_4$CH$_2$
  3-MeOC$_6$H$_4$CH$_2$

Scheme 3.26. Enantioselective reaction of oxindoles with ethyl α-ketoesters.

However, *in situ* reaction with acetic anhydride and Zn(ClO$_4$)$_2$ blocked the hydroxy function (thus preventing epimerization) and removed *N*-Boc protecting group (Scheme 3.26, Eq. (3.16)). (*S,S*)-Enantiomers could be

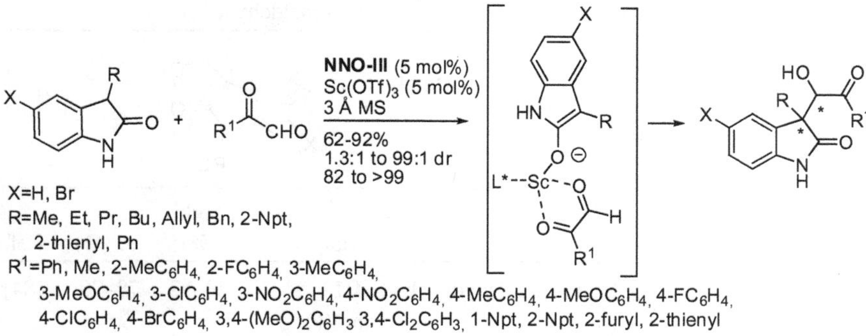

Scheme 3.27. Chiral Sc(III)-catalyzed direct aldol-type reaction.

recovered in comparable yield, diastereo- and enantioselectivity with **(DHQD)₂PHAL**. Two features have to be mentioned, the very low catalyst loading and the use of commercially available polymeric ethyl glyoxylate, which release low concentration of monomer under the reaction conditions controlling its high reactivity.

The reaction of glyoxals was performed under chiral scandium complexes (Scheme 3.27) [83]. Unfortunately, the configuration of the recovered products was not reported. However, the reaction with ethyl trifluoropyruvate was reported to give the (*R*,*R*)-product in 93% yield with 3:1 d.r. and 91% ee. The reaction worked even with 1 mol% catalyst loading in a 10 mmol-scale without affecting enantioselectivity. The source of enantioselectivity was ascribed to a chiral scandium enolate of oxindole chelated to ketoester.

Paraformaldehyde, which released low concentration of the high reactive formaldehyde, is another polymeric commercially available source of carbonyl compounds. *N*-Boc protection of oxindoles was crucial for the enantioselectivity. In addition, the hydroxymethylation in water was extensively studied by using commercially available formalin solution (Table 3.5).

Bifunctional catalysts allowed the deprotonation of the oxindole resulting in a structured chiral ion pair, which then activated formaldehyde by hydrogen-bonding interactions. On the other hand in the metal catalyzed reactions, the transition state is expected to have both the carbonyl

Table 3.5. Aldol reactions with formaldehyde.

X, PG, R	Formaldehyde source	Catalyst	Yield (%)	ee (%)	Ref.
X=H, 5-Br, 5-Me, PG=Boc, R=Bn, Bu, Pr, Me, $4\text{-ClC}_6\text{H}_4\text{CH}_2$, $4\text{-FC}_6\text{H}_4\text{CH}_2$, $4\text{-MeC}_6\text{H}_4\text{CH}_2$, $4\text{-MeOC}_6\text{H}_4\text{CH}_2$, $3\text{-MeOC}_6\text{H}_4\text{CH}_2$, $3,4\text{-(MeO)}_2\text{C}_6\text{H}_3\text{CH}_2$, 1-NptCH_2, 2-thienylCH_2	Paraformaldehyde	**TAK-III** (5 mol%)	80–98	79–90[a]	[84]
X=H, 5-MeO, PG=Boc, $R=\text{TBSO(CH}_2)_2$, $t\text{-BuO}_2\text{CCH}_2$, EtO_2CCH_2, Ts(Bn) $\text{N(CH}_2)_2$, Alloc(Bn) $\text{N(CH}_2)_2$, Moc(Bn) $\text{N(CH}_2)_2$, Ts(Me) $\text{N(CH}_2)_2$, Moc(Me) $\text{N(CH}_2)_2$, Cbz(Me) $\text{N(CH}_2)_2$, $\text{SucN(CH}_2)_2$, $\text{PhthN(CH}_2)_2$	Paraformaldehyde	**Pseudo-CIN-II c** (10 mol%)	81–95	77–97[b]	[85]
X=H, 5-Me, 5-MeO, 5-F, 7-F, PG=Boc, R=Me, Pr, *i*-Bu, Ph, Bn, $4\text{-MeC}_6\text{H}_4\text{CH}_2$, $4\text{-FC}_6\text{H}_4\text{CH}_2$, $4\text{-MeOC}_6\text{H}_4\text{CH}_2$	Paraformaldehyde	**PHOS-VI** (5 mol%)	81–98	30–94	[86]

(*Continued*)

Table 3.5.   (*Continued*)

X, PG, R	Formaldehyde source	Catalyst	Yield (%)	ee (%)	Ref.
X=H, PG=Me, R= 1-pyrrolyl	Paraformaldehyde	**CIN-VIIIa** (5 mol%)	97	89[c]	[87]
X=H, Br, PG=H, R=Me, Et, Pr, *i*-Pr, Bu, Allyl, Propargyl, Ph, Bn, 2-MeC$_6$H$_4$CH$_2$, 2-ClC$_6$H$_4$CH$_2$, 3-PhOC$_6$H$_4$CH$_2$, 4-MeOC$_6$H$_4$CH$_2$, 4-ClC$_6$H$_4$CH$_2$, 4-BrC$_6$H$_4$CH$_2$, 3,4-(OCH$_2$O)C$_6$H$_3$CH$_2$ 2,4-Cl$_2$C$_6$H$_3$CH$_2$, 1-NptCH$_2$, 2-NptCH$_2$, 2-PyCH$_2$, 2-thienylCH$_2$	Formalin	**NNO-IVa** (2 mol%), Nd(OTf)$_3$ (2 mol%), 3 Å MS	60–93	87≥99[d]	[88]
X=H, PG=Boc, R=Me	Formalin	**BOX-IV** (10 mol%), Sc(C$_{12}$H$_{25}$SO$_4$)$_3$ (10 mol%) C$_{12}$H$_{25}$SO$_4$Na (1.5 equiv)	59	82[c]	[89]

Notes: [a]3-Phenyloxindole gave an almost racemic mixture. (+)-Esermethole and (+)-physostigmine were prepared. [b]Spiro(pyrrolidinyloxindole) and hexahydropyrrolo[2,3-b]indole alkaloids were prepared. [c]Configuration was not reported. [d]The catalyst was easily recovered and reused with no loss in catalytic activity and enantioselectivity. Physostigmine and coerulescine were prepared.

and *N*-oxide oxygens of the ligand, the oxindole enolate oxygen and the formaldehyde coordinated to the metal ion. Then predominant attack from the *Re* face led to the *S*-configured product in all the reported procedures.

Finally, benzyl-protected isatin gave aldol condensation with 3-prostereogenic oxindoles in 81–88% yields as a single diastereomer, but only with less than 5% ee [90].

3.5. Mannich Reactions

The Mannich reaction of 3-prostereogenic oxindoles is often a valuable alternative to the aldol reaction. The first example of an asymmetric Mannich reaction was reported by the Chen research group in 2008 (Scheme 3.28, Eq. (3.17)) [91]. The source of the stereochemistry should be four hydrogen-bonding interactions between the two N–H of thiourea and the carbonyl of protected imine and between the protonated tertiary amine and the two carbonyls of the oxindole. More acidic 3-phenyloxindole gave rise to a racemic product owing to a fast uncatalyzed reaction. However, under phase-transfer catalysis with chiral **PHOS-VII** salt the Mannich product from 3-aryloxindoles could be obtained in 96–99% yields as practically only a diastereomer with 75–88% ee [92], but the configuration was not determined and it cannot be deduced by comparison with other papers cited in this section.

Other *syn*-Mannich reactions have been reported by Kobayashi and Ooi using a chiral calcium catalyst (Scheme 3.28, Eq. (3.18)) [93], and chiral ammonium betaine catalyst (Scheme 3.28, Eq. (3.19)) [94], respectively. In the former reaction the catalyst being non-basic, oxindole deprotonation was performed by triethylamine and triethylammonium salt protonated the intermediate that led to the product regenerating the catalyst. Moreover, the products could be readily converted to spirooxindoles without loss of optical purity. The latter reaction was only applied to 3-aryloxindoles.

Then two Mannich reactions of 3-substituted-2-oxindoles with *N*-Boc ketimines of isatin were reported in the same paper, already cited for aldol condensation [87, 90].

In particular, the Mannich reaction of 3-pyrrolyl-*N*-methyloxindole with *N*-Boc ketimine of *N*-benzylisatin led to the corresponding product in 93% yield with 95% ee [87]. The reaction under **(DHQD)₂PHAL** catalysis was studied in more detail (Scheme 3.29) [90].

Similar products were obtained by a four-component cascade reaction involving C–H functionalization/Mannich reaction of indoles co-catalyzed by a rhodium(II)/chiral phosphoric acid couple (Scheme 3.30) [95].

The metal complex gave a carbenoid species catalyzing the first step, and then the chiral-PA allowed the formation of an *E*-configured imino ester.

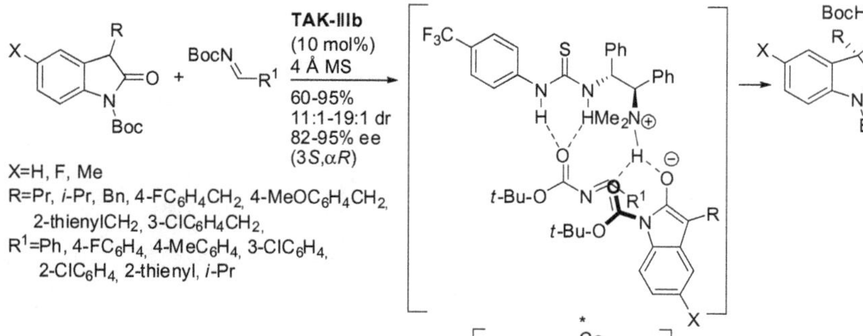

TAK-IIIb
(10 mol%)
4 Å MS
60-95%
11:1-19:1 dr
82-95% ee
(3S,αR)
X=H, F, Me
R=Pr, i-Pr, Bn, 4-FC6H4CH2, 4-MeOC6H4CH2, 2-thienylCH2, 3-ClC6H4CH2,
R1=Ph, 4-FC6H4, 4-MeC6H4, 3-ClC6H4, 2-ClC6H4, 2-thienyl, i-Pr

(3.17)

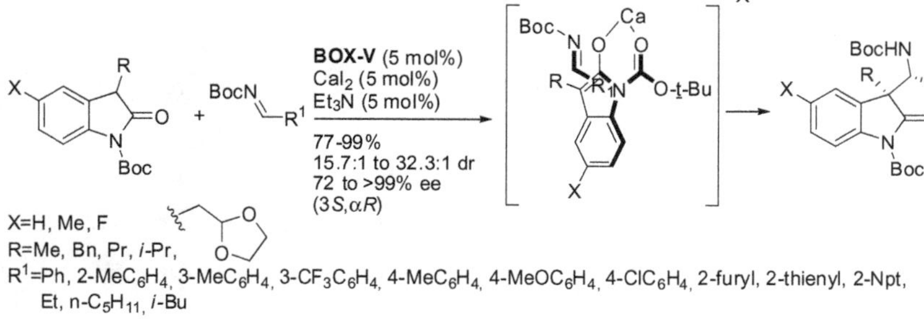

BOX-V (5 mol%)
CaI2 (5 mol%)
Et3N (5 mol%)
77-99%
15.7:1 to 32.3:1 dr
72 to >99% ee
(3S,αR)
X=H, Me, F
R=Me, Bn, Pr, i-Pr,
R1=Ph, 2-MeC6H4, 3-MeC6H4, 3-CF3C6H4, 4-MeC6H4, 4-MeOC6H4, 4-ClC6H4, 2-furyl, 2-thienyl, 2-Npt, Et, n-C5H11, i-Bu

(3.18)

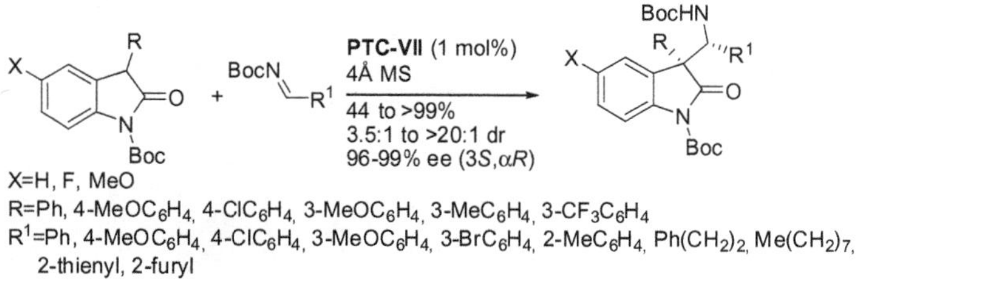

PTC-VII (1 mol%)
4Å MS
44 to >99%
3.5:1 to >20:1 dr
96-99% ee (3S,αR)
X=H, F, MeO
R=Ph, 4-MeOC6H4, 4-ClC6H4, 3-MeOC6H4, 3-MeC6H4, 3-CF3C6H4
R1=Ph, 4-MeOC6H4, 4-ClC6H4, 3-MeOC6H4, 3-BrC6H4, 2-MeC6H4, Ph(CH2)2, Me(CH2)7, 2-thienyl, 2-furyl

(3.19)

Scheme 3.28. Organocatalytic asymmetric *syn*-Mannich reaction.

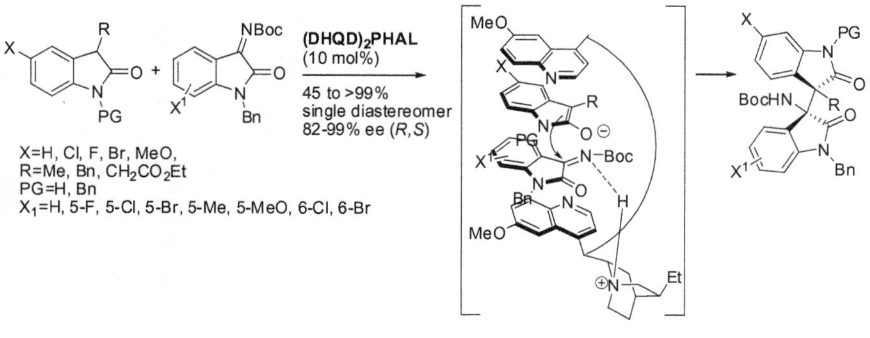

(DHQD)2PHAL
(10 mol%)
45 to >99%
single diastereomer
82-99% ee (R,S)
X=H, Cl, F, Br, MeO,
R=Me, Bn, CH2CO2Et
PG=H, Bn
X1=H, 5-F, 5-Cl, 5-Br, 5-Me, 5-MeO, 6-Cl, 6-Br

Scheme 3.29. Enantioselective addition of 3-substituted-2-oxindoles to *N*-Boc-ketimines from isatin.

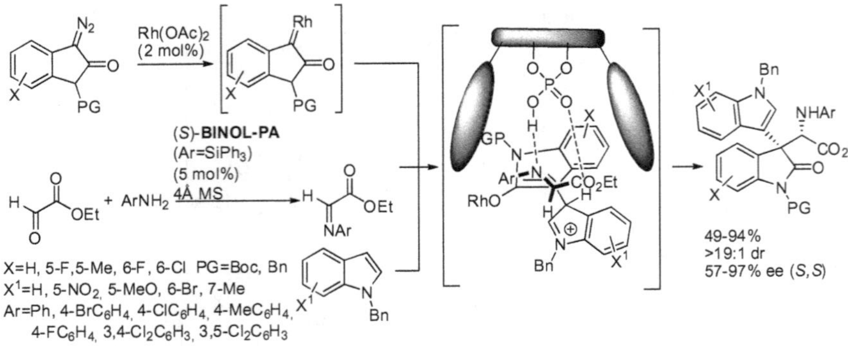

Scheme 3.30.   Four-component C–H functionalization/Mannich reaction.

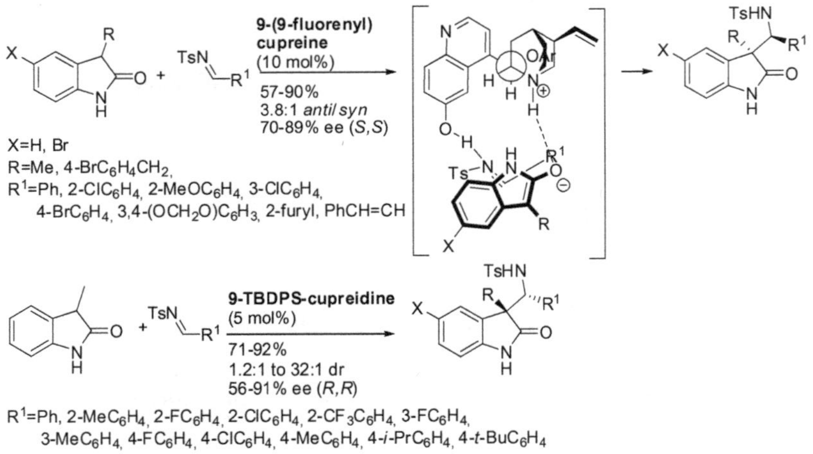

Scheme 3.31.   Organocatalytic asymmetric *anti*-Mannich reaction of oxindoles.

Finally, the chiral-PA assembled a transition state in which the sterically favored configuration had the ester group between the two indole nuclei.

The less common *anti*-Mannich products were obtained from *N*-Ts-imines (Scheme 3.31, Eq. (3.20)) [96]. High selectivity was ensured by a bulky group on C-9 position of the catalyst, but *N*-protected indoles crowded the transition state and prevented selectivity. On the other hand, Zhang and co-workers reported the asymmetric *anti*-Mannich reaction to prepare the (*R*,*R*)-enantiomer (Scheme 3.31, Eq. (3.21)) [97]. Besides the already described influence of the 9-OH group in assembling the transition state, this paper also described, but did not explain an influence of a free hydroxy group at the 6′-position on the stereoinduction.

3.6. Michael Reactions

3.6.1. *Indole as the nucleophile*

Michael additions of oxindoles to α,β-unsaturated carbonyl compounds, such as aldehydes, ketones, esters, maleimides, and quinones, are involved in many reactions leading to the formation of quaternary stereocenters at the C-3-position of the oxindole. Examples are described almost all over this book. In particular, this section will be devoted to the construction of an open-chain dialkyl derivative.

The year 2009 was very fruitful for the discovery of the Michael addition to aldehydes. In that year, Maruoka reported the reaction in the already above cited paper for the Mannich reaction (Scheme 3.32, Eq. (3.22)) [92]. In the same year, Bartoli and Melchiorre reported the catalytic asymmetric Michael reaction of oxindoles to substituted cinnamaldehydes by iminium-ion catalysis (Scheme 3.32, Eq. (3.23)) [98]. Moreover, Moyano and Rios extended the reaction to aliphatic aldehydes by the use of **JHC** catalyst, which worked badly with cinnamaldehydes. Unfortunately, they did not determine the configuration (Scheme 3.32, Eq. (3.24)) [99]. Authors also found N-Boc oxindoles gave a fast equilibrium with the product thus leading to a racemic mixture.

After these pioneering works, many other additions of 3-prostereogenic oxindoles to unsaturated aldehydes and ketones were reported in the literature and they are collected in Table 3.6.

In this table, the enantioselective tandem Michael additions to 1,4-quinones are also reported. It should be noted that excess of quinone immediately oxidized the intermediate biphenol, which was never detected. "Phase-transfer", "covalent" or "non-covalent" organocatalytic strategies as well as chiral metal complex catalysts are known. Generally, organocatalysis is foreseen as responsible for the enantio-discrimination hydrogen-bond interactions between the tight ion-pair oxindole enolate-catalyst and carbonyl or iminium ion of the carbonyl groups.

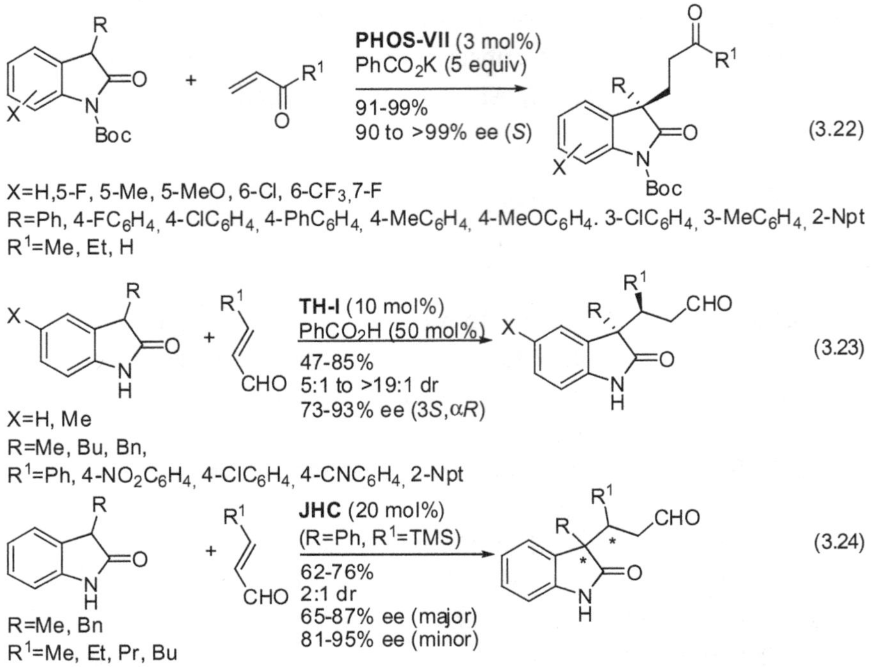

PHOS-VII (3 mol%)
PhCO2K (5 equiv)
91-99%
90 to >99% ee (S)
(3.22)
X=H,5-F, 5-Me, 5-MeO, 6-Cl, 6-CF3,7-F
R=Ph, 4-FC6H4, 4-ClC6H4, 4-PhC6H4, 4-MeC6H4, 4-MeOC6H4. 3-ClC6H4, 3-MeC6H4, 2-Npt
R1=Me, Et, H
TH-I (10 mol%)
PhCO2H (50 mol%)
47-85%
5:1 to >19:1 dr
73-93% ee (3S,αR)
(3.23)
X=H, Me
R=Me, Bu, Bn,
R1=Ph, 4-NO2C6H4, 4-ClC6H4, 4-CNC6H4, 2-Npt
JHC (20 mol%)
(R=Ph, R1=TMS)
62-76%
2:1 dr
65-87% ee (major)
81-95% ee (minor)
(3.24)
R=Me, Bn
R1=Me, Et, Pr, Bu

Scheme 3.32. Asymmetric Michael addition of unsaturated vinyl ketones to oxindoles.

On the other hand, the metal ion is able to link in its coordination sphere both oxindole enolate and the carbonyl compound. Since 3-alkyl-substituted oxindoles have lower pK_a with respect to 3-aryloxindoles, they are less reactive and scarcely represented. In most cases, 3-benzyloxindoles are employed as a model for 3-alkyl-substituted oxindoles and only few catalytic systems can be successfully applied to both aryl- and alkyl-substituted substrates.

Moreover, Michael addition of 3-prostereogenic oxindoles to (*E*)-1,4-diaryl-2-buten-1,4-diones was also performed (Scheme 3.33) [100]. The products were efficiently converted into furans by Paal–Knorr reaction without affecting the enantioselectivity.

The Michael addition to electron-deficient alkynes has been also reported under chiral *N,N′*-dioxide-Sc(III) complexes. In particular, pyrrolidinyl and pyperidinyl catalysts were better catalysts in the 3-aryl- and

Table 3.6.   Asymmetric Michael additions of 3-prostereogenic indoles to carbonyl derivatives.

(*a*)

R, X, PG	R^1, R^2	Catalyst	Yield (%)	d.r.	ee (%)	Ref.
X=H, R=Ph, 4-MeOC$_6$H$_4$, 4-MeC$_6$H$_4$, 3,5-Me$_2$C$_6$H$_3$, Me PG=Boc	R^1=Me, Et, Ph, R^2=H	**TAK-III** (20 mol%), 4Å MS	90–99[a]	—	46–88 (R=Ar) 17–48 (R=Me) (*S*)	[101]
X=H, 6-Cl, R=Bn, Me, 3-ClC$_6$H$_4$CH$_2$, Ph, PG=Boc	R^1-R^2=(CH$_2$)$_2$, (CH$_2$)$_3$, (CH$_2$)$_4$, CH$_2$C(Me$_2$)-CH$_2$, R^1=Me, R^2=Ph	**CIN-IXa** (10 mol%), PhCO$_2$H (20 mol%)	30–95	1.5:1–6:1	46–98 (3*R*,α*S*)	[102]
X=PG=H, R=Bn, 2-MeOC$_6$H$_4$CH$_2$, 3-ClC$_6$H$_4$CH$_2$, 4-ClC$_6$H$_4$CH$_2$, Pr	R^1-R^2=(CH$_2$)$_4$, R^1=Me, Et, R^2=Ph, 2-MeOC$_6$H$_4$, 3-MeOC$_6$H$_4$, 3-ClC$_6$H$_4$, 4-MeOC$_6$H$_4$, 4-FC$_6$H$_4$, 4-ClC$_6$H$_4$, Pr, *n*-C$_5$H$_{11}$	**CIN-IXb** (10 mol%), ***N*-Boc-d-Phg** (10 mol%)	71–99	2:1	94 to >99 (*S*,*S*)[b]	[103]
X=H, R=H, Me, Bn, 4-BrC$_6$H$_4$CH$_2$, 4-NO$_2$C$_6$H$_4$CH$_2$, 4-MeOC$_6$H$_4$CH$_2$, MeO$_2$CCH$_2$, PG=H, Me, Ph, Cbz	R^1-R^2=(CH$_2$)$_2$, (CH$_2$)$_3$	**L-Proline** (30 mol%) *trans*-2,5-Me$_2$piperazine (30 mol%)	63–99	1:1–2:1	5–91[b,c] (3*R*,α*S*)	[104]

(*Continued*)

Table 3.6.   (*Continued*)

R, X, PG	R^1, R^2	Catalyst	Yield (%)	d.r.	ee (%)	Ref.
X=H, 5-F, R=Ph, 4-MeC$_6$H$_4$, 2-Npt, PG=Boc	R^1=Me, R^2=H	**PHOS-VIII** (2.5 mol%)	95–97	—	90–95 (*S*)	[105]
X=H, R=Bn, Ph, Me, *i*-Bu, allyl, 4-MeOC$_6$H$_4$CH$_2$, 4-FC$_6$H$_4$CH$_2$, 4-ClC$_6$H$_4$CH$_2$, 4-BrC$_6$H$_4$CH$_2$, 3-MeC$_6$H$_4$, 3-MeOC$_6$H$_4$, 4-MeC$_6$H$_4$, 4-FC$_6$H$_4$, 2-Npt, PG=Boc	R^1=Me, Et, R^2=H	**TH-II** (5 mol%)	76–92	—	52–97 (*R*)[d]	[106]
X=H, 5-Me, 5-F, 7-F, 7-Cl, 5,7-Me$_2$, R=Ph, 4-FC$_6$H$_4$, 4-MeOC$_6$H$_4$, 4-*t*-BuC$_6$H$_4$, 4-PhC$_6$H$_4$, 3-MeC$_6$H$_4$, 2-Npt, Bn, *i*-Bu, *n*-C$_7$H$_{15}$ 3-MeOC$_6$H$_4$CH$_2$, 2-NptCH$_2$, PG=Boc	R^1=Me, Et, Ph, H, R^2=H	**PHOS-IX a-c** (10 mol%)[e]	75–97	—	71–96 (*R*)	[107]
X=H, 5-Me, 5-MeO,5-F, 5-Cl, R=Ph, Bn, 4- MeC$_6$H$_4$, CNCH$_2$, EtO$_2$CCH$_2$, *t*-BuO$_2$CCH$_2$, EtO$_2$CNH(CH$_2$)$_2$, PG=Boc	R^1=Me, H[f], R^2=H	**PHOS-X** (10 mol%) K$_2$CO$_3$ (5 equiv)	74–99	—	48–94 (*R*)[g]	[108]
X=H, 5-Me, 5-F, 5-MeO, R=Ph, 4-FC$_6$H$_4$, 4-MeC$_6$H$_4$, 3-FC$_6$H$_4$, 3-CF$_3$C$_6$H$_4$, 2-Npt, PG=Boc	R^1=H, Me, Et R^2=H	**PHOS-XI** (0.1 mol%) for acrolein, (1 mol%) for ketones	89–99	—	88–91 (*S*)	[109]

X=H, 5-F, 5-OMe, 6-Br, 7-F, R=Bn, 4-CNC$_6$H$_4$CH$_2$, 4-FC$_6$H$_4$CH$_2$, 4-BrC$_6$H$_4$CH$_2$, 4-MeOC$_6$H$_4$CH$_2$, 4-MeC$_6$H$_4$CH$_2$, 3-FC$_6$H$_4$CH$_2$, 3-MeOC$_6$H$_4$CH$_2$, 2-MeOC$_6$H$_4$CH$_2$, 2-FC$_6$H$_4$CH$_2$, 3-BrC$_6$H$_4$CH$_2$, 3,5-(CF$_3$)$_2$C$_6$H$_3$CH$_2$, 2,4-Cl$_2$C$_6$H$_3$CH$_2$, PG=4-FC$_6$H$_4$CH$_2$	R^1-R^2=(CH$_2$)$_2$, (CH$_2$)$_3$	**GU-I** (20 mol%), Et$_3$N (20 mol%)	65–99	4:1 to >20:1	73–98 (*S*,*S*)	[110]
X=H, 5-Me, 5-MeO, 5-F, R=Ph, Me, Et, *i*-Bu, Bn, *n*-C$_6$H$_{13}$, 4-MeOC$_6$H$_4$, 4-MeC$_6$H$_4$, 4-FC$_6$H$_4$, 3-MeOC$_6$H$_4$, 3-ClC$_6$H$_4$, 4-MeOC$_6$H$_4$CH$_2$, EtO$_2$CCH$_2$, NCCH$_2$, allyl, 1,3-dioxolan-2-ylCH$_2$, PG=Boc	R^1=C(OH)Me$_2$, C(OH)Bn$_2$, R^2=H	**(DHQD)2 PYR** (10 mol%)	76–95	—	30–98[h]	[111]
X=H, R=Me, Bn, Ph, 4-CNC$_6$H$_4$CH$_2$, 4-FC$_6$H$_4$CH$_2$, 4-AcOC$_6$H$_4$CH$_2$, 3-BrC$_6$H$_4$CH$_2$, PG=Ac	R^1=Me, Et, R^2=Ph, Pr, Bu, *i*-amyl, 4-MeC$_6$H$_4$, 2-Npt 4-MeOC$_6$H$_4$, 4-CNC$_6$H$_4$, 4-ClC$_6$H$_4$, 4-FC$_6$H$_4$, 4-BrC$_6$H$_4$, 3-MeC$_6$H$_4$, 3-FC$_6$H$_4$, 3-MeOC$_6$H$_4$, 3-BrC$_6$H$_4$	**TAK-IV L-tartaric acid** (20 mol%, each)	58–98	8:1 to >19:1	81–95 (3*R*,α*S*)	[112]

(*Continued*)

Table 3.6.   (*Continued*)

R, X, PG	R^1, R^2	Catalyst	Yield (%)	d.r.	ee (%)	Ref.
X=H, 6-Cl, R=Me, Bn, Bu, 2-FC$_6$H$_4$CH$_2$, 4-FC$_6$H$_4$CH$_2$, 4-BrC$_6$H$_4$CH$_2$, 4-OHC$_6$H$_4$CH$_2$, 4-CNC$_6$H$_4$CH$_2$, PG=H	R^1=Me, R^2=Ph, Me, 4-MeC$_6$H$_4$, 4-ClC$_6$H$_4$, 4-MeOC$_6$H$_4$, 4-FC$_6$H$_4$, 3-MeC$_6$H$_4$, 3-BrC$_6$H$_4$, 3-FC$_6$H$_4$, 3-MeOC$_6$H$_4$, 3-ClC$_6$H$_4$, 2-ClC$_6$H$_4$, 2-MeOC$_6$H$_4$, 2-FC$_6$H$_4$, 2-Npt	**TAK-V** (10 mol%) **N-Boc-D-Phg** (40 mol%)	50–98	3:1 to >19:1	90–99 (*R,R*)	[112]
X=H, 4-Br, 5-F, 5-Br, 5-Me, 5-MeO, 5-Cl, 6-Cl, 6-Br, 7-Cl, R=CO$_2$Et, PG=H, Me, allyl, PMB	R^1=CO$_2$Et, CO$_2$-*t*-Bu, CN	**Pseudo CIN-IIa** (10 mol%)	63–96	1.2:1–13:1	40–90 (*S,S*)[a]	[113]
X=H, 5-Me, 5-F, R=Bn, Ph, Me, Bu, 4-MeOC$_6$H$_4$CH$_2$, 4-BrC$_6$H$_4$CH$_2$, 3-FC$_6$H$_4$CH$_2$, 2-ClC$_6$H$_4$CH$_2$, 3,4-(OCH$_2$O)C$_6$H$_3$CH$_2$, 3,5-Me$_2$C$_6$H$_3$CH$_2$, MeO$_2$CCH$_2$, EtO$_2$CCH$_2$, 1-NptCH$_2$, 2-thienylCH$_2$, 2-pyridylCH$_2$, 4-MeC$_6$H$_4$, 4-PhC$_6$H$_4$, 4-ClC$_6$H$_4$, PG=Boc, Ts	R^1=PO(OMe)$_2$, PO(OEt)$_2$, PO(O-*i*-Pr)$_2$, PO(OBu)$_2$, R^2=Me, Pr, Ph, 2-furyl	**Pseudo CIN-VIIIb** (20 mol%)	51–98[i]	2:1 to >99:1	60–98[j]	[114]

X=H, 5-Me,5-MeO, 5-F, R=Ph, Me, 4-MeC$_6$H$_4$, 4-MeOC$_6$H$_4$, 4-CF$_3$C$_6$H$_4$, 4-PhC$_6$H$_4$, PG=Boc	R^1=Me, H, Ph[j,k]	**PHEN** (12 mol%), Ni(OAc)$_2$ (10 mol%)	25–99	—	30–87 (R)	[115]
X=H, 5-MeO, R=Me, Et, Pr, Bn, Allyl, Ph, 4-MeOC$_6$H$_4$, 4-MeC$_6$H$_4$, 4-BuC$_6$H$_4$, 4-FC$_6$H$_4$, 4-ClC$_6$H$_4$, 4-EtOC$_6$H$_4$, 4-PhC$_6$H$_4$, 3-MeOC$_6$H$_4$, 3,5-Me$_2$C$_6$H$_3$, indol-3-yl, 1-Bn-indol-3-yl, 1-Boc-indol-3-yl, PG=Boc	R^1=CN, R^2=Cl	**TAK-VI** (10 mol%), 4Å MS	81–97[l,m]	6:1 to >30:1	86–99 (3R,βS)	[116]
X=H, 5-Me, 5-F, 5-Br,5-CF$_3$O, 6-Cl, 6-Br, 7-F, R=Ph, 4-FC$_6$H$_4$, 4-MeC$_6$H$_4$, 2-Npt, Bn, PG=Boc	R^1=CO$_2$Et, R^2=PhthN	**CIN-IIc** (10 mol%)	90–98	8:1–15.7:1	93 > 99 (3S,βR)[a]	[117]

(b)

(Continued)

Table 3.6.   (*Continued*)

(c)

R, X, PG	R^1, R^2	Catalyst	Yield (%)	d.r.	ee (%)	Ref.
X=H, R=3,4-(OCH_2O) $C_6H_3CH_2$ PG=Boc	R^1=Ph	**JHC** (R=n-$C_{12}H_{25}$, R^1=TMS) (20 mol%)	98	>99:1	89 (nr)	[118]
X=H, R=Me, Bn, 4-$MeOC_6H_4CH_2$, 4-$MeC_6H_4CH_2$, 3-$FC_6H_4CH_2$, PG=Boc	R^1=Ph, 4-BrC_6H_4	**JHC** (R=n-$C_{12}H_{25}$, R^1=TMS) (10 mol%)	86–93	>99:1	86–91 (*S,S*)	[119]
X=H, 5-F, 5-Me, R=Ph, 4-MeC_6H_4, 4-$MeOC_6H_4$, 4-PhC_6H_4, 4-FC_6H_4, 3,4-$Me_2C_6H_3$, *i*-Pr, Bn, Bu, PG=Boc	R^1=Ph, 4-BrC_6H_4, 4-ClC_6H_4, 4-MeC_6H_4, 3-BrC_6H_4, 3-ClC_6H_4, 3-$MeOC_6H_4$, 3,4-$Cl_2C_6H_3$,Cy, *i*-Pr, Allyl[n]	**Ent-TAK-IIIa** (20 mol%)	45–92	1.7:1–99:1	85–99 (*S,S*)	[120]
X=H, 5-Cl, 5-Me, 6-Br, 7-F, R=Bn, 4-$BrC_6H_4CH_2$, 4-$ClC_6H_4CH_2$, 3-$BrC_6H_4CH_2$, 3-$MeOC_6H_4CH_2$,	R^1=Bn, Ph, Et, 4-$BrC_6H_4CH_2$, 4-$MeOC_6H_4CH_2$, 2-$ClC_6H_4CH_2$,	**GU-I** (10 mol%)	75–99	18:1 to >99:1	90 to >99 (*S,S*)	[121]

2-BrC$_6$H$_4$CH$_2$, 2-MeOC$_6$H$_4$CH$_2$, 2-Npt, 2-thienyl, PG=Bn	2-MeOC$_6$H$_4$CH$_2$, 4-t-BuC$_6$H$_4$, 3-BrC$_6$H$_4$					
X=H, 6-Cl, R=Bn, Ph, Me, 4-BrC$_6$H$_4$CH$_2$, 4-ClC$_6$H$_4$CH$_2$, 4-FC$_6$H$_4$CH$_2$, 4-MeC$_6$H$_4$CH$_2$, 4-MeOC$_6$H$_4$CH$_2$, 3-BrC$_6$H$_4$CH$_2$, 3-NO$_2$C$_6$H$_4$CH$_2$, 3-MeOC$_6$H$_4$CH$_2$, 3-ClC$_6$H$_4$CH$_2$, 2-BrC$_6$H$_4$CH$_2$, 2-MeOC$_6$H$_4$CH$_2$, 2,4-Cl$_2$C$_6$H$_3$CH$_2$, 2-furylCH$_2$, 3-thienylCH$_2$, 2-NptCH$_2$, PG=H	R^1=2-t-BuC$_6$H$_4$, 2-t-Bu-4-ClC$_6$H$_3$, 2-t-Bu-4-BrC$_6$H$_3$	**NNO-IVb** (10 mol%), Sc(OTf)$_3$ (10 mol%), K$_2$HPO$_4$ (3 equiv)	91–99	3:1 to >19:1	84–99 (*M,R,R*)	[122]
X=H, R=CO$_2$Et, PG=Me	R^1=Me, Ph, Bn, Ph$_2$CH, 4-MeC$_6$H$_4$, 4-MeOC$_6$H$_4$, 4-FC$_6$H$_4$, 4-ClC$_6$H$_4$, 4-NO$_2$C$_6$H$_4$, 3-FC$_6$H$_4$, 3-NO$_2$C$_6$H$_4$, 2-FC$_6$H$_4$	**Cinchonidine** (5 mol%)	93–99	5.5:1–10.1:1	63–85 (*S,S*)	[123]

(Continued)

Table 3.6. (*Continued*)

R, X, PG	R^1, R^2	Catalyst	Yield (%)	d.r.	ee (%)	Ref.
X=H, 5-F, 5-MeO, 5-Me, 6-Br, 5,7-Me$_2$, R=Ph, 4-MeC$_6$H$_4$, 4-MeOC$_6$H$_4$, 3-MeC$_6$H$_4$, PG=Boc[o]	R^1=2-t-BuC$_6$H$_4$, 2-t-Bu-4-ClC$_6$H$_3$, 2-t-Bu-4-ClC$_6$H$_3$, 2,5-(t-Bu)$_2$-4-BrC$_6$H$_2$, 2,5-(t-Bu)$_2$C$_6$H$_3$, 2-t-Bu-5-NO$_2$C$_6$H$_3$, 2-t-Bu-5-NHCbzC$_6$H$_3$	**CIN-VIIIb** (10 mol%)	43–98	6:1 to >19:1	84 to >99 (*M*,3*R*, αS)	[124]
X=H, 5-Br, 5-Cl, 5-F, 5-Me, R=Ph, 4-BrC$_6$H$_4$, 3-ClC$_6$H$_4$, 3-MeC$_6$H$_4$, 2-Npt, 2-thienyl, PG=H	1,4-naphtho-quinone[p]	**(DHQD)$_2$PYR** (10 mol%)	45–97	—	54–83 (*S*)	[125]
X=H, 5-Cl, 5-Me, 5-MeO, 5-i-Pr, 5-NMe$_2$, R=Ph, Me, 4-FC$_6$H$_4$, 4-MeOC$_6$H$_4$, 4-MeC$_6$H$_4$, 2,4-Me$_2$C$_6$H$_3$, 2-Npt, 2-pyridyl, PG=Boc, Cbz. Bz, Ac	1,4-naphtho-quinone, 5,8-(AcO)$_2$-1,4-naphtho-quinone, 8-MeO-1,4- naphtho-quinone	**TAK-VII** (10 mol%)[q]	75–94	—[r]	73–97 (*R*)	[126]

(*d*)

| X=H, 5-Me, 5-F, R=Ph, Bn, Me, i-Pr, 4-MeC$_6$H$_4$, 4-MeOC$_6$H$_4$, 4-FC$_6$H$_4$, 4-ClC$_6$H$_4$, 2-MeOC$_6$H$_4$, 4-MeOC$_6$H$_4$CH$_2$, 4-FC$_6$H$_4$CH$_2$, 3-MeOC$_6$H$_4$CH$_2$, 3-FC$_6$H$_4$CH$_2$, 2-MeOC$_6$H$_4$CH$_2$, 1-Npt, 2-thienyl, PG=Boc, Cbz, Me | benzoquinone, 2-Cl-benzo-quinone, 2,6-Cl$_2$benzo-quinone[o] | **CIN-VIIIb** (10 mol%) | 32–87 | —[s] | 12–96 $(-\alpha_D)$ | [127] |

Notes: [a]The pseudo-enantiomer afforded the enantiomer in comparable yield, diastereo- and enantio-selectivity. 3-Substituted cyclic enones did not react under these conditions. [b]3-Phenylindole gave racemic adduct. [c]Open-chain enones remained unreacted. [d]The absolute configuration is reported only in Ref. [106b] It is very likely that it is the same for products reported in Ref. [106a]. [e]Benzyl indoles preferred R=OTMS, phenyl vinyl ketone R=OH, respectively, otherwise R=Me was the best catalyst. [f]Products arising from acrolein were characterized after *in situ* Wittig reaction with Ph$_3$PCHCO$_2$Et. [g]An (S)-configuration is attributed to these compounds, but the schemes of the paper reported (R)-configured products and experimental section α_D values in agreement with those reported in Ref. [107], in which the (R)-configuration is attributed to the products. [h]A stereochemistry with R group out of plane was attributed because one of the products could be used to prepare (–)-esermethole in 78% overall yield and 90% ee. [i]Yields calculated as methyl or ethyl ester, benzyl amide or morpholinamide. [j]Only α_D values are reported. [k]Methyl acrylate, acrylonitrile, 3-acryloyloxazolidin-2-one and pent-3-en-2-one did not react. [l]The reaction was scalable up to 1 mmol scale. [m]These compounds were converted into pyrroloindoles or C$^\gamma$-tetrasubstituted α-amino acids. [n]Two addition products to maleic anhydride were recovered in 71% and 76% yield, 16:1 and 19:1 dr, 85% ee. [o]4-substituted oxindoles did not react. [p]Benzoquinone and 2,6-dichlorobenzoquinone did not react. 1,4-Naphthodiols can be obtained by hydrogenation without loss of stereoselectivity. 3-Methyl-5-bromooxindole gave only 19% ee. [q]The only reaction in which the quinone is stoichiometric and the air acts as the oxidant. [r]The regioisomeric ratio was 1.4:1 in favor of the 2-substituted naphthoquinone. The (1R,2S) catalyst gave the enantiomeric (S)-product in 63% yield and 78% ee and increased regioisomeric ratio up to 20:1. [s]The main isomer was not reported, but it was obtained in 49:1 regioisomeric ratio.

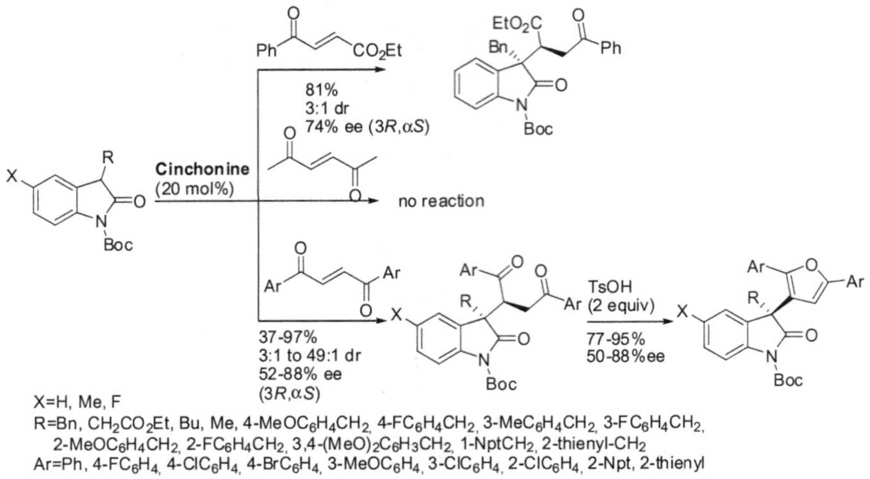

Scheme 3.33.   Enantioselective furanylation of 3-alkyloxindoles.

3-alkylindoles reaction, respectively (Scheme 3.34, Eq. (3.25)) [49]. The products were highly Z-selective (over 8.1:1 Z/E ratio) except for 1-phenyloct-2-yn-1-one (1.2:1). The interaction between the electron-enriched π orbital of the allenoate and the electron-deficient C-2 atom of the oxindole was found to be responsible of the Z-selectivity.

The enantioselectivity of the addition to acetylenedicarboxylates was found to increase, increasing the size of the ester moiety (Scheme 3.34, Eq. (3.26)) [128]. Also enantiomeric products were recovered with the pseudo-enantiomeric quinidine derivative, but in lower yields and Z/E ratios. The Z/E selectivity and a dual activation of the catalyst accounted for the enantioselectivity.

Also, allenoates can undergo Michael addition by 3-prochiral oxindoles and chiral bifunctional N-acylaminophosphine catalysts were found to be the most efficient in this catalysis. Two papers appeared in the literature almost at the same time (Scheme 3.35, Eqs. (3.27) [129], (3.28) and (3.29) [130]).

Both research groups proposed a mechanism in which deprotonation of C-3 was accomplished by the adduct arising from the addition of the

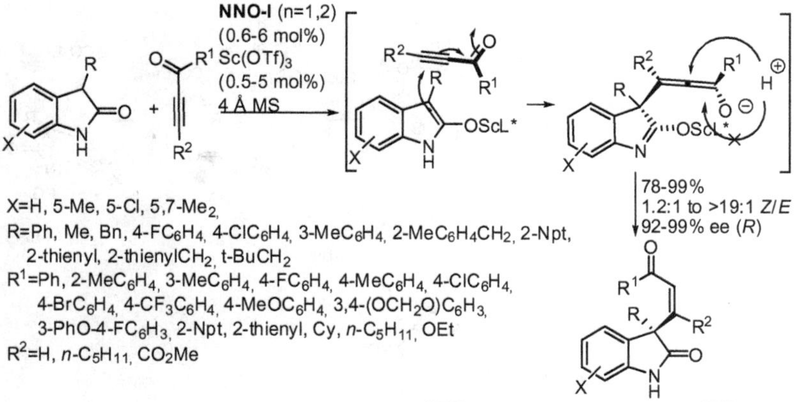
NNO-I (n=1,2)
(0.6-6 mol%)
Sc(OTf)3
(0.5-5 mol%)
4 Å MS
78-99%
1.2:1 to >19:1 Z/E
92-99% ee (R)
X=H, 5-Me, 5-Cl, 5,7-Me2,
R=Ph, Me, Bn, 4-FC6H4, 4-ClC6H4, 3-MeC6H4, 2-MeC6H4CH2, 2-Npt, 2-thienyl, 2-thienylCH2, t-BuCH2
R1=Ph, 2-MeC6H4, 3-MeC6H4, 4-FC6H4, 4-MeC6H4, 4-ClC6H4, 4-BrC6H4, 4-CF3C6H4, 4-MeOC6H4, 3,4-(OCH2O)C6H3, 3-PhO-4-FC6H3, 2-Npt, 2-thienyl, Cy, n-C5H11, OEt
R2=H, n-C5H11, CO2Me

(3.25)

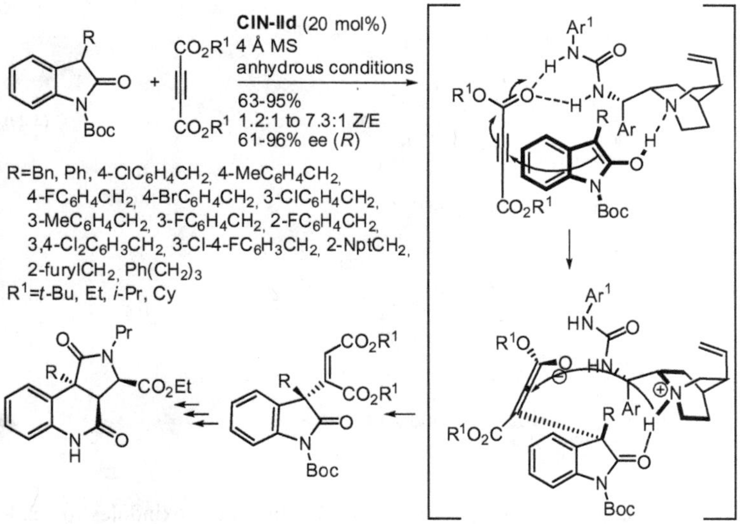
CIN-IId (20 mol%)
4 Å MS
anhydrous conditions
63-95%
1.2:1 to 7.3:1 Z/E
61-96% ee (R)
R=Bn, Ph, 4-ClC6H4CH2, 4-MeC6H4CH2, 4-FC6H4CH2, 4-BrC6H4CH2, 3-ClC6H4CH2, 3-MeC6H4CH2, 3-FC6H4CH2, 2-FC6H4CH2, 3,4-Cl2C6H3CH2, 3-Cl-4-FC6H3CH2, 2-NptCH2, 2-furylCH2, Ph(CH2)3
R1=t-Bu, Et, i-Pr, Cy

(3.26)

Scheme 3.34. Addition of 3-prochiral oxindoles to electron-deficient alkynes.

catalyst to the allenoate. Then the oxindole enolate attacks the γ-position of the phosphonium dienolate. Moreover, Cai and Zhao found that the addition of benzoic acid enhanced the selectivity although it slowed down the reaction rate because of the inhibition of the enolate formation [129].

On the other hand, Lu's group obtained the opposite stereochemistry with 3-alkyl- (Eq. (3.28)) and 3-aryloxindoles (Eq. (3.29)), but no explanation was given [130]. These products were employed as key

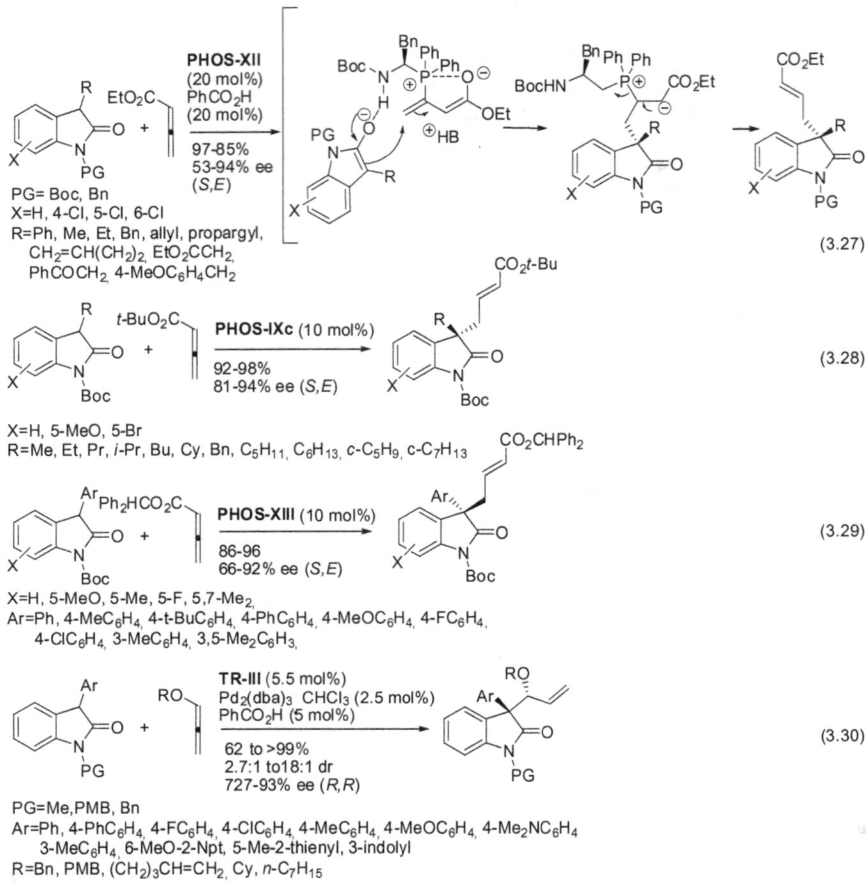

PHOS-XII (20 mol%)
PhCO$_2$H (20 mol%)
97–85%
53–94% ee
(S,E)

PG= Boc, Bn
X=H, 4-Cl, 5-Cl, 6-Cl
R=Ph, Me, Et, Bn, allyl, propargyl, CH$_2$=CH(CH$_2$)$_2$, EtO$_2$CCH$_2$, PhCOCH$_2$, 4-MeOC$_6$H$_4$CH$_2$

(3.27)

PHOS-IXc (10 mol%)
92–98%
81–94% ee (S,E)

(3.28)

X=H, 5-MeO, 5-Br
R=Me, Et, Pr, i-Pr, Bu, Cy, Bn, C$_5$H$_{11}$, C$_6$H$_{13}$, c-C$_5$H$_9$, c-C$_7$H$_{13}$

PHOS-XIII (10 mol%)
86–96
66–92% ee (S,E)

(3.29)

X=H, 5-MeO, 5-Me, 5-F, 5,7-Me$_2$,
Ar=Ph, 4-MeC$_6$H$_4$, 4-t-BuC$_6$H$_4$, 4-PhC$_6$H$_4$, 4-MeOC$_6$H$_4$, 4-FC$_6$H$_4$, 4-ClC$_6$H$_4$, 3-MeC$_6$H$_4$, 3,5-Me$_2$C$_6$H$_3$,

TR-III (5.5 mol%)
Pd$_2$(dba)$_3$ CHCl$_3$ (2.5 mol%)
PhCO$_2$H (5 mol%)
62 to >99%
2.7:1 to18:1 dr
727–93% ee (R,R)

(3.30)

PG=Me, PMB, Bn
Ar=Ph, 4-PhC$_6$H$_4$, 4-FC$_6$H$_4$, 4-ClC$_6$H$_4$, 4-MeC$_6$H$_4$, 4-MeOC$_6$H$_4$, 4-Me$_2$NC$_6$H$_4$, 3-MeC$_6$H$_4$, 6-MeO-2-Npt, 5-Me-2-thienyl, 3-indolyl
R=Bn, PMB, (CH$_2$)$_3$CH=CH$_2$, Cy, n-C$_7$H$_{15}$

Scheme 3.35. Phosphine-catalyzed γ-addition of 3-substituted oxindoles to 2,3-butadienoates.

intermediates for natural product synthesis such as (–)-esermethole and (–)-physostigmine, as well as were transformed into (S)-3-allyl-1-methyl-3-phenylindolin-2-one. In Table 3.2, the Pd-catalyzed asymmetric allylic alkylation was already described, and this reaction was also extended to the nucleophilic addition of 3-aryloxindoles to allenes (Scheme 3.35, Eq. (3.30)) [131]. It should be noted that the increased steric hindrance of *ortho*-substituent on the aromatic ring needed a more polar solvent to reach high yield, diastereo-, and enantioselectivity, but the opposite diastereomer was preferred. The reaction was then used for a new approach to gliocladins.

Finally, a three-component cascade reaction should be mentioned in which vinyl ketones or allenoates, indoles and diazooxindoles under Rh(II)/chiral phosphine combined catalysis allowed the synthesis of enantioenriched 3,3′-indolyloxindole derivatives (Scheme 3.36) [132].

The long distance between the newly formed asymmetric site and the coordination sphere of the catalyst represent an intriguing challenge to perform enantioselective 1,6-conjugate additions. However, some interesting examples, in which this problem was successfully solved, have been reported. Scandium (III) complex allowed the asymmetric addition

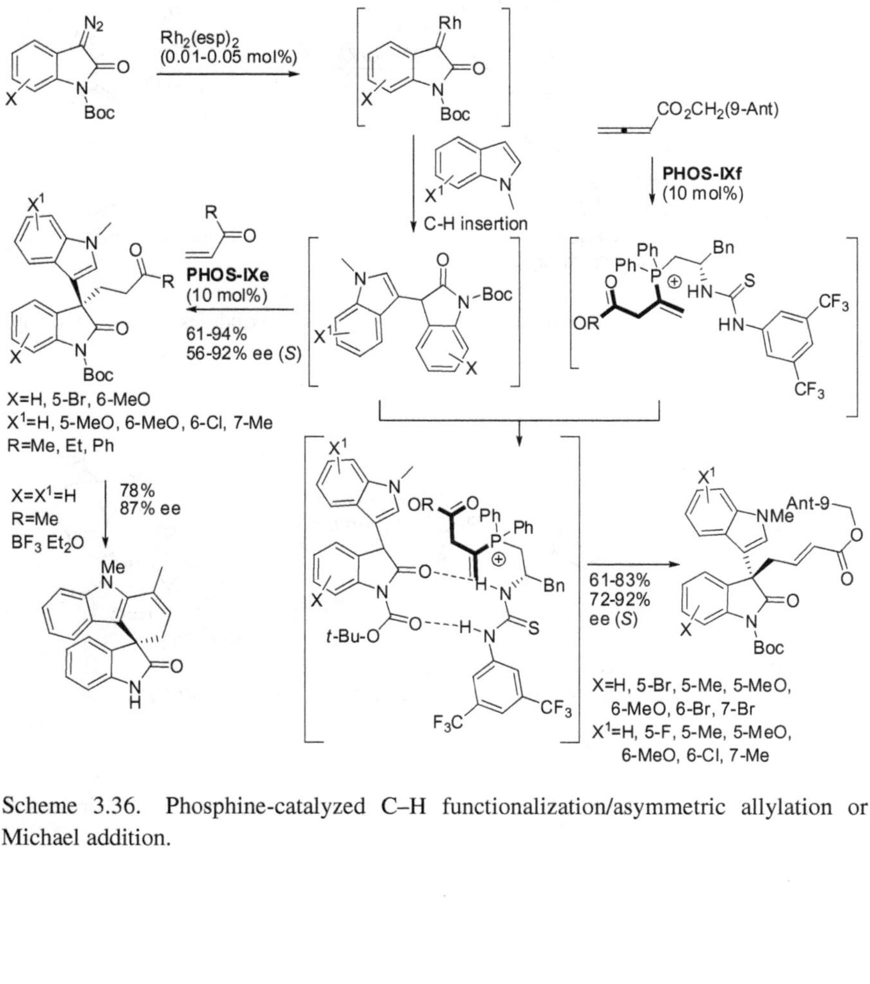

Scheme 3.36. Phosphine-catalyzed C–H functionalization/asymmetric allylation or Michael addition.

of 3-substituted oxindoles to dienyl ketones in very high enantiomeric and regioselective control, even in gram-scale reaction (Scheme 3.37) [133].

Unfortunately, authors gave only patchy information about the stereochemistry. In fact, their envisaged transition state explained the stereochemistry of the C-3, but not that of the other asymmetric carbon. Only a negative α_D value was reported.

The choice of the *N*-alkyl group of the catalyst was basic for achieving the 1,6-conjugate addition of cyclic dienones and the scarcely steric demanding cyclohexyl group was the best (Scheme 3.38) [134]. It should be noted that linear dienones afforded only 1,4-addition products under these experimental conditions.

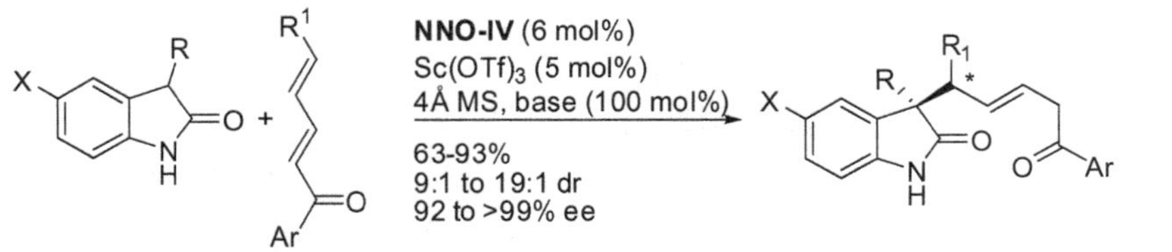

X=H, Me, Ar=Ph, 4-MeOC$_6$H$_4$

R=Bn, Me, Ph, 3-BrC$_6$H$_4$CH$_2$, 3-NO$_2$C$_6$H$_4$CH$_2$, 3-MeC$_6$H$_4$CH$_2$, 4-BrC$_6$H$_4$CH$_2$, 4-CNC$_6$H$_4$CH$_2$, 4-MeC$_6$H$_4$CH$_2$, 2,4-Cl$_2$C$_6$H$_3$CH$_2$, 4-ClC$_6$H$_4$, 2-Npt, 2-NptCH$_2$, 2-thienylCH$_2$, 2-furylCH$_2$

base=Na$_2$CO$_3$ if R^1=H; K$_3$PO$_4$.7H$_2$O if R^1=Me

Scheme 3.37. Asymmetric 1,6-addition of 3-substituted oxindoles to linear dienyl ketones.

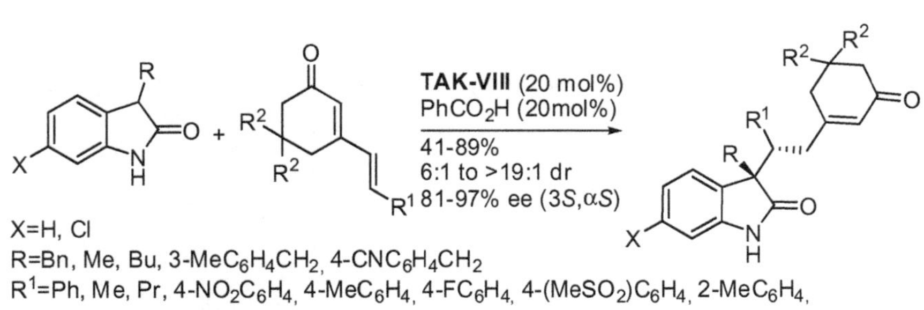

X=H, Cl

R=Bn, Me, Bu, 3-MeC$_6$H$_4$CH$_2$, 4-CNC$_6$H$_4$CH$_2$

R^1=Ph, Me, Pr, 4-NO$_2$C$_6$H$_4$, 4-MeC$_6$H$_4$, 4-FC$_6$H$_4$, 4-(MeSO$_2$)C$_6$H$_4$, 2-MeC$_6$H$_4$, 2-Npt, 3-thienyl, CO$_2$Et

R^2=H, Me

Scheme 3.38. Asymmetric 1,6-addition of 3-substituted oxindoles to cyclic dienyl ketones.

Other 1,6-addition acceptors are *para*-quinone methides. Their reaction with oxindoles was reported in very similar conditions, which also allowed scale up, by two independent research groups at the beginning of 2016 [135,136]. Some interesting features have to be outlined:

(i) Each research group obtained one of the two enantiomers, thus making complementary the two methodologies.
(ii) Fan group (Scheme 3.39, Eq. (3.31)) [135] discarded the pseudo enantiomer of the catalyst used by Enders group (Scheme 3.39, Eq. (3.32)) [136].
(iii) Enders did not use a basic cocatalyst.
(iv) Fan accounted for the stereochemistry observed affirming that the *Re*-face of *para*-quinone methide was attacked by the *Re*-face of the enolate of oxindole.

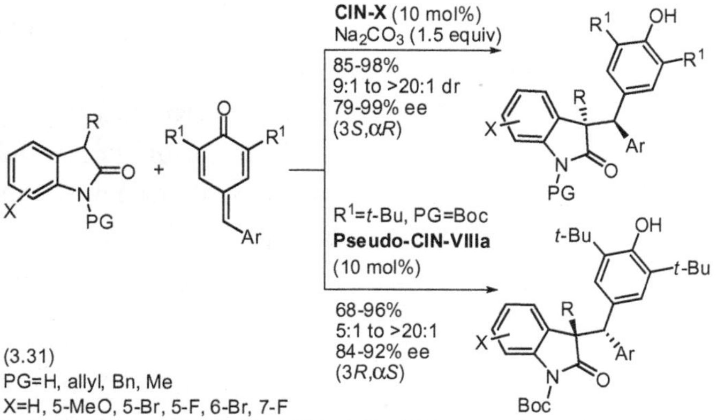

$$\text{(3.31)}$$

$$\text{(3.32)}$$

(3.31)
PG=H, allyl, Bn, Me
X=H, 5-MeO, 5-Br, 5-F, 6-Br, 7-F
R=Ph, 4-MeC$_6$H$_4$, 4-MeOC$_6$H$_4$, 4-NMe$_2$C$_6$H$_4$, 4-FC$_6$H$_4$, 4-ClC$_6$H$_4$, 3-MeC$_6$H$_4$, 3-MeOC$_6$H$_4$, 2-MeOC$_6$H$_4$, 3,5-(MeO)C$_6$H$_3$, 3,5-F$_2$C$_6$H$_3$, 2-Npt, 2-thienyl, 3-Me-2-thienyl, *N*-pyrrolyl
Ar=Ph, 4-MeC$_6$H$_4$, 4-MeOC$_6$H$_4$, 4-FC$_6$H$_4$, 4-ClC$_6$H$_4$, 4-BrC$_6$H$_4$, 3-BrC$_6$H$_4$, 2-BrC$_6$H$_4$, 2-ClC$_6$H$_4$, 2-CO$_2$MeC$_6$H$_4$, 2-AcOC$_6$H$_4$, 3,4-(OCH$_2$O)C$_6$H$_3$ 3,5-(MeO)$_2$C$_6$H$_3$, 1-Npt, 2-Npt, Me
R^1=*t*-Bu, Me, Ph, TMS
(3.32)
X=H, 5-Me
R=Ph, 4-FC$_6$H$_4$, 4-PhC$_6$H$_4$, 4-*t*-BuC$_6$H$_4$, 4-MeC$_6$H$_4$
Ar=Ph, 4-MeOC$_6$H$_4$, 4-MeC$_6$H$_4$, 4-BrC$_6$H$_4$, 4-NO$_2$C$_6$H$_4$, 2-ClC$_6$H$_4$, 2-BrC$_6$H$_4$, 2-MeOC$_6$H$_4$, 3,4-(MeO)$_2$C$_6$H$_3$, 3,4-(OCH$_2$O)C$_6$H$_3$ 1-Npt, 2-thienyl, 2-furyl

Scheme 3.39.   Asymmetric 1,6-conjugate addition/aromatization of *para*-quinone methides with oxindoles.

(v) Enders affirmed that the strongest hydrogen bonds enhanced the stereoselectivity of the reaction.

(vi) Enders used *N*-protected indoles and was able to deprotect them without any decrease of optical purity; Fan generally employed unprotected indoles.

3.6.2. *Indole as the acceptor*

In the previuos section we described subsituted oxindoles as Michael donors. However, properly substituted oxindoles, such as alkylidene oxindoles, can undergo conjugated nucleophilic attack, acting as the Michael acceptor. Pay attention, if another EWG is present at the out-of-ring terminus of the double bond the negative charge will accommodate near to the C-3 and not on it. High electron-withdrawing external groups, for example, favors the C-3 attack.

The reaction of acetylacetone with propyl oxoindolylidene acetate was one of the first reports (Scheme 3.40) [137]. The absolute configuration was not reported and diasteroselectivity was low. The reaction with nitromethane as the nucleophile was also carried out obtaining a high yield (90%) and enantioselectivity (96% ee), but no diasteroselectivity.

Later, Xiao demonstrated with anthrone that the crowding close to C-3 and bulkiness of the *N*-groups drove the attack to both termini of the double bond in oxoindolylidene acetate or ketones (Scheme 3.41) [138]. The biggest group should accommodate far from the quinuclidine nucleus of the catalyst, leading to different transition states, which favored C-α or C-3 attack, respectively.

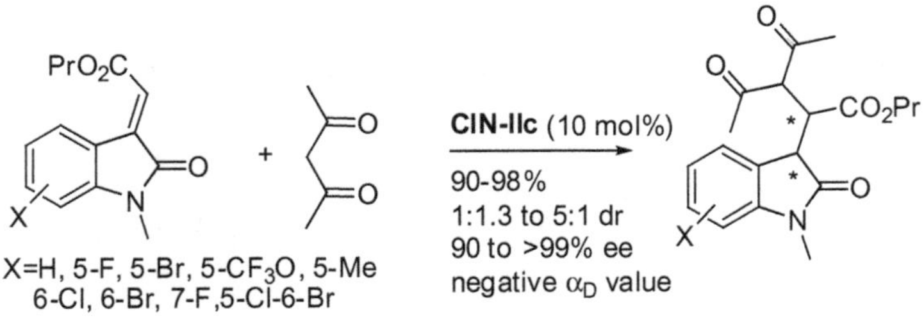

Scheme 3.40. Organocatalytic addition of acetylacetone to alkylidene oxindoles.

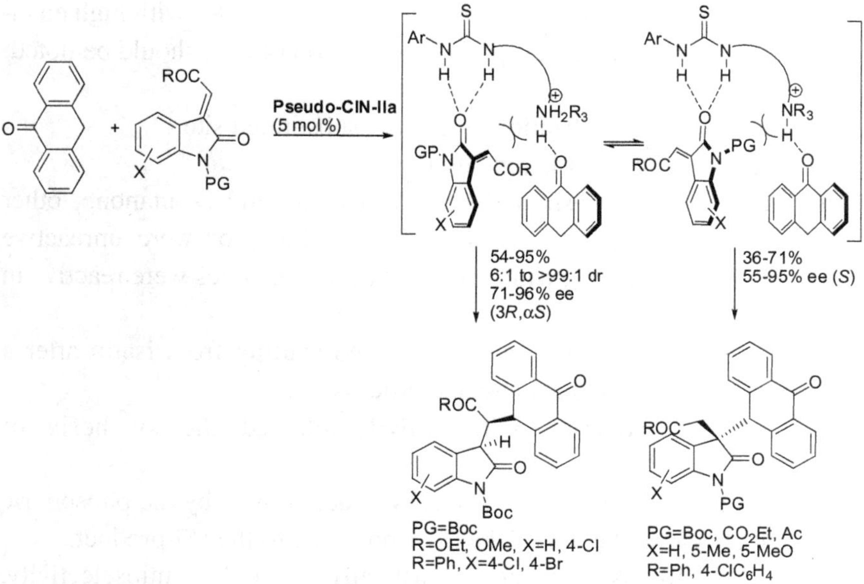

Scheme 3.41.   Addition of anthrone to oxoindolylidene acetate or ketones.

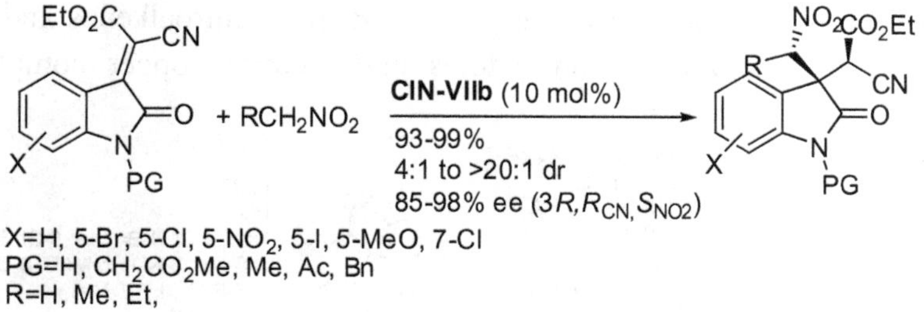

Scheme 3.42.   Organocatalytic Michael addition of nitroalkanes to indolylidenecyano-acetates.

Higher electron-withdrawing external groups such as oxoindolylidene cyanoacetates or isatylidene malononitriles were completely regioselective, as, for instance in the reaction of nitroalkanes with oxoindolylidene cyanoacetates (Scheme 3.42) [139]. The reduction of the nitro group to amine led to a spontaneous cyclization to spirolactams, which also allowed determining the exact configuration of the stereocenters.

Ketone enolates can add to isatylidene malononitriles with high enantioselectivity (Scheme 3.43, Eqs. (3.33)–(3.35)) [140]. It should be noted:

(i)   The exclusive enolization at the less substituted site.

(ii) The large excess of ketone used.

(iii) The reaction is restricted only to acetone and 2-butanone, other ketones were not tested (Eq. (3.33)) [140a] or were unreactive (Eq. (3.34)) [140b], while different acyclic ketones were reactive in Ref. [140c].

(iv) The reaction could be also performed starting from isatin after a Knoevenagel–Michael cascade process.

(v)   That the reduction with $NaBH_4$ allowed the synthesis of spirooxindole.

(vi) In Ref 140a, the enantioselectivity is determined by the phosphoric acid, because the use of the (S)-isomer led to the (S)-product.

(vii) In Ref 140a, N-substitution drastically lowered enantioselectivity, while in Ref 140c both N-unprotected and N-protected oxindole reacted.

Two Friedel–Crafts reactions of isatin-derived nitroalkenes and β,γ-unsaturated α-keto esters on indoles under chiral copper complexes

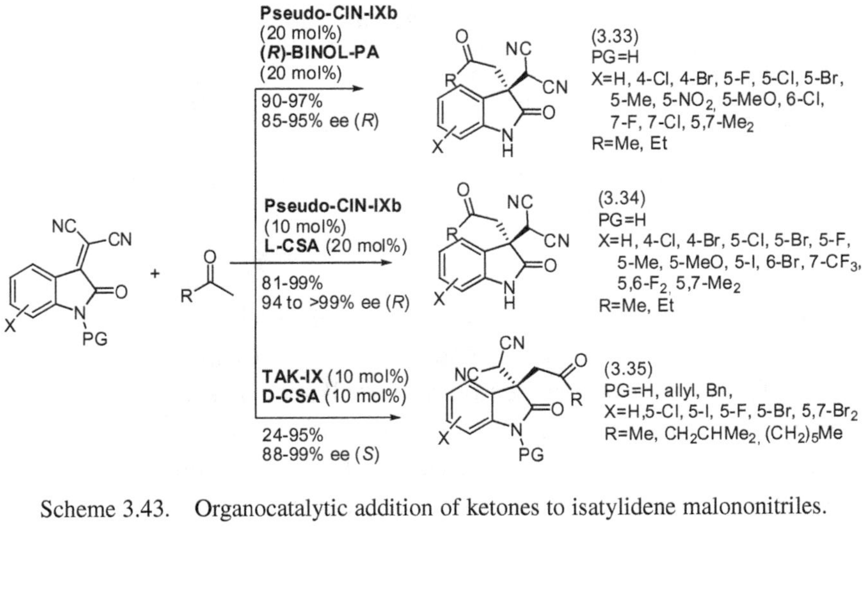

Scheme 3.43. Organocatalytic addition of ketones to isatylidene malononitriles.

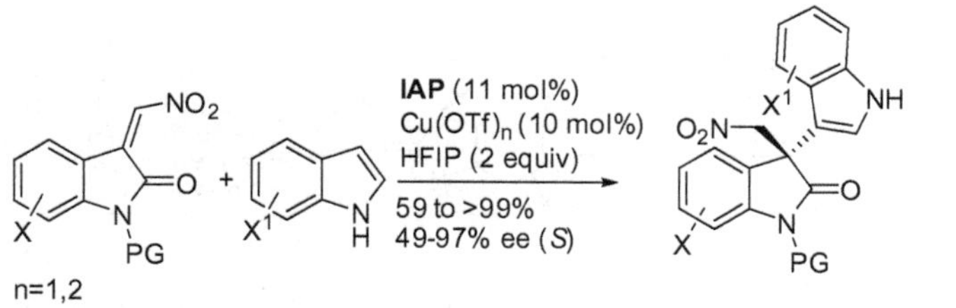

IAP (11 mol%)
Cu(OTf)$_n$ (10 mol%)
HFIP (2 equiv)

59 to >99%
49-97% ee (*S*)

(3.36)

n=1,2
X=H, 5-F, 5-Cl, 5-Br, 5-MeO, 5-Me, 7-Me
PG=Me, Bn, allyl
X^1=H, 5-MeO, 5-Me, 5-Br, 7-Me

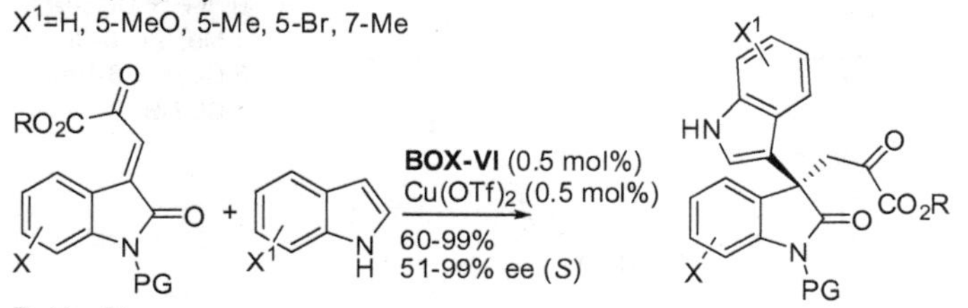

BOX-VI (0.5 mol%)
Cu(OTf)$_2$ (0.5 mol%)

60-99%
51-99% ee (*S*)

(3.37)

R=Me, Et
X=H, 5-F, 5-Cl, 5-Br, 5-MeO, 5-Me, 6-Cl, 6-Br, 7-F, 7-Cl, 7-Me, 5,7-Me$_2$, 5,6-F$_2$
PG=H, Me, MOM, Boc, allyl
X^1=H, 4-MeO, 5-F, 5-I, 5-MeO, 5-Me, 5-Br, 5-Ph, 5-CN, 6-F, 6-Cl,
     6-Br, 6-I, 6-MeO, 7-Me

Scheme 3.44. Copper-catalyzed asymmetric synthesis of 3,3-*bis*-indoles.

catalysis have been reported. In the former reaction (Scheme 3.44, Eq. (3.36)), IAP/Cu(I) or (II) complexes had to be selected time by time, depending on the substrate, and the reaction of *N*-acetyloxindole, 2-methyl- and 4-methylindoles resulted in low yields [141]. The latter reaction (Scheme 3.44, Eq. (3.37)) used very low amounts of a chiral BINOL-derived *bis*-oxazoline copper complex, was also scaled-up on a gram scale, and the stereochemical outcome was rationalized by DFT calculations [142]. Products arising from both reactions were further manipulated without loss of selectivity.

Isatylidene-3-acetaldehyde underwent conjugate addition with functionalization of C-3 rather than out-of-ring position through iminium catalysis, which allows a LUMO lowering of the substrate.

Under these reaction conditions, malonates [143], indoles [144], and nitromethane [145] gave the expected C-3 functionalization in reactions easily scalable up to the gram scale (Scheme 3.45). From these reactions, some key intermediates for the synthesis of natural products

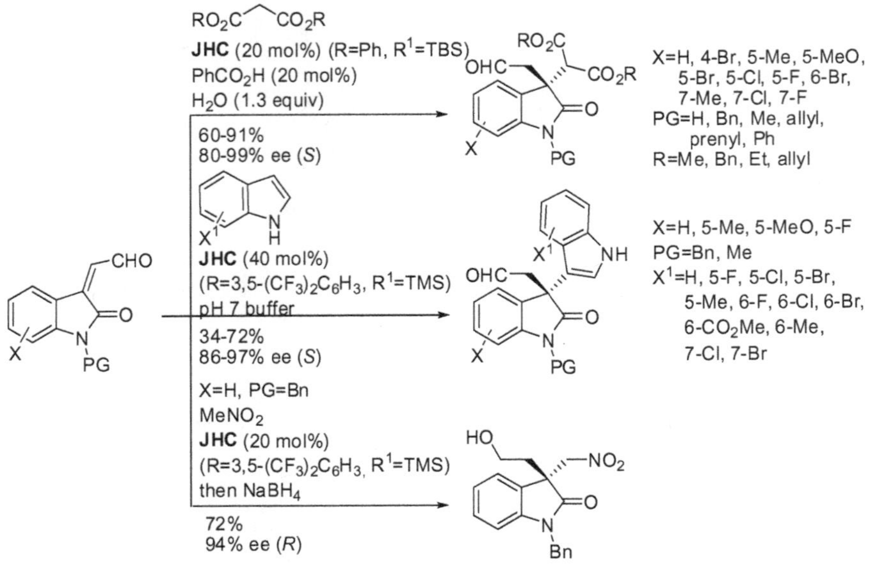

Scheme 3.45.   Michael addition of isatylidene-3-acetaldehydes.

were achieved. In particular, (−)-debromoflustramine E (seven steps, 29% overall yield) from malonate addition; (+)-gliocladin C (27% overall yield in three steps) and (−)-chimonanthine (47% overall yield in seven steps, by using the enantiomer of the catalyst depicted in Scheme 3.49) from indole addition; (−)-horsfiline (33% overall yield) and (−)-coerulescine (46% overall yield), in a three-step, one-pot reaction from nitromethane.

3.6.3. *Addition to activated double bonds*

Besides classical Michael addition to unsaturated carbonyl derivatives, the 3-prostereogenic oxindole unit can also add to a variety of activated double bonds, such as nitroalkenes, sulfones, selenones and phosphonates.

Nitroalkenes are certainly the most studied class of such compounds, and in Table 3.7 are collected all the examples of this reaction. The catalysts have to be basic to enolize the oxindole, and then the tight ion pair enolate-catalyst hides one face of the oxindole allowing the attack to the nitroalkene from the other face. At the same time, H-bonding or

Table 3.7.   Asymmetric Michael additions of 3-prostereogenic indoles to nitroalkenes.

X, R, PG	R^1	Catalyst	Yield (%)	d.r.	ee (%)	Ref.
X=H, 5-MeO, R=Me, Et, Allyl, (E)-PhCH=CH, 4-BrC$_6$H$_4$CH$_2$, PG=Boc	Ph, 4-BrC$_6$H$_4$, 4-MeOC$_6$H$_4$, 4-OHC$_6$H$_4$, 3-BrC$_6$H$_4$, 3,4-Cl$_2$C$_6$H$_3$, 2-furyl, n-C$_7$H$_{15}$, CH(OMe)$_2$	**TAK-X** (10 mol%)	65–97	3:1 to >20:1	88–99[a] (3R,αS)	[148]
X=H, 5-MeO, 6-Cl, R=Ph, 2-Npt, 2-benzothienyl PG=Boc[b]	Ph, 2-furyl, Ph(CH$_2$)$_2$, Pr	**PTC-VIII** (1 mol%), water	90–93	3:1–19:1	80–95 (−α$_D$)	[149]
X=H, 5-MeO, 5-F, R=Me, Bn, PG=Boc	Ph, 4-ClC$_6$H$_4$, 4-MeOC$_6$H$_4$, 4-BrC$_6$H$_4$, 3-BrC$_6$H$_4$, 2-BrC$_6$H$_4$, 2-furyl, 2-thienyl, PhCH=CH, Me(CH$_2$)$_3$CH=CH	**MN** (2.5 mol%)	83–99	5:1 to >30:1	85–96 (3R,αS)	[150]
X=H, R=Me, PG=Ph[c]	Ph, 4-NO$_2$C$_6$H$_4$, 4-PhC$_6$H$_4$, 4-ClC$_6$H$_4$, 4-MeOC$_6$H$_4$, 4-MeC$_6$H$_4$, 3-NO$_2$C$_6$H$_4$, 2-ClC$_6$H$_4$, 2-BrC$_6$H$_4$, 2,4-(MeO)$_2$C$_6$H$_3$, 3,4-(OCH$_2$O)-C$_6$H$_3$, 2-Npt, Ph(CH$_2$)$_2$, i-Bu	**TAK- VI** (10 mol%)	90–99	2:1–5:1	78–98 (S,S)	[151]

(*Continued*)

Table 3.7. (*Continued*)

X, R, PG	R^1	Catalyst	Yield (%)	d.r.	ee (%)	Ref.
X=H, 5-F, 5-Br, R=Ph, Bn, 2-Npt, 4-ClC$_6$H$_4$, PG=H	i-Bu, Pr, Ph, 2-Npt, 4-CF$_3$C$_6$H$_4$, 4-ClC$_6$H$_4$, 2,4-Cl$_2$C$_6$H$_3$, 2-thienyl, 2-furyl	**Quinidine** (10 mol%)	61–96	1.6:1–5:1	n.r.[d]	[152]
X=H, 5-F, 5-Cl, 5-Br, 5-MeO R=Ph, 2-Npt, 4-ClC$_6$H$_4$, prenyl, 4-ClC$_6$H$_4$CH$_2$, 2-ClC$_6$H$_4$CH$_2$, Me, PG=H	Ph, Pr, i-Bu, 4-BrC$_6$H$_4$, 4-CF$_3$C$_6$H$_4$, 4-MeOC$_6$H$_4$, 2,4-Cl$_2$C$_6$H$_3$, 2-furyl, 2-thienyl	**CIN-XIa** (10 mol%), 5Å MS[e]	76–96	3:1–21:1	91–99 (3S,αR)	[153]
X=H, 5-F, 5-Me, R=Bn, 4-MeC$_6$H$_4$CH$_2$, 4-MeOC$_6$H$_4$CH$_2$, 3-MeC$_6$H$_4$CH$_2$, 3-MeOC$_6$H$_4$CH$_2$, 2-FC$_6$H$_4$CH$_2$, 2-MeOC$_6$H$_4$CH$_2$, 3,5-Me$_2$C$_6$H$_3$CH$_2$, 3,4-(MeO)$_2$C$_6$H$_3$CH$_2$, 2-NptCH$_2$, 2-thienyl CH$_2$, EtO$_2$CCH$_2$, Pr, Bu, PG=Boc, Ph, CO$_2$Et	NHAc, NHBz	**IN** (10 mol%)	85–99[f]	6:1 to >99:1	55–89 (3R,αS)	[146]
X=H, 5-Me, 5-F, R=Me, Et, Bn, PG=CO$_2$Et, Boc, Cbz	Ph, 4-FC$_6$H$_4$, 4-BrC$_6$H$_4$, 4-ClC$_6$H$_4$, 4-NO$_2$C$_6$H$_4$, 4-MeOC$_6$H$_4$, 4-MeC$_6$H$_4$, 2-BrC$_6$H$_4$, 2-MeOC$_6$H$_4$, 2-fyryl, 2-thienyl	**TAK-XI** (12 mol%), Ni(OAc)$_2$ (10 mol%)	74–95	1.5:1–99:1	76–97[g] (3S,αR)	[147]

X=H, 5-MeO, 5-Me R=Ph, Bn, Me, 3-$FC_6H_4CH_2$, 4-$MeOC_6H_4CH_2$, 4-$MeC_6H_4CH_2$, piperonylmethyl PG=Boc	Ph, 4-MeC_6H_4, 4-FC_6H_4, 2-furyl[j]	**JHC** (R=n-$C_{12}H_{25}$, R^1=TMS) (10 mol%)	88–98	49:1 to >99:1	83 to >99 (3*S*,α*R*)	[118]
X=H, 5-Br, 5-Me, 5-F, 5-MeO, 5-CF_3O, 6-Cl, 7-CF_3 PG=Boc, Cbz, CO_2Me[h], R= 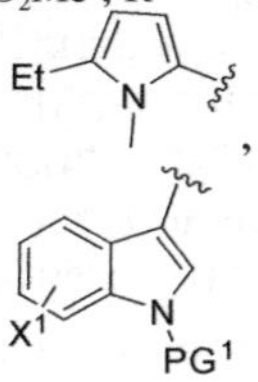 (X^1=H, 4-Me, 5-Me, 5-Br, 6-Me, 7-Me, 7-Cl, 7-I; PG^1=H, Me, Boc)	H	**IM-I** (11 mol%), $Ni(OAc)_2 \cdot 4H_2O$ (5 mol%)	30–95	—	44–93 (*S*)	[154]
X=H, 5-MeO, 5-Br, 5-F, 7-Cl R=CO_2Et, CO_2-*i*-Bu PG=Me, Bn, MOM, CO_2Et, H	Ph, 4-$MeOC_6H_4$, 4-MeC_6H_4, 4-BrC_6H_4, 4-ClC_6H_4, 4-FC_6H_4, 4-$CF_3C_6H_4$, 4-$NO_2C_6H_4$, 3-BrC_6H_4, 2-BrC_6H_4, 2-ClC_6H_4, 2,4-$Cl_2C_6H_3$, 2-Npt, 2-furyl, Pr	**CIN-IIa** (10 mol%), $PhCO_2H$ (10 mol%)[i]	48–99	1.5:1–19:1	72–98 (*S*,*S*)	[155]

(*Continued*)

Table 3.7.   (*Continued*)

X, R, PG	R^1	Catalyst	Yield (%)	d.r.	ee (%)	Ref.
X=H, 5-MeO, 5-F R=Me, Et, Bn, Ph PG=Boc[k]	Ph, 4-MeOC$_6$H$_4$, 4-MeC$_6$H$_4$, 4-BrC$_6$H$_4$, 4-ClC$_6$H$_4$, 4-FC$_6$H$_4$, 2-ClC$_6$H$_4$, 3,4-(MeO)$_2$C$_6$H$_3$, 2-thienyl, 2-furyl, Et[l] Ph(CH$_2$)$_2$	**TAK-XII** (2 mol%)	33–97	3:1–49:1	44–88 (3*S*,α*R*)	[156]
X=PG=H R=Cl, Bn, 4-NO$_2$ C$_6$H$_4$CH$_2$, 4-BrC$_6$H$_4$CH$_2$, 4-MeOC$_6$H$_4$CH$_2$	Ph	**CIN-IIIb** (10 mol%)	44–99	1.6:1–6.7:1	61–82[m] (3*S*,α*R*)	[157]
X=H, 5-Me, 5-MeO, 5-F, 6-Cl R=Me, Ph, 2-MeC$_6$H$_4$, 3-MeC$_6$H$_4$, 3-CF$_3$C$_6$H$_4$, 4-FC$_6$H$_4$, 2-Npt, 2-benzothienyl PG=Boc	H, 2-MeOC$_6$H$_4$, 3-MeOC$_6$H$_4$, 2-FC$_6$H$_4$, 2-furyl, 2-thienyl	**PTC-VIII** (1 mol%)	34–99	1.8:1 –99:1[n]	20–95 (*S*,*S*)	[158]
X=H R=Ph, Bu, CyCH$_2$, Bn, 4-MeOC$_6$H$_4$CH$_2$, 4-*i*-PrC$_6$H$_4$CH$_2$, 4-FC$_6$H$_4$CH$_2$, 4-ClC$_6$H$_4$CH$_2$, 3-MeOC$_6$H$_4$CH$_2$, 2-furyl, 2-thienyl PG=Boc	Ph, Bn[o], Pr[o], 4-FC$_6$H$_4$, 4-ClC$_6$H$_4$, 4-BrC$_6$H$_4$, 4-MeC$_6$H$_4$, 4-*i*-PrC$_6$H$_4$, 4-MeOC$_6$H$_4$, 2-ClC$_6$H$_4$, 3,4-Cl$_2$C$_6$H$_3$, 2-furyl, 2-thienyl	**GU-II** (10 mol%)	82–99	11.5: to >99:1	87 to >99[p] (3*S*,α*R*)	[159]
X=H, 5-F, 5-Cl, 5-Me, 6-Cl PG=H, Me, Ac, Ph, Bn, R= ⟨pyrrolyl⟩	Ph, Bn[l], Pr, 4-MeOC$_6$H$_4$, 4-FC$_6$H$_4$, 4-BrC$_6$H$_4$, 4-NO$_2$C$_6$H$_4$, 4-CNC$_6$H$_4$, 3-MeOC$_6$H$_4$, 3-FC$_6$H$_4$, 2-MeOC$_6$H$_4$, 2-FC$_6$H$_4$, 2-ClC$_6$H$_4$, 2-NO$_2$C$_6$H$_4$, 3,4-(MeO)$_2$C$_6$H$_4$, 2,4-Cl$_2$C$_6$H$_4$, 1-Npt, 2-furyl, 2-thienyl	**CIN-VIIIa** (5 mol%)	91–98	2.1:1 –24:1	74–99[p] (3*S*,α*R*)	[87]

X=H, 5-Cl, 5-Br, 5-Me, 5-MeO, 5-F, 6-Cl R=Ph, Me, Bn, 4-FC$_6$H$_4$, 4-MeC$_6$H$_4$, 3-FC$_6$H$_4$, 3-MeC$_6$H$_4$, 2-Npt PG=H, Boc, Me, Bn, Ac, Cbz, CO$_2$Me	CF$_3$[q]		**CIN-VIIIa** (5 mol%), 4Å MS	27–99	6:1 to >20:1	69–99 (R,R)[r]	[160]
X=H R=Me, Et PG=Boc	Ph, 4-MeC$_6$H$_4$, 4-MeOC$_6$H$_4$, 4-BrC$_6$H$_4$, 4-NO$_2$C$_6$H$_4$, 3-NO$_2$C$_6$H$_4$, 2-ClC$_6$H$_4$, 2-furyl, *i*-Pr		**IM-IIa** (5 mol%), AgBF$_4$ (10 mol%)	42 to >99	1.5:1–16:	88–97 (S,S)	[161]
			IM-IIb (5 mol%), AgBF$_4$ (10 mol%)	60 to >99	1 2:1–9:1	57–94 (3S,αR)	

Notes: [a](+)-Esermethole was then prepared in 72% overall yield over three steps. [b]N-Boc-3-methyloxindole gave only 34% yield, with 7:3 d.r. and 20% ee. 1-Nitro-1-pentene led to about 1:1 dr. [c]The 3-phenyl indole gave product only with 53% ee. N-Boc protected oxindole improved selectivity (19:1 d.r., 89% ee). It should be noted that Barbas III [148] and Luo and Cheng [151] used catalysts with opposite stereochemistry and obtained diastereoisomeric and not enantiomeric products. [d]n.r.= not reported, however, the major products were demonstrated to have (R,R) or (S,S) configuration. Only with catalyst **CIN-IIa**, the ee was calculated (85% ee with 2.2:1 d.r.) (PG= H, R = Ph, R^1 = 2,4-Cl$_2$C$_6$H$_3$). [e]3-Alkyloxindoles reacted at room temperature with 20 mol% of catalyst. [f]The reaction was scaled up to 5 mmol without decrease of yield and stereoselectivity. [g]PG=H, Bn did not work, PG=Boc, Cbz ee not reported. [h]PG=H, Ac did not work. The synthetic utility of the reaction was demonstrated by the construction of an optically active indole-containing cyclotryptamine motif. [i]An amount of 1.62 g was obtained with 96% yield, 6.1:1 d.r., and 95% ee. One of the products was transformed into spirooxindoles and into a CPC-1 derivative. [j]R^1=alkyl did not work. [k]Other protecting groups or NH-indole gave worse results. [l]R^1=Et gave the worst yield, R^1=*i*-Pr, Cy did not work. [m]The absolute configuration of the product with chlorine was not determined, but the *syn* diastereomer prevails anyway. (R,R)-**TAK-I** (C=S instead of C=O) afforded the enantiomeric (3R,αS) product. [n]R^1=Pr, Ph(CH$_2$)$_2$ gave 1.2 (94% ee):1 (84% ee) and 6.7(91% ee):1(n.r.) prevalence of the diastereomer (3S,αR). [o]These nitroalkenes need much longer reaction times. [p]The product in which R=R^1=Ph gave the opposite enantiomer, but no explanation was given. [q]The synthetic utility of the reaction was demonstrated by the construction of an optically active heptacyclic spirooxindole and of a pyrrolidinoindoline derivative. The reaction was also performed at a gram scale. The reaction with PG=Ac is faster. NH indole works well, although authors did not explore this reaction better. [r]A trifluoromethyl-substituted 2,3,3a,8-tetrahydropyrrolo[2,3-b]indole derivative was prepared in 60% overall yield (three steps) without affecting the ee.

coordination to the NO_2 oxygen atoms activates the nitroalkene and masks one face. Some reactions among those reported in Table 3.7 worked through a mechanism different from this general one. For instance, (1S,2R)-1-(pyrrolidin-1-yl)-2,3-dihydro-1H-inden-2-ol (**IN**) can form only a single hydrogen-bonding interaction between its hydroxyl group and the nitro group, but it was enough to assemble a transition state, in which the *Si*-face of the protected (Z)-2-amino-1-nitroethene was preferably attacked by the *Re*-face of the oxindole enolate [146].

The mechanism of the nickel catalysts were studied in detail. Yuan surmised that the nitro group occupied the open apical position on the nickel [147].

To minimize steric interactions, the bulky binaphthalene moiety had to be oriented away from the aromatic group of the nitroalkenes, and also π-π stacking interactions between the aromatic rings of oxindoles and nitroolefins are important because aliphatic nitroalkenes did not work. Then, the oxindole metal enolate approached from its *Si*-face accounting for the absolute configuration of the final products.

Imidazoline-aminophenol (IAP, Appendix A) nickel catalyst introduced by Arai [154] generated the enolate by exchanging one of the two IAP ligand of the nickel complex. The subsequent 1,4-addition to nitroalkenes afforded a zwitterion, which in turn exchanged the ligand with the starting 3-indolyl-3-oxindole leading to the product with regeneration of the enolate.

Mechler and Peters [161] demonstrated that the axially chiral *bis*-imidazolium salt determined the configuration of the stereocenter generated at the nitroalkene, while the oxindole stereocenter configuration is governed by the imino alcohol side arms.

Finally, in contrast to common opinion, Görling and Tsogoeva [157] demonstrated by DFT calculations that the thiourea moiety and the sulfonamide nitrogen of their catalysts coordinated only an oxygen atom of the nitro group, and also van der Waals interactions narrowed the reacting complex, enhancing stereoselectivity.

The reactions with nitroalkenes were employed in some cascade reactions. In the synthesis of indolyl-indanes (Scheme 3.46, Eq. (3.38)), [162] 3-methyloxindole (R=Me) and oxindole (R=H) gave only 39% and 44% yield with 33% and 73% ee, respectively.

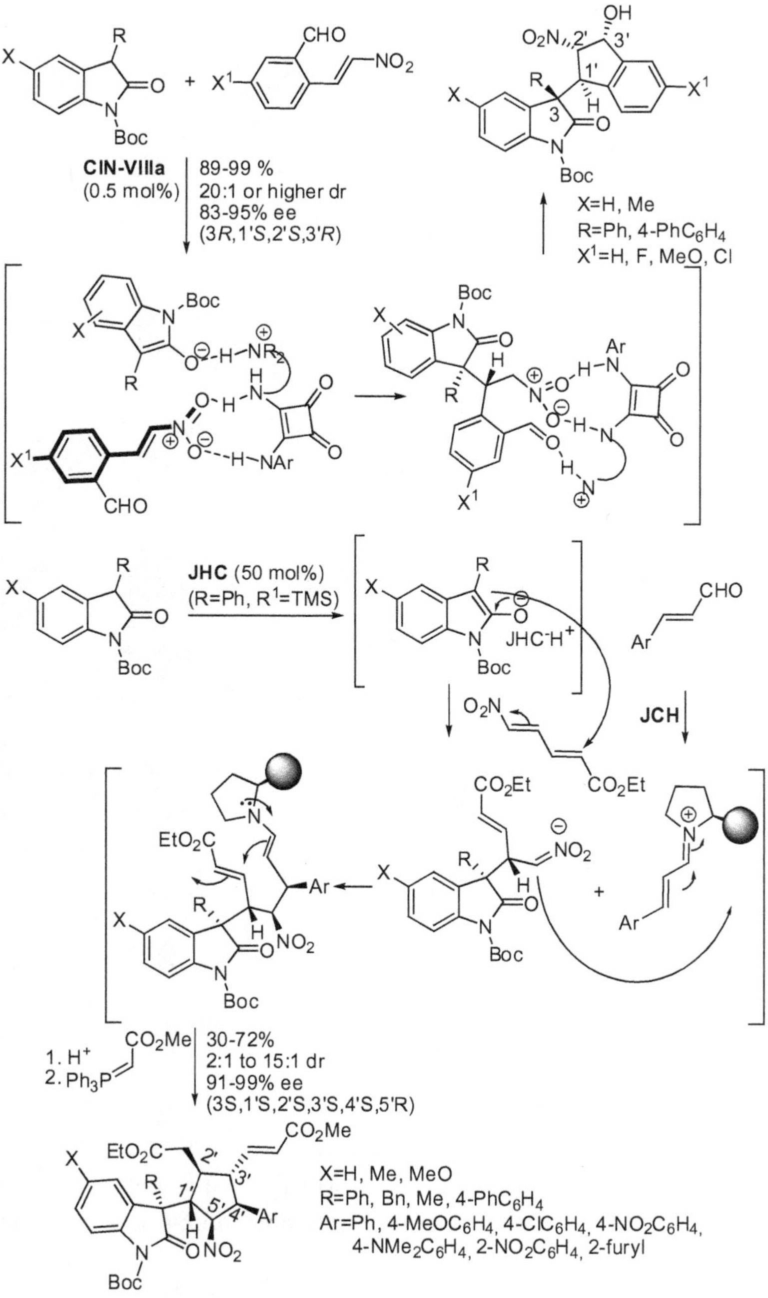

CIN-VIIIa
(0.5 mol%)

89-99 %
20:1 or higher dr
83-95% ee
(3*R*,1'*S*,2'*S*,3'*R*)

X=H, Me
R=Ph, 4-PhC$_6$H$_4$
X^1=H, F, MeO, Cl

(3.38)

JHC (50 mol%)
(R=Ph, R^1=TMS)

(3.39)

1. H$^+$
2. Ph$_3$P

30-72%
2:1 to 15:1 dr
91-99% ee
(3S,1'S,2'S,3'S,4'S,5'R)

X=H, Me, MeO
R=Ph, Bn, Me, 4-PhC$_6$H$_4$
Ar=Ph, 4-MeOC$_6$H$_4$, 4-ClC$_6$H$_4$, 4-NO$_2$C$_6$H$_4$,
4-NMe$_2$C$_6$H$_4$, 2-NO$_2$C$_6$H$_4$, 2-furyl

Scheme 3.46.   Cascade reactions involving nitroalkenes and oxindoles.

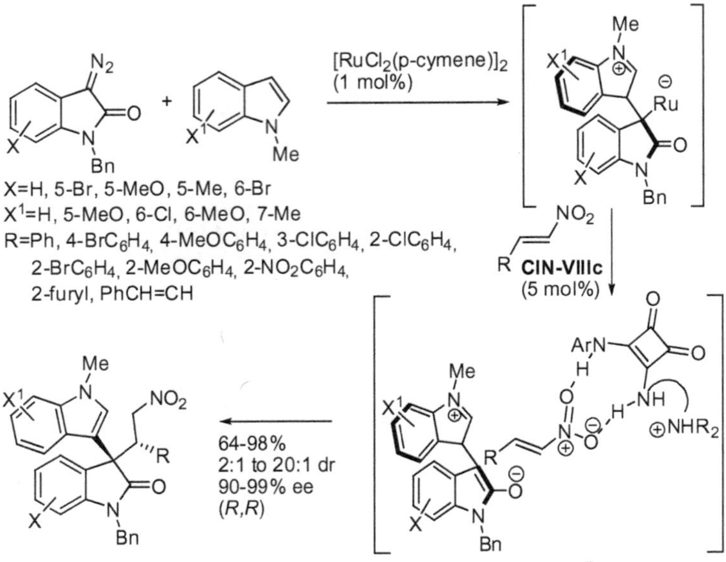

X=H, 5-Br, 5-MeO, 5-Me, 6-Br
X^1=H, 5-MeO, 6-Cl, 6-MeO, 7-Me
R=Ph, 4-BrC$_6$H$_4$, 4-MeOC$_6$H$_4$, 3-ClC$_6$H$_4$, 2-ClC$_6$H$_4$,
    2-BrC$_6$H$_4$, 2-MeOC$_6$H$_4$, 2-NO$_2$C$_6$H$_4$,
    2-furyl, PhCH=CH

[RuCl$_2$(p-cymene)]$_2$ (1 mol%)

CIN-VIIIc (5 mol%)

(3.40)

64-98%
2:1 to 20:1 dr
90-99% ee
(R,R)

X=H, F, Cl, Br, Me
Ar=4-OMeC$_6$H$_4$, 4-AcNHC$_6$H$_4$, 3,4-(MeO)$_2$C$_6$H$_3$,
    3-Br-4-MeOC$_6$H$_3$, 3-I-4-MeOC$_6$H$_3$,
    3-Me-4-MeOC$_6$H$_3$, 3,4-Me$_2$-thien-2-yl
R=Ph, n-C$_5$H$_{11}$, TMS, 4-FC$_6$H$_4$, 4-BrC$_6$H$_4$

Ph$_3$PAuCl (1 mol%) **CIN-XIb**
AgOTf (1.5 mol%)   (10 mol%)
then   5Å MS
44-99%
1.5:1 to >20:1 dr
93-99% ee (R,R)

Ph$_3$PAuNTf$_2$ (20 mol%)
80% overall
>20:1 dr
>99% ee

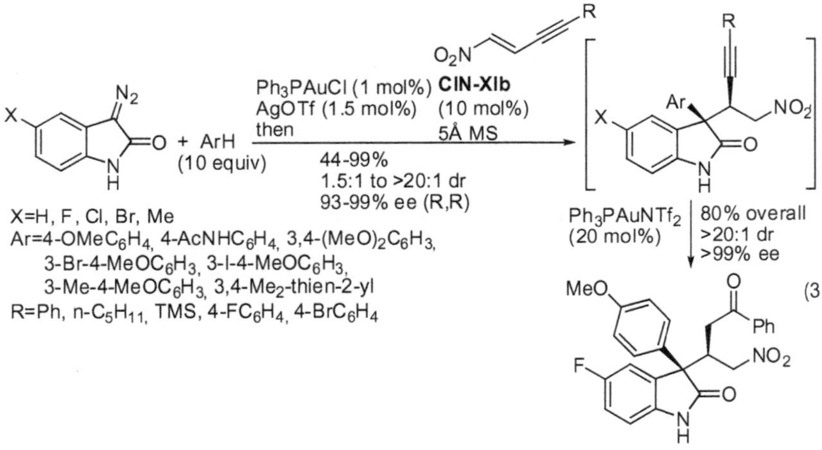

(3.41)

Scheme 3.46.   (*Continued*).

In the synthesis of fully substituted cyclopentyl-oxindoles (Scheme 3.46, Eq. (3.39)) [163], it is worth noting the exclusive 1,4-conjugate addition to the nitro group, notwithstanding four possible conjugate additions.

N-benzyl-3-diazooxindoles, *N*-methylindoles and nitroalkenes (Scheme 3.46, Eq. (3.40)) [164] reacted through a mechanism similar to that reported in Scheme 3.30 and can be applied to a seven-step total synthesis of (–)-folicanthine with 14.5% overall yield.

Then the reaction was extended to arenes other than indoles (Scheme 3.46, Eq. (3.41)) [165]. The very large excess of arene is undoubtedly a drawback. On the other hand, the resulting adducts were easily cyclized to spirooxindoles or a hydration step could be added to the one-pot sequence to give a quaternary oxindole with a ketone moiety, in which the diastereomeric ratio is even better because the minor diastereomer was more difficultly hydrated. The selective 1,4-conjugate addition found in these cascade reactions has to be outlined, being that 1,6-conjugated additions are known (see Schemes 3.37–3.39).

Vinyl sulfones are other substrates for the addition of 3-prostereogenic oxindoles. One of the first examples of this reaction was reported by

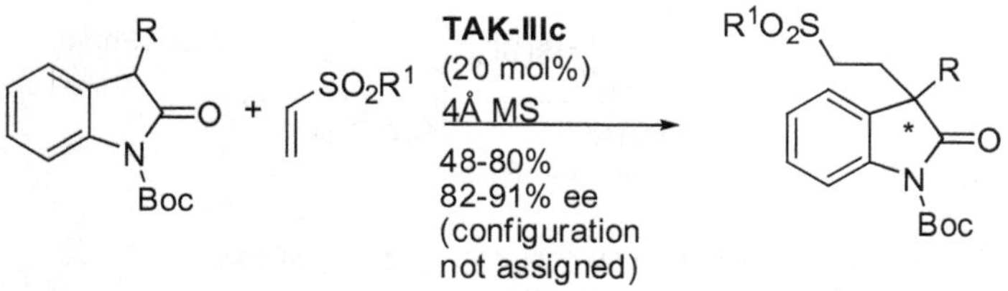

$$(3.42)$$

R=Ph, 4-MeC$_6$H$_4$, 4-MeOC$_6$H$_4$, 3,5-Me$_2$C$_6$H$_3$
R^1=Me, Ph

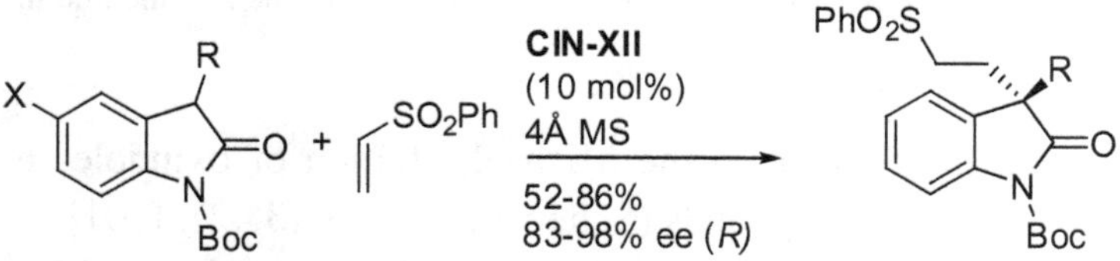

$$(3.43)$$

X=H, F, Me, MeO
R=Ph, 4-FC$_6$H$_4$, 4-MeC$_6$H$_4$, 4-MeOC$_6$H$_4$, 3-MeC$_6$H$_4$, 2-MeC$_6$H$_4$, 3,5-Me$_2$C$_6$H$_3$, 2-Npt

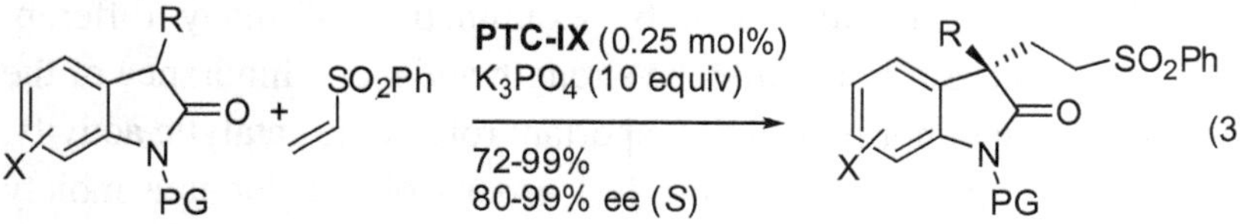

$$(3.44)$$

PG= PMB, Bn, Me, prenyl
X=H, 5-F, 5-Cl, 5-MeO, 5-Br, 6-Br, 7-F
R=Me, Et, Pr, Bn, n.C$_6$H$_{13}$, allyl, prenyl, CH$_2$=C(Me)CH$_2$, propargyl, 4-MeOC$_6$H$_4$CH$_2$, 4-ClC$_6$H$_4$CH$_2$, 2-furylCH$_2$, MeCO(CH$_2$)$_2$, CNCH$_2$, MeO$_2$CCH$_2$, EtO$_2$CCH$_2$

Scheme 3.47.   Asymmetric addition of oxindoles to vinyl sulfones.

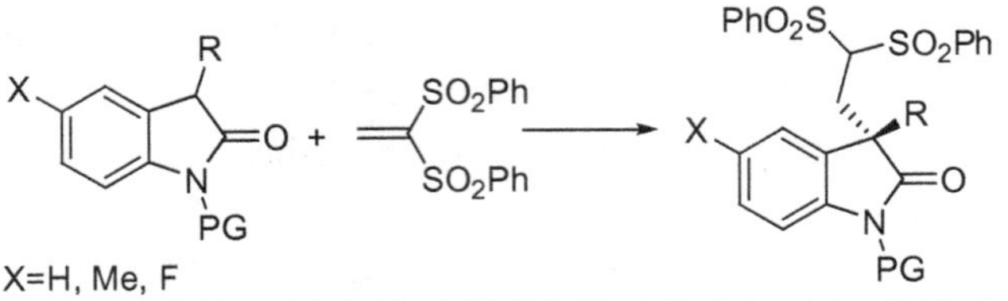

X=H, Me, F
R=Ph, 4-FC$_6$H$_4$, 4-MeC$_6$H$_4$, 4-MeOC$_6$H$_4$, 4-PhC$_6$H$_4$, 3-MeC$_6$H$_4$, 2-MeC$_6$H$_4$,
   2-MeOC$_6$H$_4$, 2-Npt, Bn, 4-FC$_6$H$_4$CH$_2$, 4-MeOC$_6$H$_4$CH$_2$, Pr, Bu,n-C$_6$H$_{13}$, n-C$_{11}$H$_{23}$
PG=Boc, H

ref. 168: R=Aryl: **Pseudo-CIN-IIa** (20 mol%), 94-98%, 90-99% ee (*R*)
         R=Alkyl: **CIN-XIII** (20 mol%),72-88%, 77-91% ee (*R*)
ref. 169: R=Aryl: **TAK-XIII** (10 mol%), 80-95%, 85-99% ee (*R*)
         R=Alkyl: **TAK-XIV** (10 mol%), 80-92%, 91-97% ee (*R*)

Scheme 3.48.   Asymmetric addition of oxindoles to vinyl *bis*-sulfones.

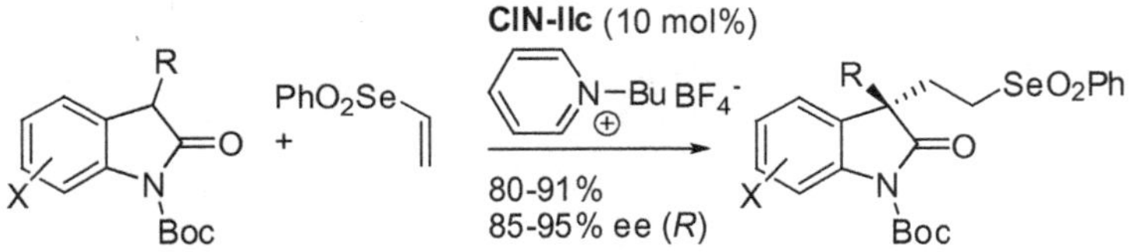

X=H, 5-Me,
R=Me, Et, Pr, Bn, 4-MeC$_6$H$_4$CH$_2$, 3-MeC$_6$H$_4$CH$_2$, 3-MeOC$_6$H$_4$CH$_2$,
   2-MeC$_6$H$_4$CH$_2$, 2-MeOC$_6$H$_4$CH$_2$, 1-NptCH$_2$

Scheme 3.49.   Enantioselective addition of oxindoles to vinyl selenone in ionic liquid.

Luo and Cheng in their paper on the Michael addition of oxindoles to
α,β-unsaturated carbonyl compounds (Scheme 3.47, Eq. (3.42)) [101].

Phenyl vinyl sulfone worked better than methyl vinyl sulfone, which,
in particular, was completely unreactive with 3-(4-methoxyphenyl)- and
3-methyl-*N*-Boc-2-oxindole. Then, by examination of many different
catalysts, Zhao and Shi found that the strength and steric hindrance of the
hydrogen-bonding donor played an important role in the catalytic activity,
creating hydrogen-bonds among the NH groups of the thiourea moiety
and the sulfone oxygen atoms, as well as between the nitrogen of the quinuclidine framework and the carbonyls of indole and Boc group [166].
Under their conditions, however, protecting groups other than Boc and
3-alkyl substituents afforded the corresponding in very low yields or did
not react at all (Scheme 3.47, Eq. (3.43)). The phase-transfer pentanidium
catalyst allowed a version of the reaction easily scalable, highly

enantioselective, and useful to convert them into pyrroloindoline derivatives with very low catalyst loading (Scheme 3.47, Eq. (3.44)) [167].

1,1-*Bis*(benzenesulfonyl)ethylene was also employed as activated olefins for the addition of 3-substituted oxindoles. Two different catalysts were found necessary for 3-aryl and 3-alkyl substituted oxindoles, respectively (Scheme 3.48) [168, 169, 106b]. The different pKa values between aryl and alkyl substituents could account for this feature.

The surmised mechanism, once more, contemplated strong hydrogen bonds among the NH groups and the sulfone oxygen atoms, as well as between the nitrogen of the quinuclidine framework and the carbonyls of indole and Boc group, which was essential to ensure high selectivity. Moreover, π–π stacking interactions could be also involved in 3-aryl-substituted oxindoles.

Just one example employed α,β-unsaturated selenones (Scheme 3.49) [170]. Only 3-alkyl-2-oxindoles were tested. Ionic liquids were found to influence yields and enantioselectivity that is imidazole-type ionic decreased yields than pyridine-type, while more stable anions lowered enantioselectivity, because they interfered with the classical chiral ion-pair quinuclidine nitrogen oxindole enolate.

Since selenones can be more easily transformed into other functional group than sulfur analogues, products were further manipulated into iodide, azides and pyrroloindoline.

Vinylphosphonates are other valuable activated alkenes (Scheme 3.50) [171, 172]. The stereochemistry of these reactions is in agreement with

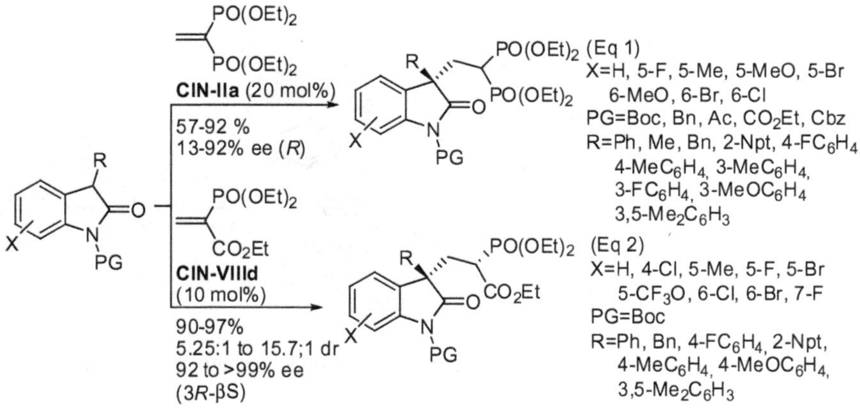

Scheme 3.50. Enantioselective addition of oxindoles to vinyl phosphonates.

the classical mechanism. Interestingly, carrying out a Horner–Wadsworth–Emmons reaction with formaldehyde on products from both reactions, the enantioselectivity drastically decreased, owing to a retro-Michael addition/aldol reaction. Furthermore, under thiourea catalysis [171], 3-(*ortho*-substituted phenyl)-, *N*-methyl and *N*-unprotected oxindoles did not react, and *N*-acetyl-oxindoles led to low yield and selectivity. Both pseudo-enantiomers of the squaramide catalyst gave the corresponding products in comparable yields and selectivities [172].

3.7. Rearrangements

The pericyclic rearrangements such as [3,3]sigmatropic ones are surely among the most stereoselective reactions in organic chemistry. Among them, rearrangement of 2-indolyl esters represents an example of the O-to-C rearrangement discovered by Black in 1986 with benzofuranones [173]. Unfortunately, the starting materials are difficult to prepare due to the competing reaction at O and C atoms and only *O*-acetylation of oxindoles followed by *N*-protection can lead to the 2-indolyl esters in acceptable yield.

Only in 2003 the first examples of this reaction appeared in the literature, when Fu was able to obtain the O-to-C rearrangement of 2-indolyl carbonates by means of a chiral ferrocene (Scheme 3.51) [174]. The bulkiness of the carbonate group increased the enantioselectivity and trichloro-*t*-butyl group was found the best. The less reactive 3-alkyl substituted oxindoles required higher loading of catalyst. The true mechanism was not simultaneous, but stepwise.

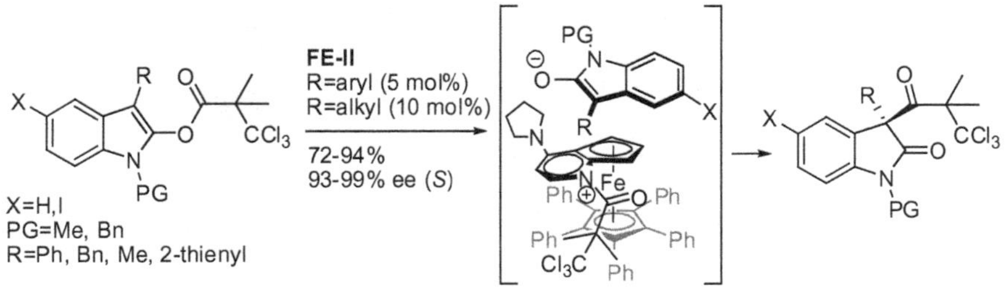

Scheme 3.51. Enantioselective rearrangement of indolyl acetates and carbonates by ferrocene catalysis.

Vedejs and coworkers introduced some chiral pyridine catalysts and tested them in this rearrangement reaction [175]. They found the *N*-diphenylacetyl group as the most efficient protecting group to promote rearrangement and two different catalysts gave best results for indolyl acetates and carbonates, respectively (Scheme 3.52). The two catalysts gave the opposite configuration, but the authors did not give any explanation. The reaction can be scaled up to gram scale without affecting yield and selectivity with only 1 mol% of catalyst.

2-Amino allyl vinyl ethers rearrange to γ,δ-unsaturated amides (Meerwein–Eschenmoser–Claisen rearrangement) as well as 2-amino propargyl vinyl ethers to allenyl amides (Saucy–Marbet–Claisen rearrangement), but only stable hemiaminals can undergo asymmetric catalysis and indole derivatives are surely an interesting candidate. Actually, they facilely rearranged under **BINAP** or **PHOS**-**I** palladium catalysis [176, 177]. In the reaction of 2-allyloxyindoles (Scheme 3.53, Eq. (3.45)) [176], the two catalysts were complementary, in fact, **BINAP** catalyst gave higher selectivity with bulkier ester groups and smaller R groups, whereas **PHOS**-**I** catalyst provided best results with small ester groups

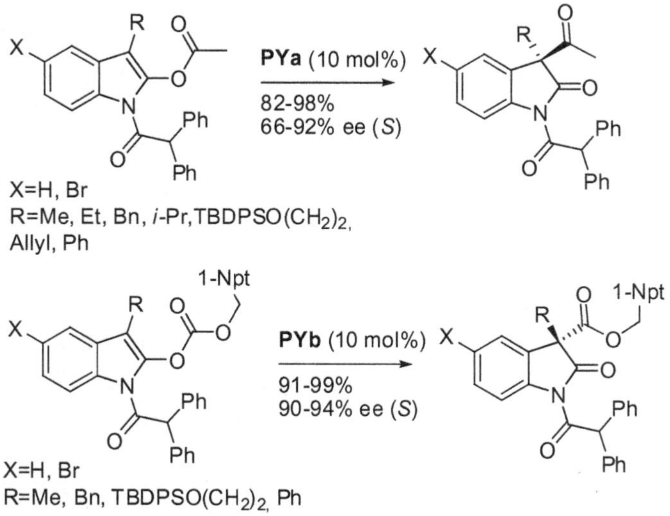

Scheme 3.52. Enantioselective rearrangements of indolyl acetates and carbonates by pyridine catalysis (note the different configuration but the same Cahn–Ingold–Prelog descriptor).

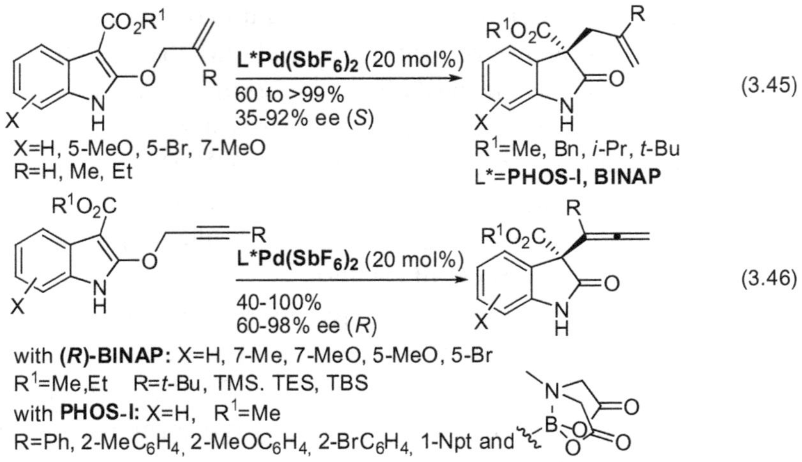

CO2R1
L*Pd(SbF6)2 (20 mol%)
60 to >99%
35-92% ee (S)
X=H, 5-MeO, 5-Br, 7-MeO
R=H, Me, Et
R1=Me, Bn, i-Pr, t-Bu
L*=PHOS-I, BINAP
(3.45)
R1O2C
L*Pd(SbF6)2 (20 mol%)
40-100%
60-98% ee (R)
(3.46)
with (R)-BINAP: X=H, 7-Me, 7-MeO, 5-MeO, 5-Br
R1=Me,Et R=t-Bu, TMS. TES, TBS
with PHOS-I: X=H, R1=Me
R=Ph, 2-MeC6H4, 2-MeOC6H4, 2-BrC6H4, 1-Npt and

Scheme 3.53. Meerwein-Eschenmoser-Claisen (Eq. (3.45)) and Saucy–Marbet–Claisen (Eq. (3.46)) rearrangements of indole derivatives (note the same configuration but the opposite Cahn–Ingold–Prelog descriptor).

and large R groups. The stereochemical outcome was explained by a palladium complex in which the Lewis acidic palladium ion coordinated the ether and the ester oxygen atoms with the oxazolidine ring far from the C-3. In the reaction of 2-propargyloxyindoles (Scheme 3.53, Eq. (3.46)) [177], a bulky substituent on the triple bond is always necessary for high stereoselectivity (R=H, Me gave only 8% and 4% ee, respectively), and **PHOS-I** catalyst did not work well with non-aryl R groups. The products can be converted into spirocyclic lactones through a tandem rearrangement, as well as into other derivatives, by hydration or desilylation.

Trost and Osipov applied the asymmetric allylic alkylation (see Table 3.2) to the decarboxylative rearrangement of *N,N′*-diBoc-2,2′-*bis*(allyloxycarbonyloxy)-3,3′-biindole (Scheme 3.54) [178], and the reaction was efficiently scaled up to more than nine grams of the product. Then the key intermediates for the synthesis of (–)-chimonanthine, (–)-folicanthine, (+)-calycanthine, (–)-WIN 64821 and (–)-ditryptophenaline were prepared. Authors carried out some experiments to elucidate the mechanism and found that the two allylation steps are not synchronous, but the first was enantio-determining with about 93% ee, while the second allylation step decreased the selectivity.

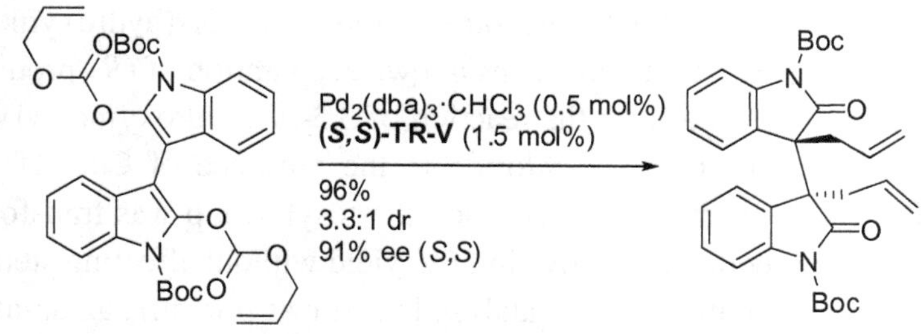

Scheme 3.54.   Asymmetric Pd-catalyzed decarboxylative asymmetric allylic alkylation.

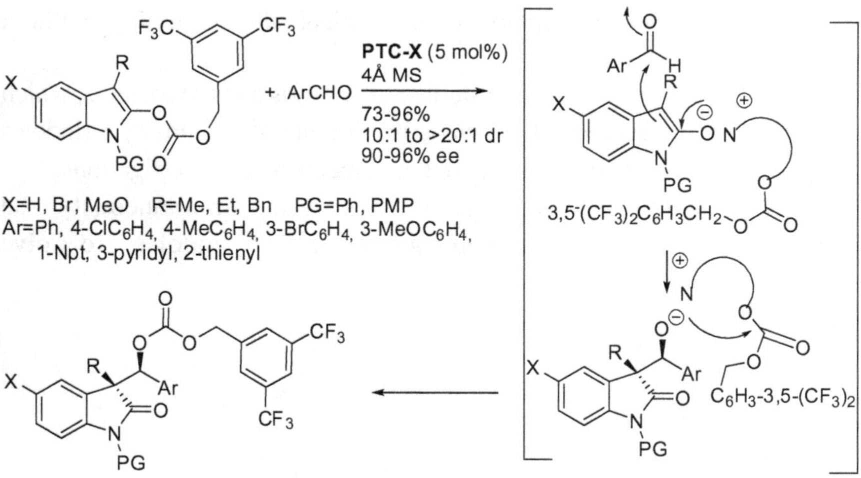

X=H, Br, MeO   R=Me, Et, Bn   PG=Ph, PMP
Ar=Ph, 4-ClC$_6$H$_4$, 4-MeC$_6$H$_4$, 3-BrC$_6$H$_4$, 3-MeOC$_6$H$_4$,
   1-Npt, 3-pyridyl, 2-thienyl

Scheme 3.55.   Asymmetric Steglich-type rearrangement.

The Steglich-type rearrangement was also performed (Scheme 3.55) [179]. The surmised mechanism contemplated a disruption of the carbonate moiety by the betaine negatively charged oxygen and the formation of a tight chiral ion-pair with the indole enolate.

Then an insertion of an aldehyde from the less hindered face and an acyl transfer occurred.

3.8. Desymmetrization

Kinetic resolution of 3-prochiral oxindole is another method to obtain enantiomerically enriched 3,3-dialkyloxindoles. The first example was the copper-catalyzed asymmetric monotosylation of 3,3-*bis*(hydroxymethyl) oxindoles [180]. The best conditions allowed separation of (*R*)-product in 81% yield and 99% ee from the reaction of 3,3-*bis*(hydroxymethyl)oxindole and 1.2 equiv. of tosyl chloride in the presence of Cu(OTf)$_2$ and (***R,R***)-**BOX-III** (10 mol%, each). Then the tosyl group was transformed into the 3-cyanomethyl derivative in 76% yield without affecting stereoselectivity. More recently, the Cu-catalyzed kinetic asymmetric azide-alkyne cycloaddition was reported (Scheme 3.56) [181]. Cyclohexyl azide gave only 13% yield, albeit with 86% ee. Some chemical transformations, such as [3+2] cycloaddition, Sonogashira coupling, and full or partial hydrogenation, were then attempted at the residual alkyne moiety, without affecting the selectivity.

Later, another research group studied the ligands and the solvents, which allowed achieving reproducible kinetic resolutions of quaternary oxindoles [182], but these reactions still need detailed mechanistic investigations.

Also, racemic spiroepoxyoxindoles can simultaneously be enantiomerically enriched and alkylated by indoles to give

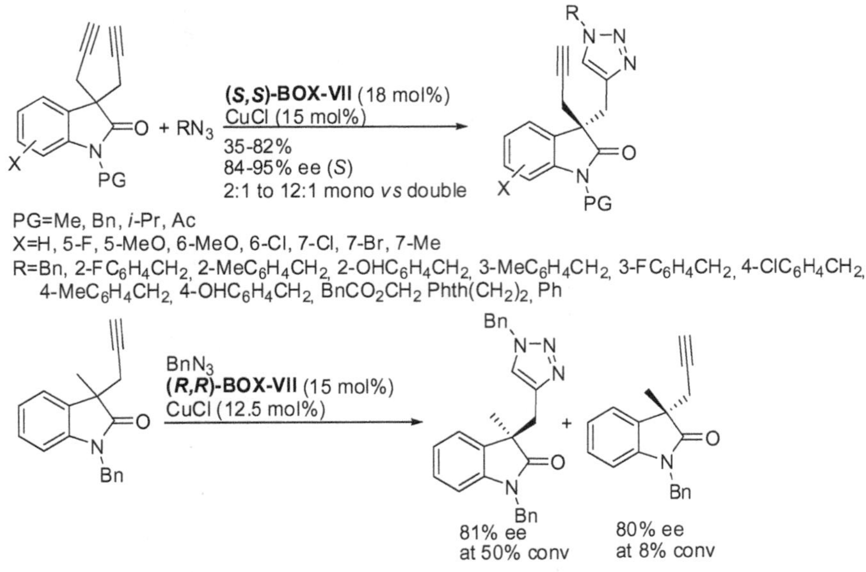

Scheme 3.56. Asymmetric Cu(I)-catalyzed azide-alkyne cycloaddition.

PHOS-XIV (20 mol%)
Amberlite CG50

30-50%
88-99% ee (*R*)

33-49%
93 to >99% (*R*)

X=H, 5-Me, 5-MeO, 5-F, 5-Cl, 5-Br,
  6-Cl, 6-Br, 7-Cl,
X^1=H, 4-Me, 4-Cl, 4-MeO. 5-Me, 5-Br, 5-MeO,5-Cl, 6-MeO, 6-F, 7-Me, 7-F, 7-Cl

Scheme 3.57.   Alkylative kinetic resolution of spiroepoxyoxindoles.

CIN-XIV (5 mol%)
R^{2}Br (0.55 equiv)
K$_2$CO$_3$
then
NEt$_3$, Me(CH$_2$)$_7$OCOCl

47-79%
18-95% ee (*S*)

21-48% conv
>99:1 dr
87-99% ee (3*S*,α*R*)

R=H, Me, Et, F, R^1=Me, Bn
R^2=Bn, 4-CNC$_6$H$_4$CH$_2$, 4-FC$_6$H$_4$CH$_2$, 4-CF$_3$C$_6$H$_4$CH$_2$,
  4-MeC$_6$H$_4$CH$_2$, 3,5-Br$_2$C$_6$H$_3$CH$_2$, allyl, CH$_2$=CMeCH$_2$,
TMS—≡—CH$_2$

Scheme 3.58.   Alkylative kinetic resolution of 3-prochiral oxindoles.

CIN-VIIa
(10 mol%)

65-95%
8:1 to >20;1 dr
74-94% ee
(3*R*,1'*R*,3'*S*)

X=H, Me, Cl
R=Ph, 4-PhC$_6$H$_4$, 4-FC$_6$H$_4$, 4-MeC$_6$H$_4$
R^1=Bn, 4-*t*-BuC$_6$H$_4$CH$_2$, 4-ClC$_6$H$_4$CH$_2$, 4-BrC$_6$H$_4$CH$_2$,
  3-ClC$_6$H$_4$CH$_2$, 2-ClC$_6$H$_4$CH$_2$, 2-Npt, Allyl

Scheme 3.59.   Organocatalytic desymmetrization of cyclopent-4-ene-1,3-diones.

3-(3-indolyl)-oxindole-3-methanols (Scheme 3.57) [183]. The reaction was interestingly applied to a gram-scale synthesis of (+)-gliocladin C. Moreover, enantioenriched (*R*)-spiroepoxyoxindole was also regioselectively opened by amines from the less-hindered side leading to 3-hydroxy-3-aminomethyoxindole without affecting enantiopurity.

Sorrentino and Connon described another alkylative kinetic resolution of 3-prochiral oxindoles to give benzylated, allylated, and propargylated 2-oxindoles by bifunctional phase-transfer catalysis (Scheme 3.58) [184]. If R=H, the resolved starting materials underwent racemization via a retro-Michael–Michael process.

Finally, desymmetrization of prochiral cyclopent-4-ene-1,3-diones has been performed with 3-aryloxindoles under organocatalytic Michael reaction (Scheme 3.59) [185]. A gram-scale reaction led to the product in 92% yield, >20:1 d.r., 94% ee.

Appendix A. Catalysts Cited in This Chapter

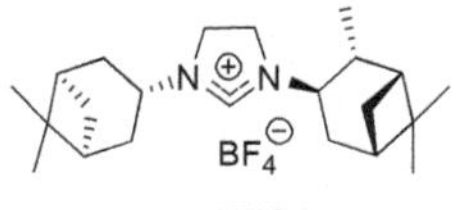

NHC-I
(Table 3.1)

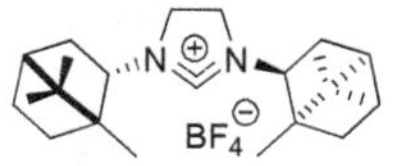

NHC-II
(Table 3.1)

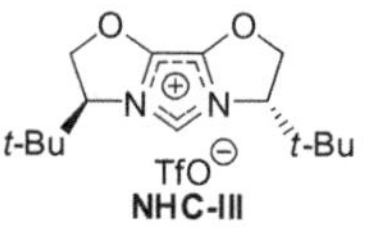

NHC-III
(Table 3.1)

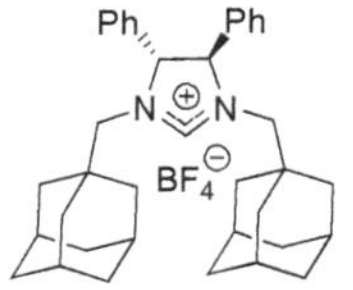

NHC-IV
(Table 3.1)

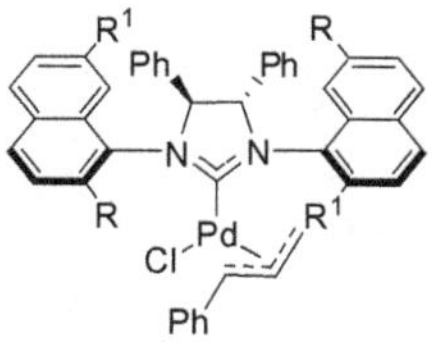

NHC-V
(Table 3.1)

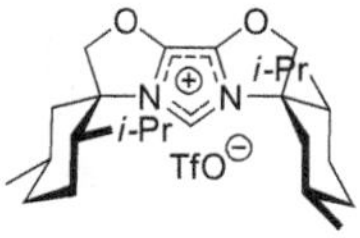

NHC-VI
(Table 3.1)

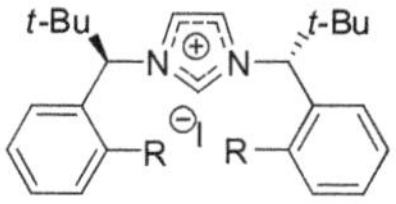

NHC-VII
(Table 3.1)

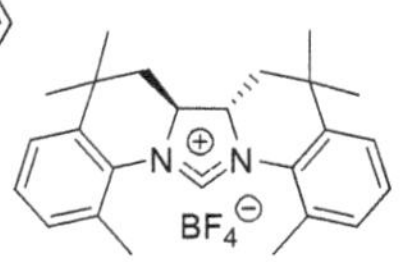

NHC-VIII
(Table 3.1)

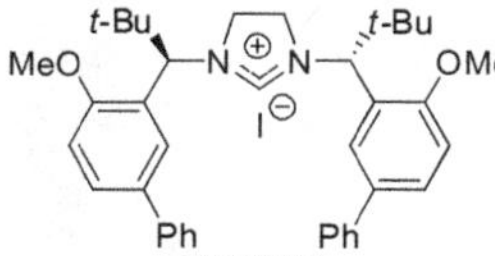

NHC-IX
(Table 3.1)

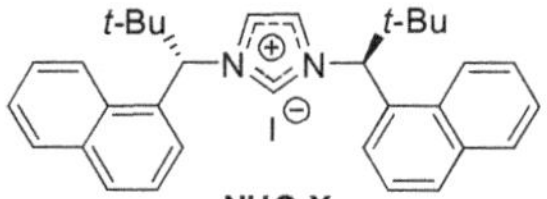

NHC-X
(Table 3.1)

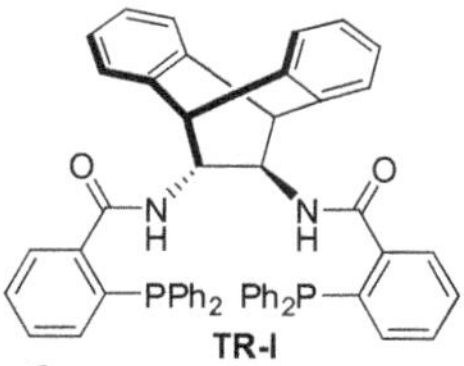

TR-I
(Scheme 3.10)
(Table 3.2)

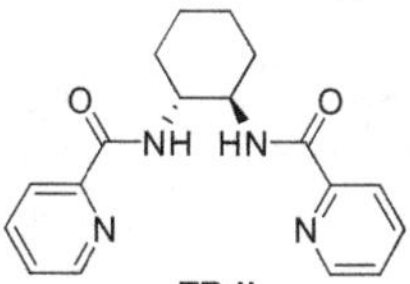

TR-II
(Table 3.2)

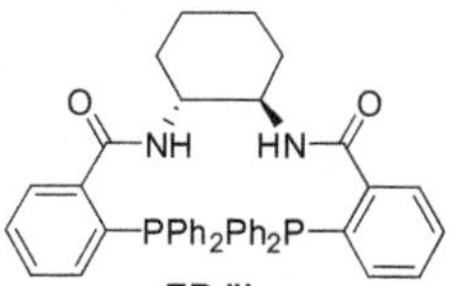

TR-III
(Table 3.2)
(Scheme 3.35)

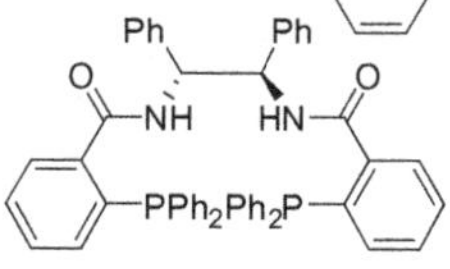

TR-V
(Table 3.2)
(Scheme 3.54)

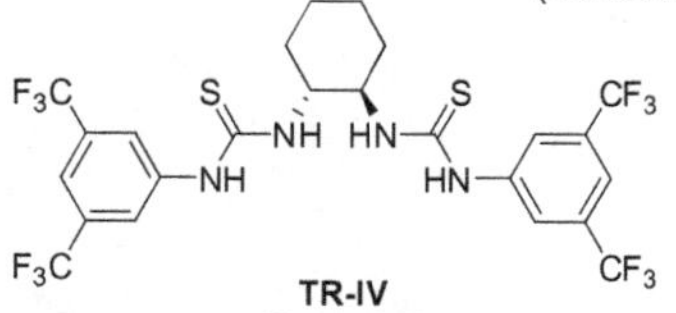

TR-IV
(Table 3.2)

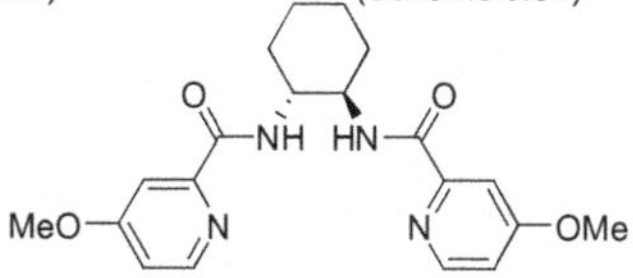

TR-VI
(Table 3.2)

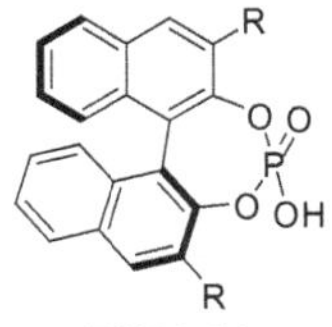

BINOL-PA
(Schemes 3.10, 3.30)
(Table 3.4)

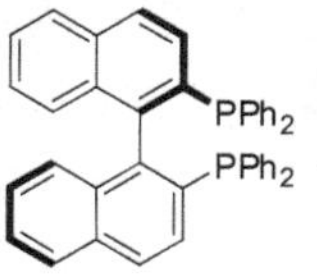

BINAP
(Table 3.1)
(Schemes 3.1, 3.53)

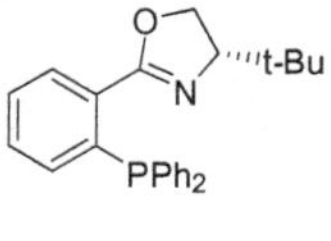

PHOS-I
(Table 3.1)
(Schemes 3.5, 3.53)

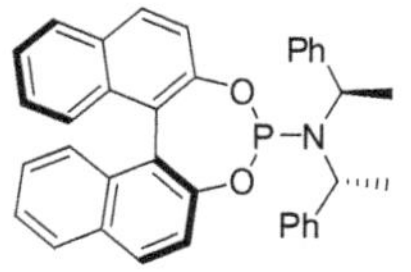

PHOS-II
(Scheme 3.7)

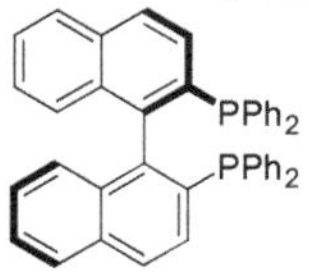

PHOS-III
(Table 3.2)

PHOS-IV
(Scheme 3.16)

PHOS-V
(Table 3.4)

PHOS-VI
(Table 3.5)

PHOS-VII
(Scheme 3.32)

PHOS-VIII
(Table 3.6)

PHOS-IX
a: R=R^1=Me, X=O (Table 3.6)
b: R=OH, R^1=Me, X=O (Table 3.6)
c: R=OTMS, R^1=Me, X=O (Table 3.6)
d: R=R^1=H, X=O (Scheme 3.35)
e: R=Et, R^1=Me, X=O (Scheme 3.36)
f: R=Bn, R^1=Me, X=S (Scheme 3.36)

PHOS-X
(Table 3.6)

PHOS-XI
(Table 3.6)

PHOS-XII
(Scheme 3.35)

PHOS-XIII
(Scheme 3.35)

PHOS-XIV
(Scheme 3.57)

PTC-I
(Scheme 3.11)

PTC-II
(Scheme 3.12)

PTC-III
(Scheme 3.16)

PTC-IV
(Scheme 3.17)

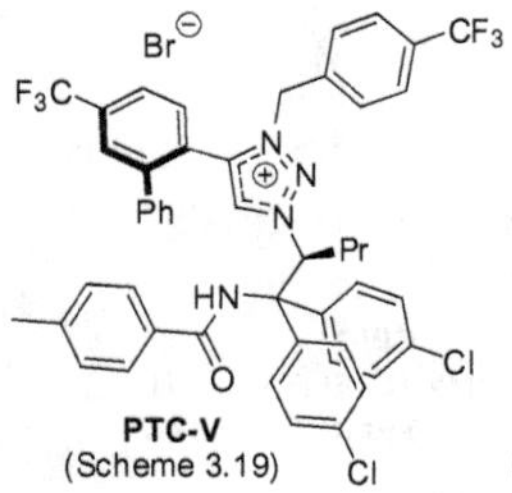

PTC-V
(Scheme 3.19)

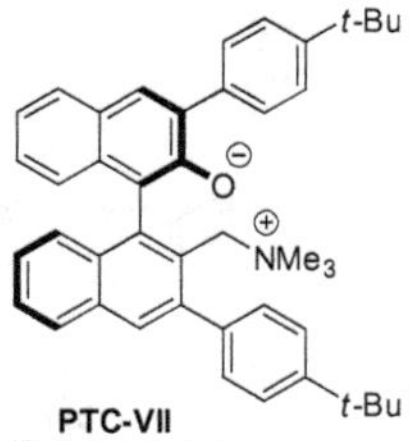

PTC-VII
(Scheme 3.28)

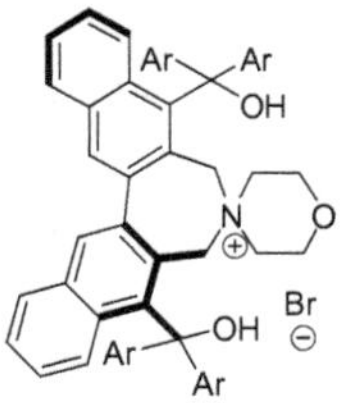

PTC-VIII
[Ar=3,5-$(CF_3)_2C_6H_3$]
(Table 3.7)

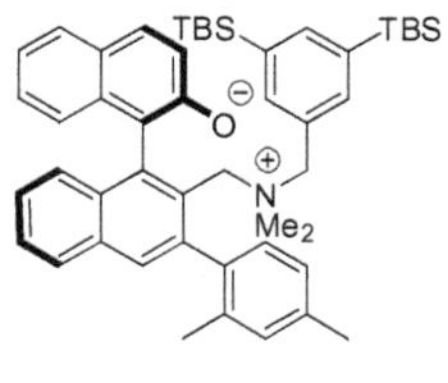

PTC-X
(Scheme 3.55)

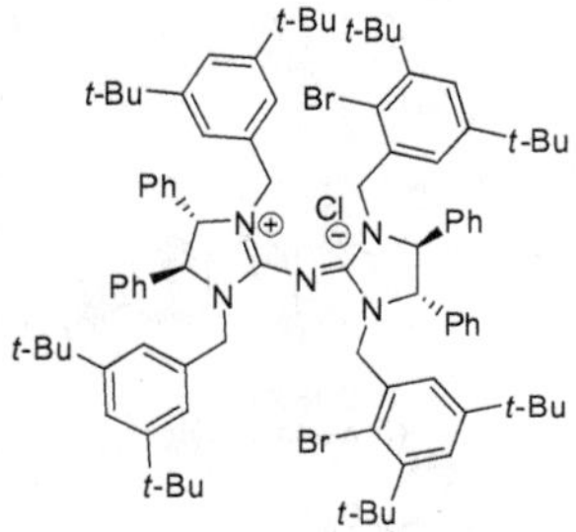

PTC-IX
(Scheme 3.47)

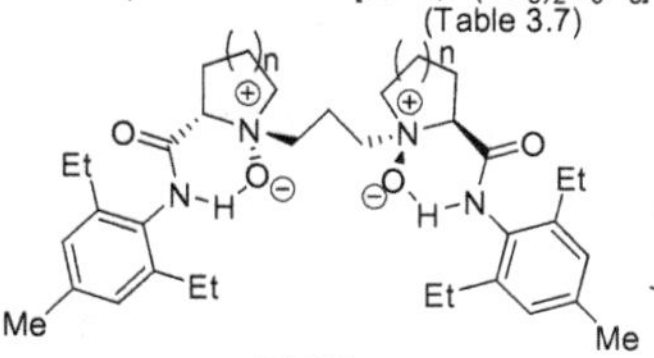

NNO-I
(Scheme 3.34)

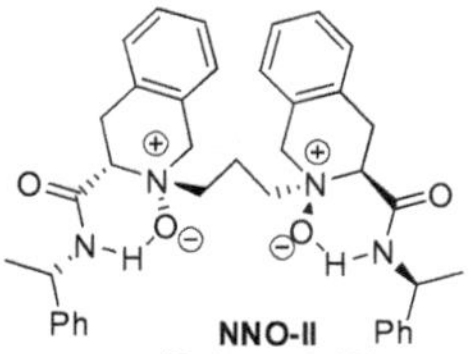

NNO-II
(Scheme 3.16)

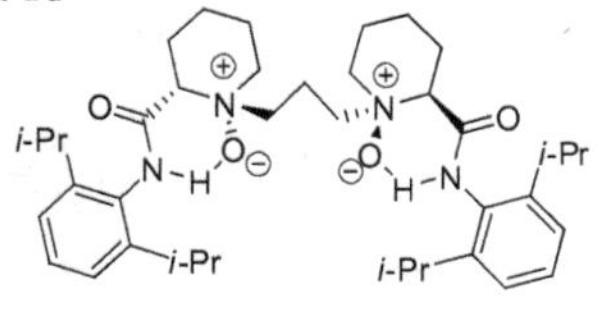

NNO-III
(Scheme 3.27)

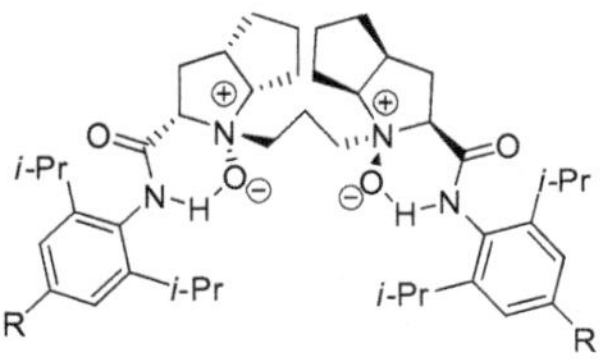

NNO-IV
a: R=H (Table 3.5) (Scheme 3.37)
b: R=i-Pr (Table 3.6)

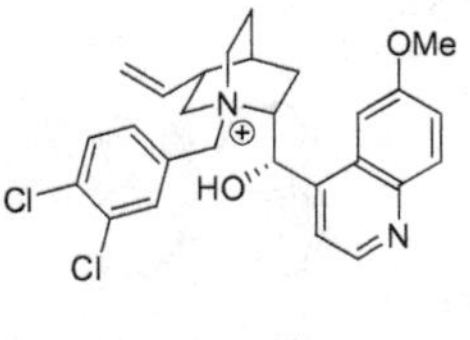

CIN-I
(Scheme 3.11)

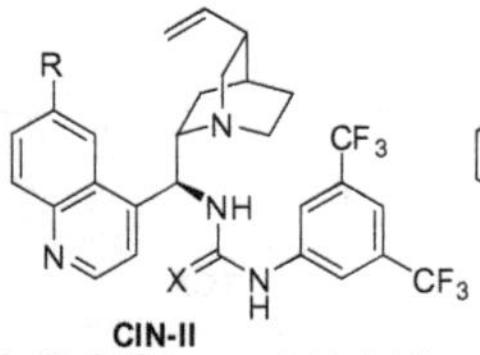

CIN-II
a: R=OMe, X=S (Schemes 3.14, 3.41)
(Table 3.6, 3.7)
b: R=OH, X=S (Table 3.4)
c: R=H, X=S (Schemes 3.40, 3.49) (Tables 3.5, 3.6)
d: R=OMe, X=O (Scheme 3.34)

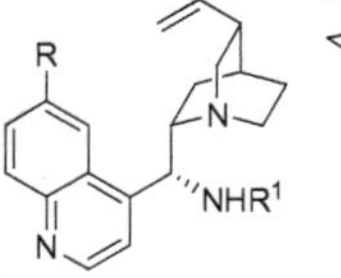

CIN-III
a: R=R^1=H (Table 3.2)
b: R=OMe, R^1=Ts (Table 3.7)

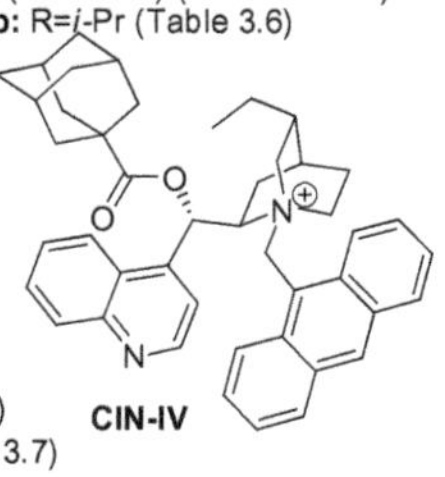

CIN-IV

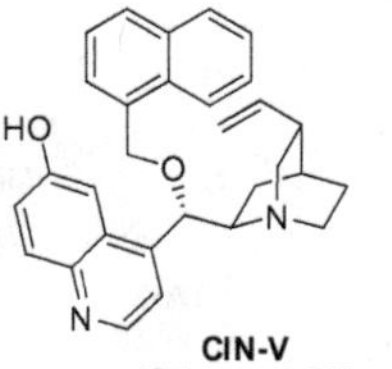

CIN-V
(Scheme 3.20)

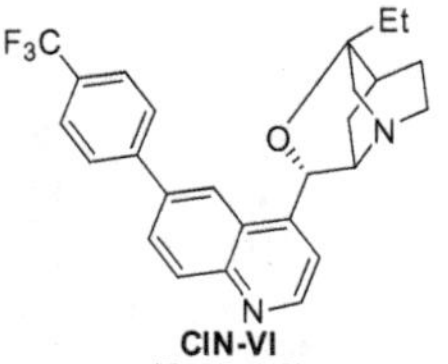

CIN-VI
(Table 3.3)

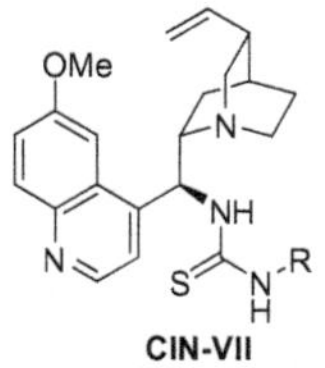

CIN-VII
a: R=Ph (Table 3.4)
b: R=t-Bu (Scheme 3.42)

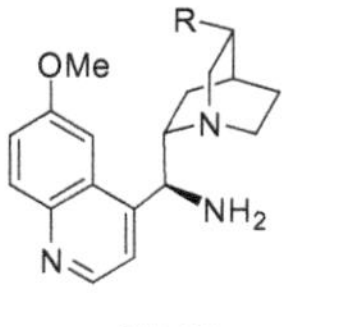

CIN-IX
a: R=Et (Table 3.6)
b: R=vinyl (Table 3.6)
(Scheme 3.43)

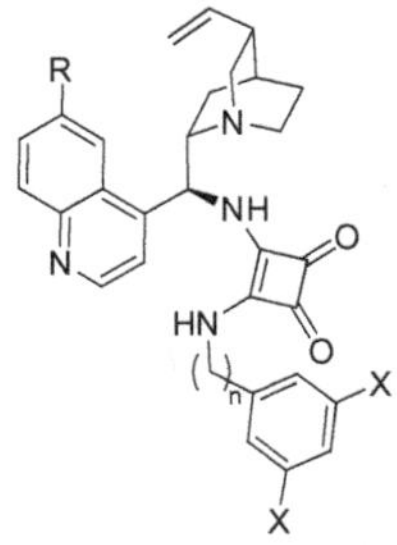

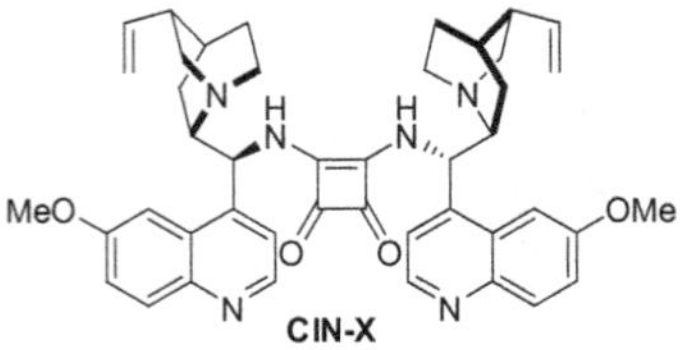

CIN-X
(Scheme 3.39)

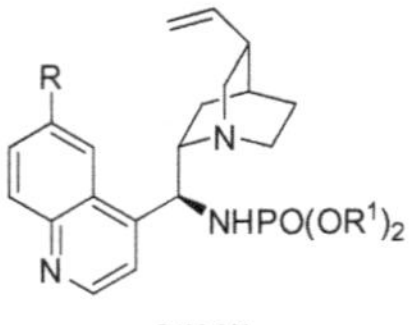

CIN-XI
a: R=OMe, R^1=Et (Table 3.7)
b: R=H, R^1=i-Pr (Scheme 3.46)

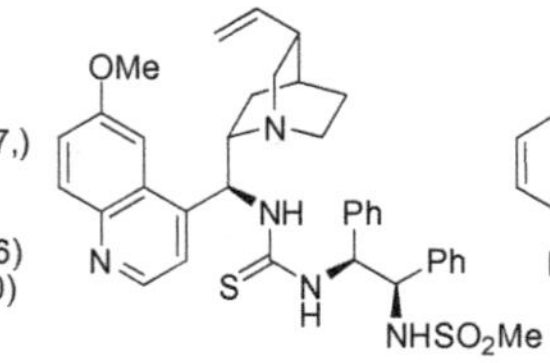

CIN-VIII
a: n=0, R=OMe, X=CF$_3$ (Table 3.5, 3.7,)
(Schemes 3.39, 3.45)
b: n=0, R=H, X=CF$_3$ (Table 3.6)
c: n=1, R=OMe, X=CF$_3$ (Scheme 3.46)
d: n=0, R=OMe, X=Me (Scheme 3.50)

CIN-XII
(Scheme 3.47)

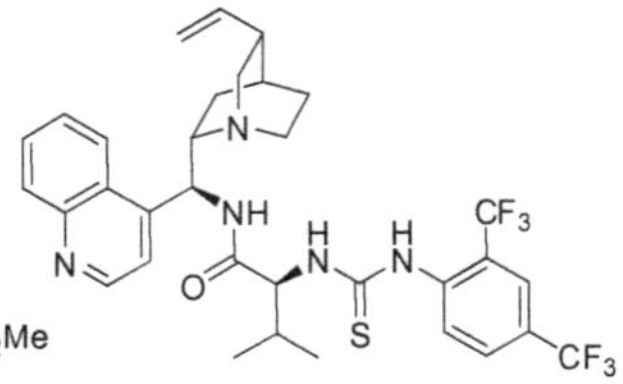

CIN-XIII
(Scheme 3.48)

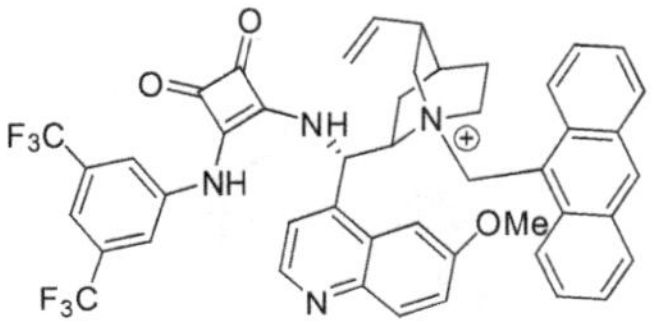

CIN-XIV
(Scheme 3.57)

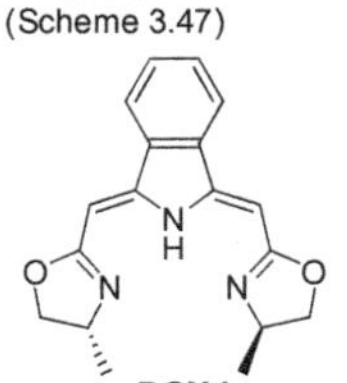

BOX-I
(Scheme 3.13)

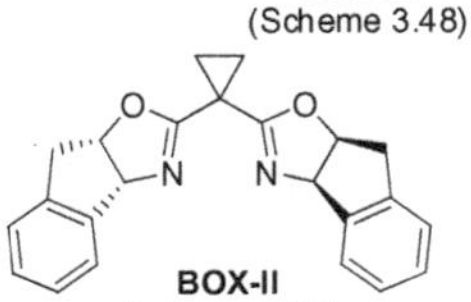

BOX-II
(Scheme 3.22)

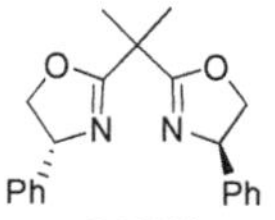

BOX-III
(Scheme 3.23)

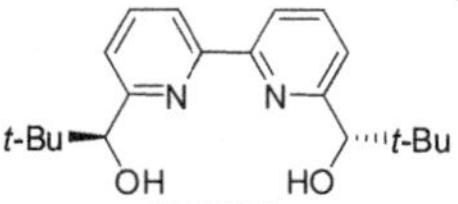

BOX-IV
(Table 3.5)

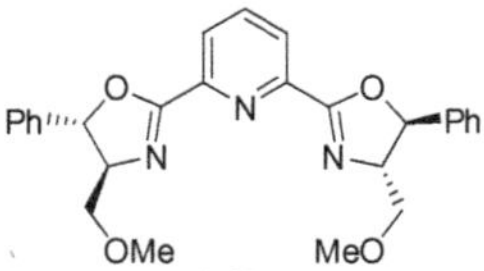

BOX-V
Scheme 3.28)

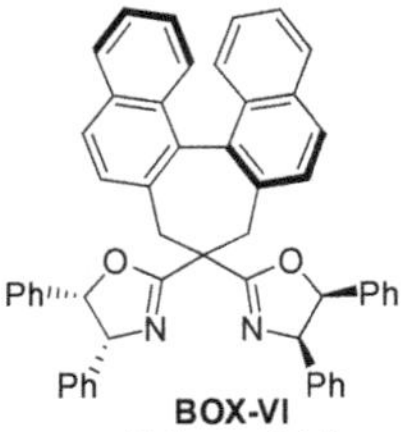

BOX-VI
(Scheme 3.44)

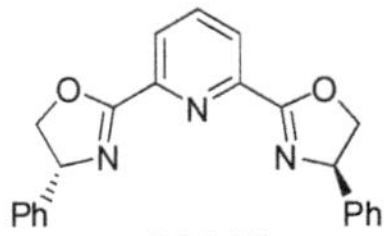

BOX-VII
(Scheme 3.56)

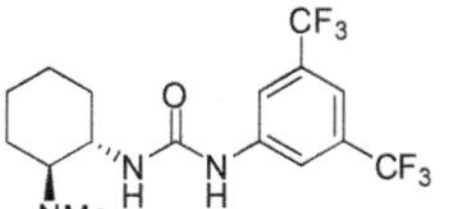

TAK-I
(Scheme 3.20)

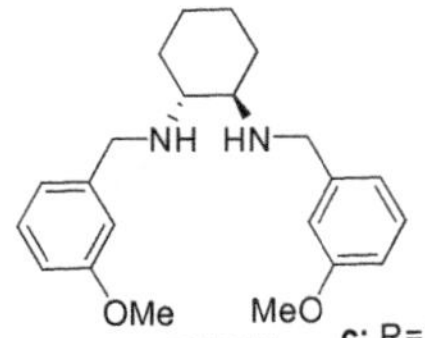

TAK-II
(Scheme 3.21)

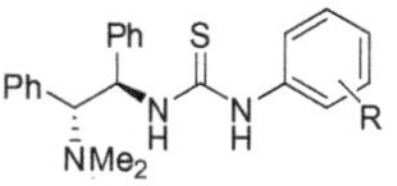

TAK-III
a: R=4-NO$_2$ (Tables 3.5, 3.6)
b: R=4-CF$_3$ (Scheme 3.28)
c: R=3,5-(CF$_3$)$_2$ (Table 3.6) (Scheme 3.47)

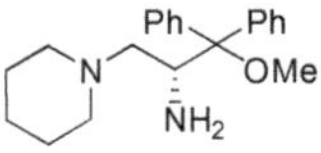

TAK-IV
(Table 3.6)

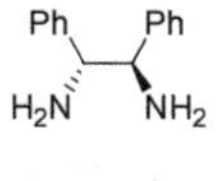

TAK-V
(Table 3.6)

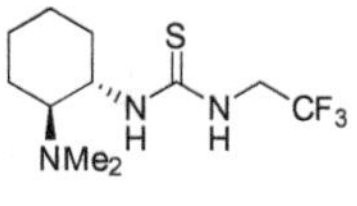

TAK-VI
(Table 3.6, 3.7)

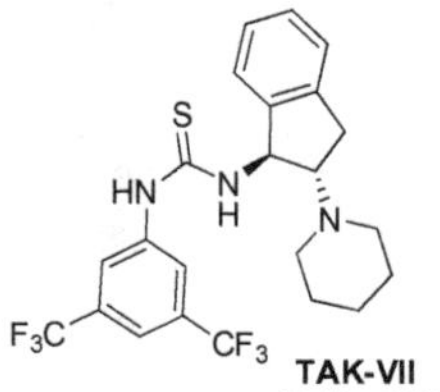

TAK-VII
(Table 3.6)

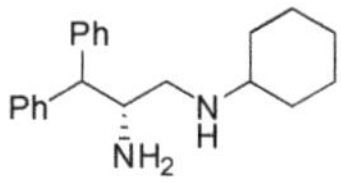

TAK-VIII
(Scheme 3.38)

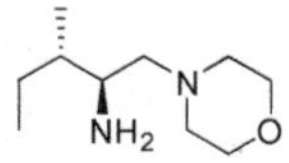

TAK-IX
(Scheme 3.43)

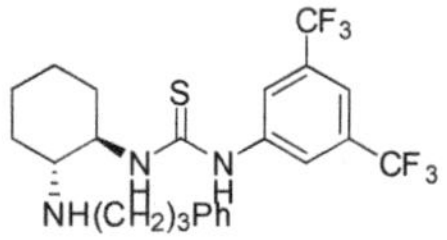

TAK-X
(Table 3.7)

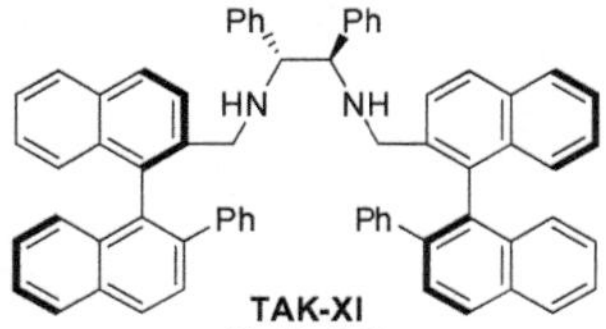

TAK-XI
(Table 3.7)

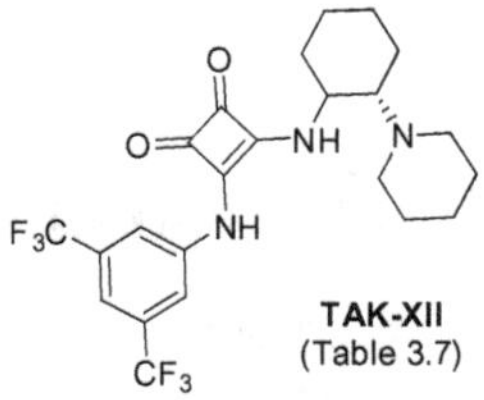

TAK-XII
(Table 3.7)

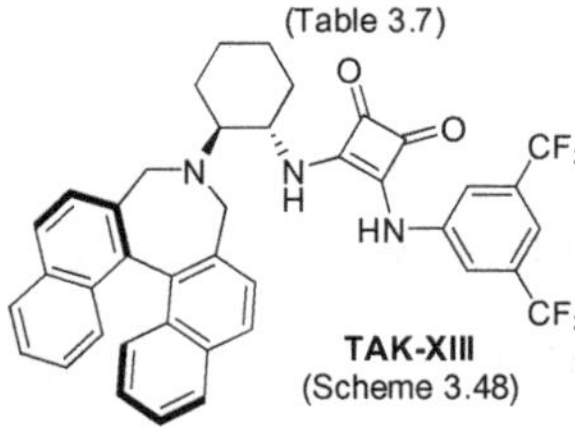

TAK-XIII
(Scheme 3.48)

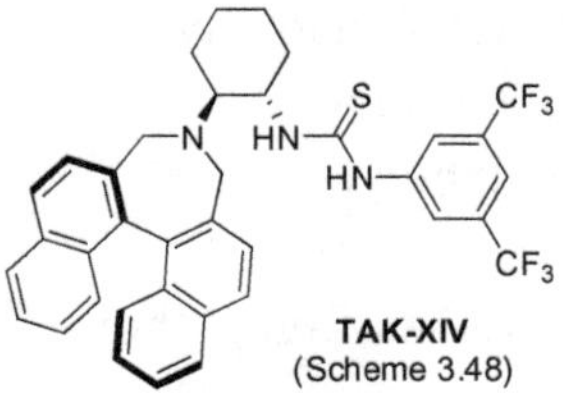

TAK-XIV
(Scheme 3.48)

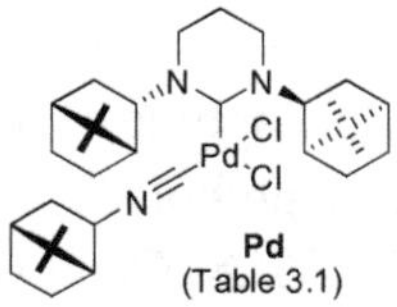

Pd
(Table 3.1)

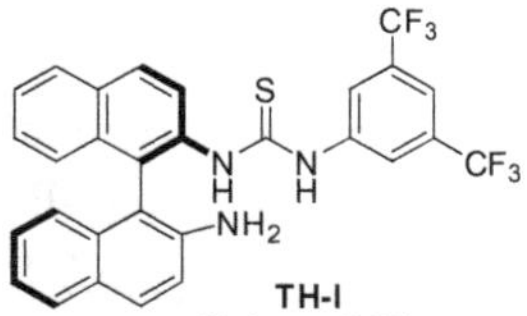

TH-I
(Scheme 3.32)

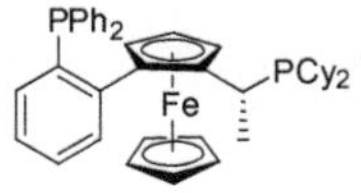

FE-I

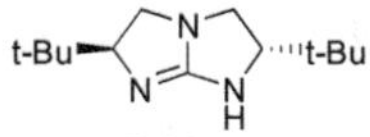

GU-I
(Table 3.6)

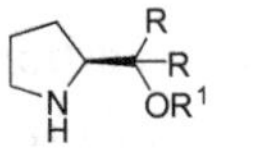

JHC
(Scheme 3.32)
(Table 3.6)

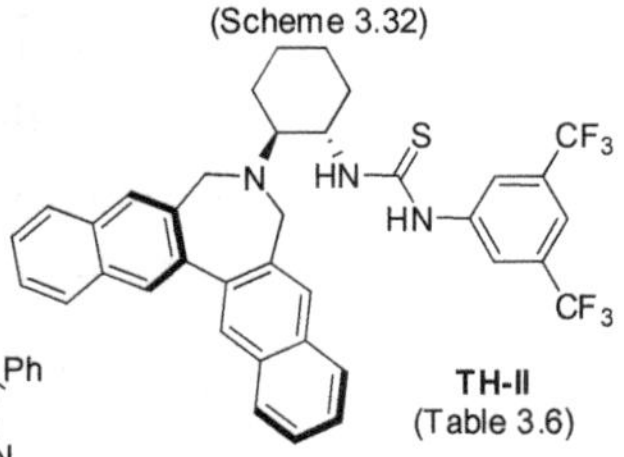

TH-II
(Table 3.6)

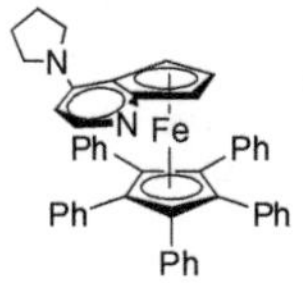

FE-II
(Scheme 3.51)

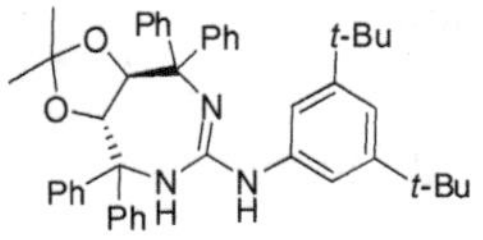

GU-II
(Table 3.7)

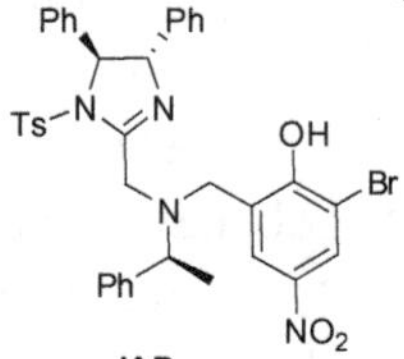

IAP
(Scheme 3.44)

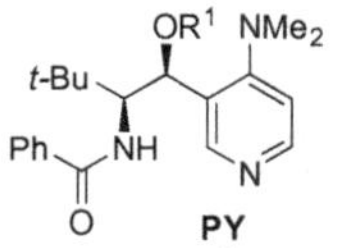

PY
a: R^1=Ac
b: R^1=Bn
(Scheme 3.52)

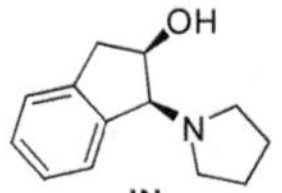

IN
(Table 3.7)

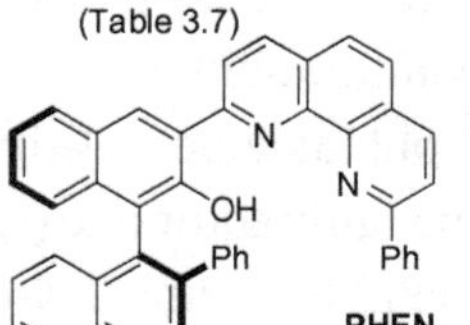

PHEN
(Table 3.6)

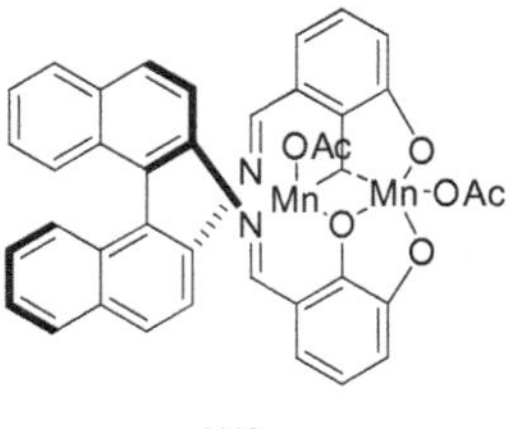

MN
(Table 3.7)

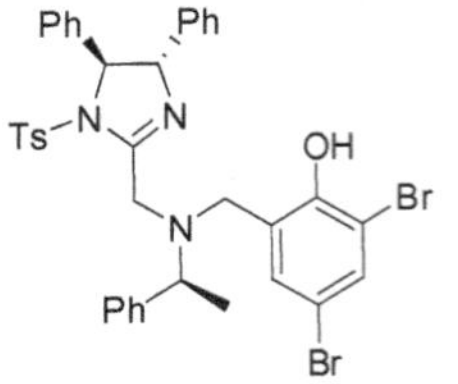

IM-I
(Table 3.7)

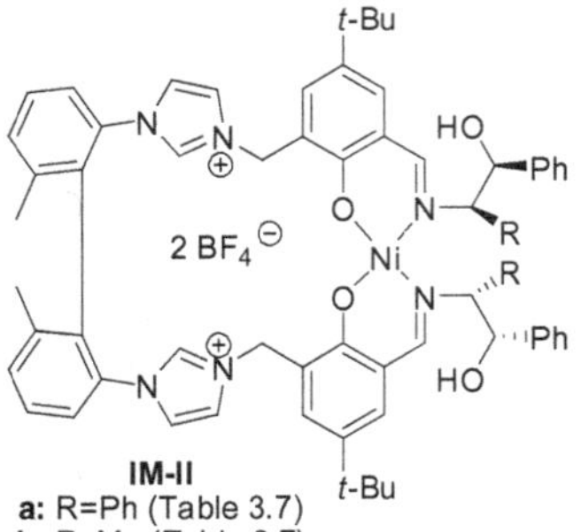

IM-II
a: R=Ph (Table 3.7)
b: R=Me (Table 3.7)

Appendix B. List of Abbreviation

Alloc	Allyloxycarbonyl
BINAP	2,2′-*Bis*(diphenylphosphino)-1,1′-binaphthyl
H_8-BINAP	5,6,7,8,5′,6′,7′,8′-Octahydro-2,2′-*bis*(diphenylphosphino)-1,1′-binaphthyl
BINOL	1,1′-Bi-2-naphthol
H_8-BINOL	5,6,7,8,5′,6′,7′,8′-Octahydro-1,1′-bi-2-naphthol
Bn	Benzyl (PhCH$_2$)
Boc	*tert*-Butoxycarbonyl
Bz	Benzoyl (PhCO)
Cbz	Benzyloxycarbonyl
CSA	Camphor sulfuric acid
Cy	Cyclohexyl
DABCO	1,4-Diazabicyclo[2.2.2]octane
dba	Dibenzylideneacetone
DDQ	2,3-Dichloro-5,6-dicyano-1,4-benzoquinone
(DHQ)$_2$AQN	Hydroquinine anthraquinone-1,4-diyl diether
(DHQ)$_2$PHAL	Hydroquinine phthalazine-1,4-diyl diether
(DHQD)$_2$AQN	Hydroquinidine anthraquinone-1,4-diyl diether
(DHQD)$_2$PHAL	Hydroquinidine phthalazine-1,4-diyl diether
(DHQD)$_2$PYR	Hydroquinidine 2,5-diphenylpyrimidine-4,6-diyl diether
DIFLUORPHOS	5,5′-*bis*(diphenylphosphino)-2,2,2′,2′-tetrafluoro-4,4′-bi-1,3-benzodioxole

DIPEA	*N,N*-Diisopropylethaneamine
DMPU	*N,N*-Dimethylpropylene urea
DUANPHOS	(1*R*,1′*R*,2*S*,2′*S*)-2,2′-Di-*tert*-butyl-2,3,2′,3′-tetrahydro-1*H*,1′*H*-(1,1′)biisophosphindolyl
esp	α,α,α′,α′-Tetramethyl-1,3-benzenedipropionic acid
HMDS	Hexamethyldisilazane
ICD	Isocupreidine
IQD	Isoquinidine
LG	Leaving group
MBH	Morita–Baylis–Hillman
Moc	Methoxycarbonyl
MOM	Methoxymethyl
MS	Molecular sieves
NHC	*N*-Heterocyclic carbene
Npt	Naphthyl
PA	Phosphoric acid
PG	Protecting group
Phth	Phthalimidyl
PMB	4-Methoxybenzyl
PMP	4-Methoxyphenyl
SEM	2-Trimethylsilylethyoxymethyl
TBAT	Tetrabutylammonium difluorotriphenylsilicate
TBDPS	*tert*-Butyldiphenylsilyl
TBS	*tert*-Butyldimethylsilyl
TFA	Trifluoroacetic acid
TfO	Triflate (trifluoromethanesulfonate)
TIPS	Tri-*iso*-propylsilyl
TMEDA	*N,N,N,N*-Tetramethylethanediamine
TMG	*N,N,N,N*-Tetramethylguanidine
TMS	Trimethylsilyl
Tol-BINAP	2,2′-*Bis*[*bis*(4-methylphenyl)phosphino)-1,1′-binaphthyl
Tol	*o,m* or *p*-Methylphenyl
Ts	Tosyl (4-methylbenzenesulfonyl)

Appendix C. Natural Products Cited in This Chapter

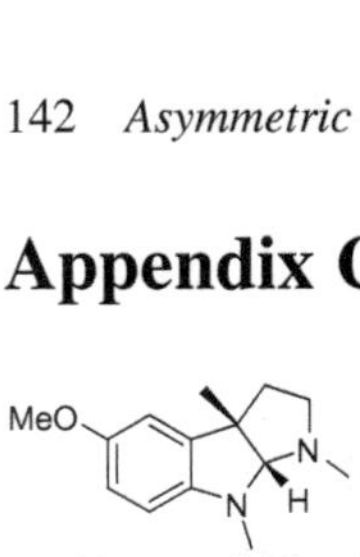

(-)-esermethole
(Schemes 3.1, 3.5,
3.11, 3.35)
(Tables 3.5, 3.6, 3.7)

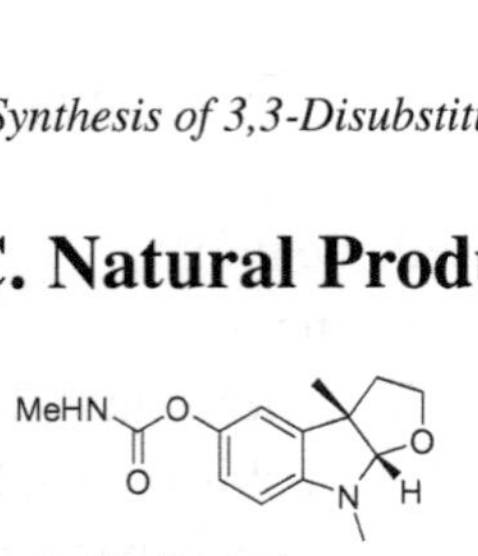

(-)- physovenine
(Scheme 3.1)

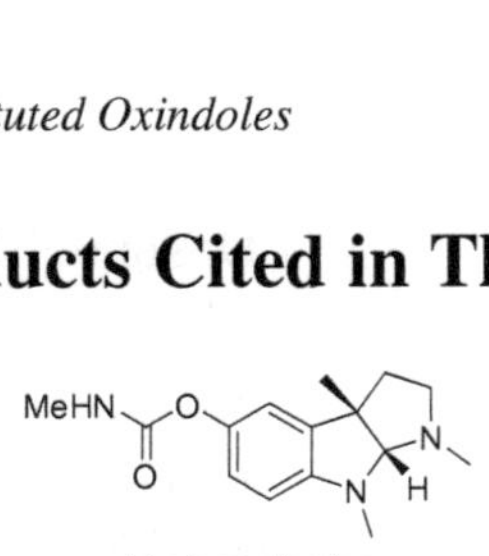

(-)-physostigmine
(Schemes 3.1, 3.4, 3.35)
(Table 3.5)

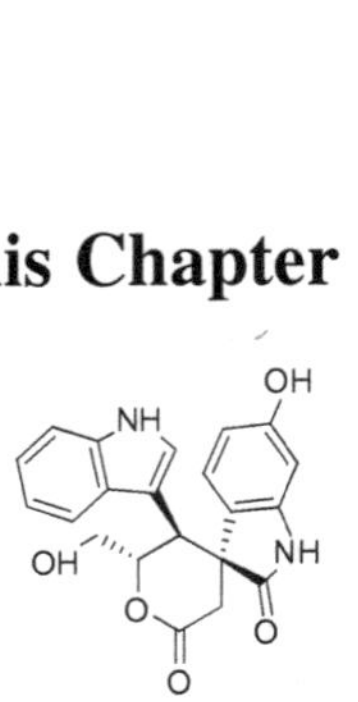

(+)-trigolute B
(Scheme 3.14)

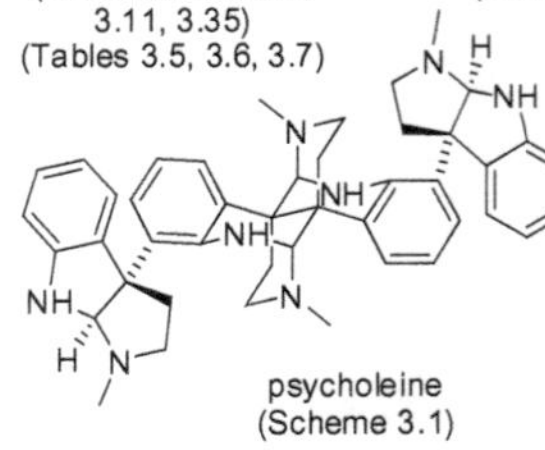

psycholeine
(Scheme 3.1)

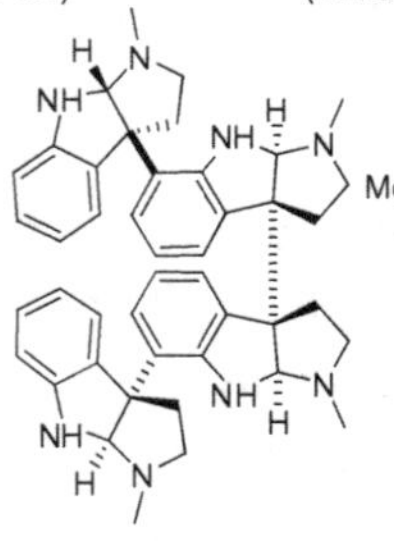

quadrigemine C
(Scheme 3.1)

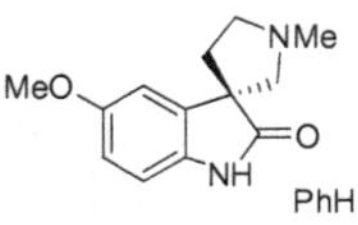

(+)-horsfiline
(Table 3.2)
(Scheme 3.45)

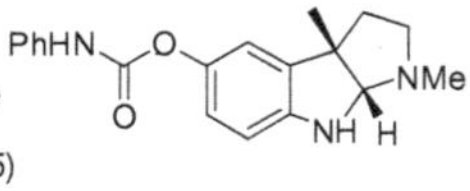

(-)-phenserine
(Table 3.2)

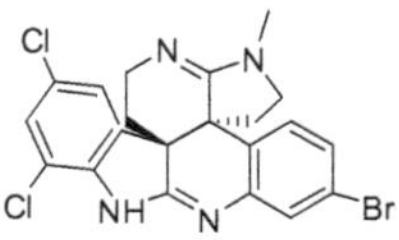

(-)-perophoramidine
(Table 3.2)
(Scheme 3.21)

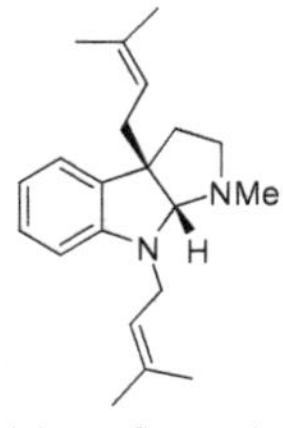

(-)-debromoflustramine B
(Table 3.2)

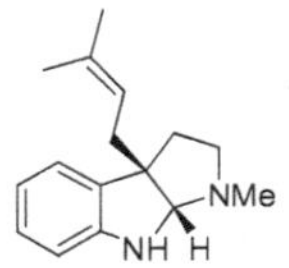

(-)-debromoflustramine E
(Scheme 3.45)

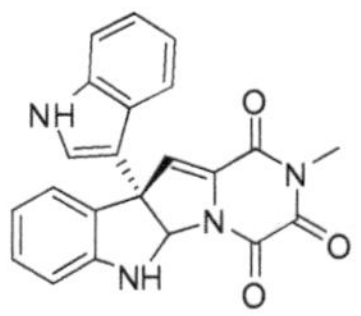

(+)-gliocladin C
(Table 3.4)
(Schemes 3.35, 3.45, 3.57)

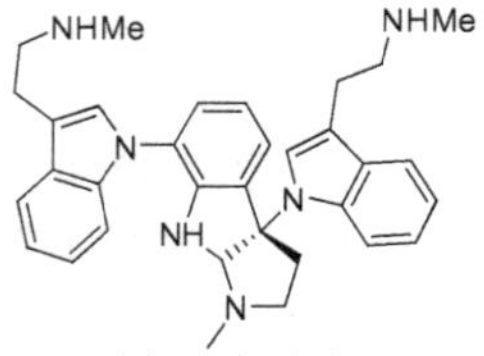

(+)-psychotrimine
(Scheme 3.22)

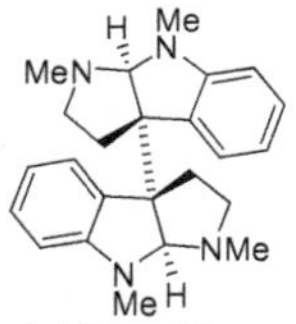

(+)-folicanthine
(Table 3.4)
(Schemes 3.46, 3.54)

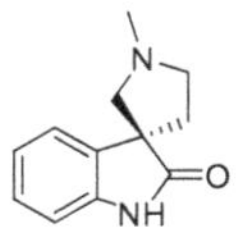

(-)-coerulescine
(Table 3.5)
(Scheme 3.45)

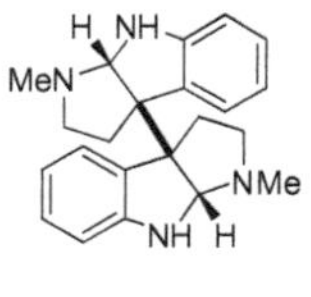

(-)-chimonanthine
Schemes 3.45, 3.54)

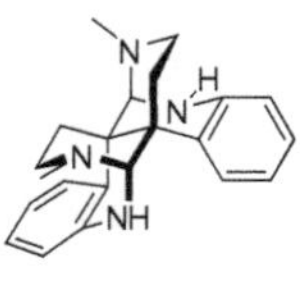

(+)-calycanthine
(Scheme 3.54)

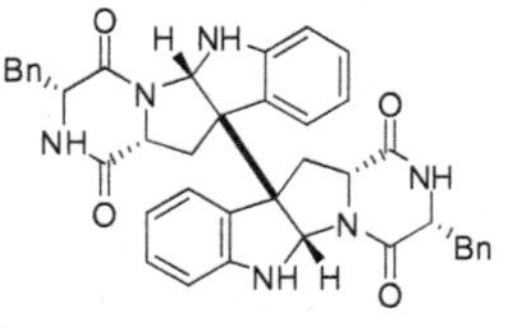

(-)-WIN 64821
(Scheme 3.54)

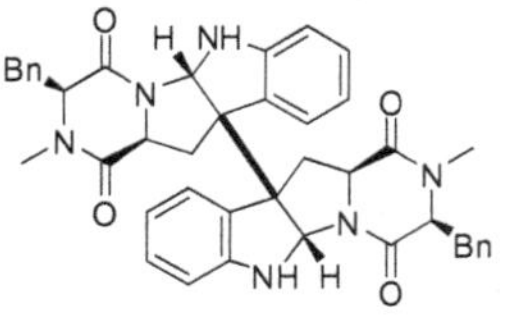

(-)-ditryptophenaline
(Scheme 3.54)

References

1. (a) A. Ashimori, T. Matsuura, L. E. Overman, D. J. Poon, *J. Org. Chem.* **1993**, *58*, 6949. (b) T. Matsuura, L. E. Overman, D. J. Poon, *J. Am. Chem. Soc.* **1998**, *120*, 6500.
2. A. L. Lebsack, J. T. Link, L. E. Overman, B. A. Stearns, *J. Am. Chem. Soc.* **2002**, *124*, 9008.
3. A. B. Dounay, K. Hatanaka, J. J. Kodanko, M. Oestreich, L. E. Overman, L. A. Pfeifer, M. M. Weiss, *J. Am. Chem. Soc.* **2003**, *125*, 6261.
4. S. Lee, J. F. Hartwig *J. Org. Chem.* **2001**, *66*, 3402.
5. F. Glorius, G. Altenhoff, R. Goddard, C. Lehmann, *Chem. Commun.* **2002**, 2704.
6. T. Arao, K. Kondo, T. Aoyama, *Chem. Pharm. Bull.* **2006**, *54*, 1743.
7. (a) T. Arao, K. Kondo, T. Aoyama, *Tetrahedron Lett.* **2006**, *47*, 1417. (b) T. Arao, K. Sato, K. Kondo, T. Aoyama, *Chem. Pharm. Bull.* **2006**, *54*, 1576.
8. X. Luan, R. Mariz, C. Robert, M. Gatti, S. Blumentritt, A. Linden, R. Dorta, *Org. Lett.* **2008**, *10*, 5569.
9. S. Würtz, C. Lohre, K. Bergander, F. Glorius, *J. Am. Chem. Soc.* **2009**, *131*, 8344.
10. (a) E. P. Kündig, T. M. Seidel, Y.-X. Jia, G. Bernardinelli, *Angew. Chem. Int. Ed.* **2007**, *46*, 8484. (b) Y.-X. Jia, D. Katayev, G. Bernardinelli, T. M. Seidel, E. P. Kündig, *Chem. Eur. J.* **2010**, *16*, 6300.
11. L. Liu, N. Ishida, S. Ashida, M. Murakami, *Org. Lett.* **2011**, *13*, 1666.
12. X. Luan, L. Wu, E. Drinkel, R. Mariz, M. Gatti, R. Dorta, *Org. Lett.* **2010**, *12*, 1912.
13. D. Katayev, E. P. Kündig, *Helv. Chim. Acta.* **2012**, *95*, 2287.
14. W. Kong, Q. Wang, J. Zhu, *Angew. Chem. Int. Ed.* **2017**, *56*, 3987.
15. M. J. Spallek, D. Riedel, F. Rominger, A. S. K. Hashmi, O. Trapp, *Organometallics*, **2012**, *31*, 1127.
16. (a) D. Katayev, Y.-X. Jia, A. K. Sharma, D. Banerjee, C. Besnard, R. B. Sunoj, E. P. Kündig, *Chem. Eur. J.* **2013**, *19*, 11916. (b) E. P. Kündig, *Chim. Oggi–Chem. Today*, **2014**, *32*, 9.
17. M. S. Joshi, F. C. Pigge, *ACS Catal.* **2016**, *6*, 4465.
18. A. Pinto, Y. Jia, L. Neuville, J. Zhu, *Chem. Eur. J.* **2007**, *13*, 961.
19. W. Kong, Q. Wang, J. Zhu, *J. Am. Chem. Soc.* **2015**, *137*, 16028.
20. W. Kong, Q. Wang, J. Zhu, *Angew. Chem. Int. Ed.* **2016**, *55*, 9714.
21. Y. Yasui, H. Kamisaki, Y. Takemoto, *Org. Lett.* **2008**, *10*, 3303.
22. N. Duguet, A. M. Z. Slawin, A. D. Smith, *Org. Lett.* **2009**, *11*, 3858.

23. T. Hashimoto, K. Yamamoto, K. Maruoka, *Chem. Commun.* **2014**, *50*, 3220.

24. H. Qiu, M. Li, L.-Q. Jiang, F.-P. Lv, L. Zan, C.-W. Zhai, M. P. Doyle, W. Hu, *Nat. Chem.* **2012**, *4*, 733.

25. Y. Yang, C. Ma, N. J. Thumar, W. Hu, *J. Org. Chem.* **2016**, *81*, 8537.

26. K. Yamamoto, Z. Qureshi, J. Tsoung, G. Pisella, M. Lautens, *Org. Lett.* **2016**, *18*, 4954.

27. A. Nakazaki, A. Mori, S. Kobayashi, T. Nishikawa, *Chem. Asian J.* **2016**, *11*, 3267.

28. T. B. K. Lee, G. S. K. Wong, *J. Org. Chem.* **1991**, *56*, 872.

29. F. Deng, S. A. Moteki, K. Maruoka, *Asian J. Org. Chem.* **2014**, *3*, 395.

30. K. Ohmatsu, M. Kiyokawa, T. Ooi, *J. Am. Chem. Soc.* **2011**, *133*, 1307.

31. J. Qiu, D. Wu, P. G. Karmaker, G. Qi, P. Chen, H. Yin, F.-X. Chen, *Org. Lett.* **2017**, *19*, 4018.

32. J.-Z. Huang, C.-L. Zhang, Y.-F. Zhu, L.-L. Li, D.-F. Chen, Z.-Y. Han, L.-Z. Gong, *Chem. Eur. J.* **2015**, *21*, 8389.

33. B. M. Trost, M. U. Frederiksen, *Angew. Chem. Int. Ed.* **2005**, *44*, 308.

34. B. M. Trost, Y. Zhang, *J. Am. Chem. Soc.* **2006**, *128*, 4590.

35. B. M. Trost, M. K. Brennan, *Org. Lett.* **2006**, *8*, 2027.

36. B. M. Trost, M. Osipov, S. Krgger, Y. Zhang, *Chem. Sci.* **2015**, *6*, 349.

37. B. M. Trost, Y. Zhang, *J. Am. Chem. Soc.* **2007**, *129*, 14548.

38. B. M. Trost, Y. Zhang, *Chem. Eur. J.* **2010**, *16*, 296.

39. A. Boucherif, S.-W. Duan, Z.-G. Yuan, L.-Q. Lu, W.-J. Xiao, *Adv. Synth. Catal.* **2016**, *358*, 2594.

40. B. M. Trost, L. C. Czabaniuk, *J. Am. Chem. Soc.* **2010**, *132*, 15534.

41. V. Franckevičius, J. D. Cuthbertson, M. Pickworth, D. S. Pugh, R. J. K. Taylor, *Org. Lett.* **2011**, *13*, 4264.

42. M. Jackson, C. Q. O'Broin, H. Müller-Bunz, P. J. Guiry, *Org. Biomol. Chem.* **2017**, *15*, 8166.

43. B. M. Trost, J. T. Masters, A. C. Burns, *Angew. Chem. Int. Ed.* **2013**, *52*, 2260.

44. K. Jiang, J. Peng, H.-L. Cui, Y.-C. Chen, *Chem. Commun.* **2009**, 3955.

45. D. Wang, Y.-L. Yang, J.-J. Jiang, M. Shi, *Org. Biomol. Chem.* **2012**, *10*, 7158.

46. Z. Qiao, N.-J. Zhong, T. Zhang, L. Liu, D. Wang, Y.-J. Chen, *Eur. J. Org. Chem.* **2013**, 2140.

47. B. M. Trost, Y. Zhang, *Chem. Eur. J.* **2011**, *17*, 2916.

48. T. B. Poulsen, L. Bernardi, J. Alemán, J. Overgaard and K. A. Jørgensen, *J. Am. Chem. Soc.* **2007**, *129*, 441.

49. Z. Wang, Z. Zhang, Q. Yao, X. Liu, Y. Cai, L. Lin, X. Feng, *Chem. Eur. J.* **2013**, *19*, 8591.

50. J. Guo, S. Dong, Y. Zhang, Y. Kuang, X. Liu, L. Lin, X. Feng, *Angew. Chem. Int. Ed.* **2013**, *52*, 10245.

51. S. Shirakawa, K. Koga, T. Tokuda, K. Yamamoto, K. Maruoka, *Angew. Chem. Int. Ed.* **2014**, *53*, 6220.

52. A. M. Taylor, R. A. Altman, S. L. Buchwald, *J. Am. Chem. Soc.* **2009**, *131*, 9900.

53. Q. Jin, C. Zheng, G. Zhao, G. Zou J. *Org. Chem.* **2017**, *82*, 4840.

54. T. Zhang, Z. Qiao, Y. Wang, N. Zhong, L. Liu, D. Wang, Y.-J. Chen, *Chem. Commun.* **2013**, *49*, 1636.

55. C.-L. Ren, T. Zhang, X.-Y. Wang, T. Wu, J. Ma, Q.-Q. Xuan, F. Wei, H.-Y. Huang, D. Wang, L. Liu, *Org. Biomol. Chem.* **2014**, *12*, 9881.

56. K. Ohmatsu, Y. Ando, T. Ooi, *J. Am. Chem. Soc.* **2013**, *135*, 18706.

57. L. Wang, D. Li, D. Yang, K. Wang, J. Wang, P. Wang, W. Su, R. Wang, *Chem. Asian J.* **2016**, *11*, 691.

58. J. Zuo, Y.-H. Liao, X.-M. Zhang, W.-C. Yuan, *J. Org. Chem.* **2012**, *77*, 11325.

59. C. Wu, G. Li, W. Sun, M. Zhang, L. Hong, R. Wang, *Org. Lett.* **2014**, *16*, 1960.

60. H. Zhang, L. Hong, H. Kang, R. Wang, *J. Am. Chem. Soc.* **2013**, *135*, 14098.

61. H. Zhang, H. Kang, L. Hong, W. Dong, G. Li, R. Wang, *Org. Lett.* **2014**, *16*, 2394.

62. S. Ma, X. Han, S. Krishnan, S. C. Virgil, B. M. Stoltz, *Angew. Chem. Int. Ed.* **2009**, *48*, 8037.

63. C. W. Lee, S.-J. Han, S. C. Virgil, B. M. Stoltz, *Tetrahedron*, **2015**, *71*, 3666.

64. L.-J. Zhou, Y.-C. Zhang, J.-J. Zhao, F. Shi, S.-J. Tu, *J. Org. Chem.* **2014**, *79*, 10390.

65. J. Peng, X. Huang, H.-L. Cui, Y.-C. Chen, *Org. Lett.* **2010**, *12*, 4260.

66. G.-Y. Chen, F. Zhong, Y. Lu, *Org. Lett.* **2012**, *14*, 3955.

67. X. Huang, J. Peng, L. Dong, Y.-C. Chen, *Chem. Commun.* **2012**, *48*, 2439.

68. X. Sun, J. Peng, S. Zhang, Q. Zhou, L. Dong, Y. C. Chen, *Acta. Chim. Sinica*, **2012**, *70*, 1682.

69. L. Song, Q.-X. Guo, X.-C. Li, J. Tian, Y.-G. Peng, *Angew. Chem. Int. Ed.* **2012**, *51*, 1899.

70. C. Guo, J. Song, J.-Z. Huang, P.-H. Chen, S.-W. Luo, L.-Z. Gong, *Angew. Chem. Int. Ed.* **2012**, *51*, 1046.

71. J. Song, C. Guo, A. Adele, H. Yin, L.-Z. Gong, *Chem. Eur. J.* **2013**, *19*, 3319.

72. Y. Zhang, S.-Y. Wang, X.-P. Xu, R. Jiang, S.-J. Ji, *Org. Biomol. Chem.* **2013**, *11*, 1933.

73. W. Tan, B.-X. Du, X. Li, X. Zhu, F. Shi, S.-J. Tu, *J. Org. Chem.* **2014**, *79*, 4635.

74. Y. Liu, H.-H. Zhang, Y.-C. Zhang, Y. Jiang, F. Shi, S.-J. Tu, *Chem. Commun.* **2014**, *50*, 12054.

75. X.-X. Sun, B.-X. Du, H.-H. Zhang, L. Ji, F. Shi, *Chem. Cat. Chem.* **2015**, *7*, 1211.

76. X.-D. Tang, S. Li, R. Guo, J. Nie, J.-A. Ma, *Org. Lett.* **2015**, *17*, 1389.

77. M.-H. Zhuo, G.-F. Liu, S.-L. Song, D. An, J. Gao, L. Zheng, S. Zhang, *Adv. Synth. Catal.* **2016**, *358*, 808.

78. J.-Z. Huang, X. Wu, L.-Z. Gong, *Adv. Synth. Catal.* **2013**, *355*, 2531.

79. Z.-H. Xing, Y. Zhang, Y. Wang, X.-P. Xu, S.-J. Ji, *Tetrahedron Lett.* **2017**, *58*, 1094.

80. S. Adhikari, S. Caille, M. Hanbauer, V. X. Ngo, L. E. Overman, *Org. Lett.* **2005**, *7*, 2795.

81. S. Ogawa, N. Shibata, J. Inagaki, S. Nakamura, T. Toru, M. Shiro, *Angew. Chem. Int. Ed.* **2007**, *46*, 8666.

82. F. Pesciaioli, P. Righi, A. Mazzanti, C. Gianelli, M. Mancinelli, G. Bartoli, G. Bencivenni, *Adv. Synth. Catal.* **2011**, *353*, 2953.

83. X. Shen, K. Liu, W. Zheng, X. Li, L. Hu, Lin, X. Feng, *Chem. Eur. J.* **2010**, *16*, 3736.

84. X.-L. Liu, Y.-H. Liao, Z.-J. Wu, L.-F. Cun, X.-M. Zhang, W.-C. Yuan, *J. Org. Chem.* **2010**, *75*, 4872.

85. S. De, M. K. Das, S. Bhunia, A. Bisai, *Org. Lett.* **2015**, *17*, 5922.

86. X. Gao, J. Han, L. Wang, *Org. Chem. Front.* **2016**, *3*, 656.

87. B.-D. Cui, Y. You, J.-Q. Zhao, J. Zuo, Z.-J. Wu, X.-Y. Xu, X.-M. Zhang, W.-C. Yuan, *Chem. Commun.* **2015**, *51*, 757.

88. K. Shen, X. Liu, W. Wang, G. Wang, W. Cao, W. Li, X. Hu, L. Lin, X. Feng, *Chem. Sci.* **2010**, *1*, 590.

89. S. Kobayashi, M. Kokubo, K. Kawasumi, T. Nagano, *Chem. Asian J.* **2010**, *5*, 490.

90. Z. Tang, Y. Shi, H. Mao, X. Zhu, W. Li, Y. Cheng, W.-H. Zheng, C. Zhu, *Org. Biomol. Chem.* **2014**, *12*, 6085.

91. X. Tian, K. Jiang, J. Peng, W. Du, Y.-C. Chen, *Org. Lett.* **2008**, *10*, 3583.

92. R. He, C. Ding, K. Maruoka, *Angew. Chem. Int. Ed.* **2009**, *48*, 4559.

93. S. Shimizu, T. Tsubogo, P. Xu, S. Kobayashi, *Org. Lett.* **2015**, *17*, 2006.

94. M. Torii, K. Kato, D. Uraguchi, T. Ooi, *Beilstein J. Org. Chem.* **2016**, *12*, 2099.

95. C. Jing, D. Xing, W. Hu, *Org. Lett.* **2015**, *17*, 4336.

96. L. Cheng, L. Liu, H. Jia, D. Wang, Y.-J. Chen, *J. Org. Chem.* **2009**, *74*, 4650.

97. Y. Jin, D. Chen, X. R. Zhang, *Chirality*, **2014**, *26*, 801.

98. P. Galzerano, G. Bencivenni, F. Pesciaioli, A. Mazzanti, B. Giannichi, L. Sambri, G. Bartoli, P. Melchiorre, *Chem. Eur. J.* **2009**, *15*, 7846.

99. N. Bravo, I. Mon, X. Companyò, A.-N. Alba, A. Moyano, R. Rios, *Tetrahedron Lett.* **2009**, *50*, 6624.

100. Y.-H. Liao, X.-L. Liu, Z.-J. Wu, X.-L. Du, X.-M. Zhang, W.-C. Yuan, *Chem. Eur. J.* **2012**, *18*, 6679.

101. X. Li, Z.-G. Xi, S. Luo, J.-P. Cheng, *Org. Biomol. Chem.* **2010**, *8*, 77.

102. F. Pesciaioli, X. Tian, G. Bencivenni, G. Bartoli, P. Melchiorre, *Synlett*, **2010**, 1704.

103. W. Sun, L. Hong, C. Liu, R. Wang, *Tetrahedron: Asymmetry*, **2010**, *21*, 2493.

104. M. H. Freund, S. B. Tsogoeva, *Synlett*, **2011**, 503.

105. W. Zheng, Z. Zhang, M. J. Kaplan, J. C. Antilla, *J. Am. Chem. Soc.* **2011**, *133*, 3339.

106. (a) H. J. Lee, D. Y. Kim, *Bull. Korean Chem. Soc.* **2012**, *33*, 3171. (b) H. J. Lee, S. B. Woo, D. Y. Kim, *Molecules*, **2012**, *17*, 7523.

107. F. Zhong, X. Dou, X. Han, W. Yao, Q. Zhu, Y. Meng, Y. Lu, *Angew. Chem. Int. Ed.* **2013**, *52*, 943.

108. X. Wu, Q. Liu, Y. Liu, Q. Wang, Y. Zhang, J. Chen, W. Cao, G. Zhao, *Adv. Synth. Catal.* **2013**, *355*, 2701.

109. S. Shirakawa, A. Kasai, T. Tokuda, K. Maruoka, *Chem. Sci.* **2013**, *4*, 2248.

110. C. Yang, W. Chen, W. Yang, B. Zhu, L. Yan, C.-H. Tan, Z. Jiang, *Chem. Asian J.* **2013**, *8*, 2960.

111. E. Badiola, B. Fiser, E. Glmez-Bengoa, A. Mielgo, I. Olaizola, I. Urruzuno, J. M. Garcìa, J. M. Odriozola, J. Razkin, M. Oiarbide, C. Palomo, *J. Am. Chem. Soc.* **2014**, *136*, 17869.

112. Y. Wei, S. Wen, Z. Liu, X. Wu, B. Zeng, J. Ye, *Org. Lett.* **2015**, *17*, 2732.

113. J. Gao, J.-R. Chen, S.-W. Duan, T.-R. Li, L.-Q. Lu, W.-J. Xiao, *Asian J. Org. Chem.* **2014**, *3*, 530.

114. L. Chen, Y. You, M.-L. Zhang, J.-Q. Zhao, J. Zuo, X.-M. Zhang, W.-C. Yuan, X.-Y. Xu, *Org. Biomol. Chem.* **2015**, *13*, 4413.

115. Y. Naganawa, H. Abe, H. Nishiyama, *Synlett*, **2016**, *27*, 1973.

116. X. Li, S. Luo, J.-P. Cheng, *Chem. Eur. J.* **2010**, *16*, 14290.

117. S.-W. Duan, J. An, J.-R. Chen, W.-J. Xiao, *Org. Lett.* **2011**, *13*, 2290.

118. C. Wang, X. Yang, D. Enders, *Chem. Eur. J.* **2012**, *18*, 4832.

119. X.-N. Yang, C. Wang, Q.-J. Ni, D. Enders, *Synthesis*, **2012**, *44*, 2601.

120. Y.-H. Liao, X.-L. Liu, Z.-J. Wu, L.-F. Cun, X.-M. Zhang, W.-C. Yuan, *Org. Lett.* **2010**, *12*, 2896.

121. L. Li, W. Chen, W. Yang, Y. Pan, H. Liu, C.-H. Tan, Z. Jiang, *Chem. Commun.* **2012**, *48*, 5124.

122. J. Zhang, Y. Zhang, L. Lin, Q. Yao, X. Liu, X. Feng, *Chem. Commun.* **2015**, *51*, 10554.

123. J. Zhou, L.-N. Jia, L. Peng, Q.-L. Wang, F. Tian, X.-Y. Xu, L.-X. Wang, *Tetrahedron*, **2014**, *70*, 3478.

124. N. Di Iorio, L. Soprani, S, Crotti, E. Marotta, A. Mazzanti, P. Righi, G. Bencivenni, *Synthesis*, **2017**, *49*, 1519.

125. J.-S. Yu, F. Zhou, Y.-L. Liu, J. Zhou, *Beilstein J. Org. Chem.* **2012**, *8*, 1360.

126. W.-Y. Siau, W. Li, F. Xue, Q. Ren, M. Wu, S. Sun, H. Guo, X. Jiang, J. Wang, *Chem. Eur. J.* **2012**, *18*, 9491.

127. Y. Zhang, X. Hu, S. Li, Y. Liao, W. Yuan, X. Zhang, *Tetrahedron*, **2014**, *70*, 2020.

128. G. Kang, Q. Wu, M. Liu, Q. Xu, Z. Chen, W. Chen, Y. Luo, W. Ye, J. Jiang, H. Wu, *Adv. Synth. Catal.* **2013**, *355*, 315–320.

129. J. Chen, Y. Cai, G. Zhao, *Adv. Synth. Catal.* **2014**, *356*, 359.

130. T. Wang, W. Yao, F. Zhong, G. H. Pang, Y. Lu, *Angew. Chem. Int. Ed.* **2014**, *53*, 2964.

131. B. M. Trost, J. Xie, J. D. Sieber, *J. Am. Chem. Soc.* **2011**, *133*, 20611.

132. D.-F. Chen, C.-l. Zhang, Y. Hu, Z.-Y. Han, L.-Z. Gong, *Org. Chem. Front.* **2015**, *2*, 956.

133. Z. Wang, T. Kang, Q. Yao, J. Ji, X. Liu, L. Lin, X. Feng, *Chem. Eur. J.* **2015**, *21*, 7709.

134. Y. Wei, Z. Liu, X. Wu, J. Fei, X. Gu, X. Yuan, J. Ye, *Chem. Eur. J.* **2015**, *21*, 18921.

135. Y.-H. Deng, X.-Z. Zhang, K.-Y. Yu, X. Yan, J.-Y. Du, H. Huang, C.-A. Fan, *Chem. Commun.* **2016**, *52*, 4183.

136. K. Zhao, Y. Zhi, A. Wang, D. Enders, *ACS Catal.* **2016**, *6*, 657.

137. S.-W. Duan, H.-H. Lu, F.-G. Zhang, J. Xuan, J.-R. Chen, W.-J. Xiao, *Synthesis*, **2011**, 1847.

138. H. Zhao, M. Xiao, L. Xu, L. Wang, J. Xiao, *RSC Adv.* **2016**, *6*, 38558.

139. L. Liu, D. Wu, S. Zheng, T. Li, X. Li, S. Wang, J. Li, H. Li, W. Wang, *Org. Lett.* **2012**, *14*, 134.

140. (a) L. Liu, D. Wu, X. Li, S. Wang, H. Li, J. Li, W. Wang, *Chem. Commun.* **2012**, *48*, 1692. (b) H. Zhao, Y.-B. Lan, Z.-M. Liu, Y. Wang, X.-W. Wang,

J.-C. Tao, *Eur. J. Org. Chem.* **2012**, 1935. (c) A. Kumar, S. S. Chimni, *Beilstein J. Org. Chem.* **2014**, *10*, 929.

141. T. Arai, Y. Yamamoto, A. Awata, K. Kamiya, M. Ishibashi, M. A. Arai, *Angew. Chem. Int. Ed.* **2013**, *52*, 2486.

142. N.-K. Li, L.-P. Kong, Z.-H. Qi, S.-J. Yin, J.-Q. Zhang, B. Wu, X.-W. Wang, *Adv. Synth. Catal.* **2016**, *358*, 3100.

143. R. Liu, J. Zhang, *Chem. Eur. J.* **2013**, *19*, 7319.

144. R. Liu, J. Zhang, *Org. Lett.* **2013**, *15*, 2266.

145. T. Mukaiyama, K. Ogata, I. Sato, Y. Hayashi, *Chem. Eur. J.* **2014**, *20*, 13583.

146. X.-L. Liu, Z.-J. Wu, X.-L. Du, X.-M. Zhang, W.-C. Yuan, *J. Org. Chem.* **2011**, *76*, 4008.

147. Y.-Y. Han, Z.-J. Wu, W.-B. Chen, X.-L. Du, X.-M. Zhang, W.-C. Yuan, *Org. Lett.* **2011**, *13*, 5064.

148. T. Bui, S. Syed, C. F. Barbas III, *J. Am. Chem. Soc.* **2009**, *131*, 8758.

149. R. He, S. Shirakawa, K. Maruoka, *J. Am. Chem. Soc.* **2009**, *131*, 16620.

150. Y. Kato, M. Furutachi, Z. Chen, H. Mitsunuma, S. Matsunaga, M. Shibasaki, *J. Am. Chem. Soc.* **2009**, *131*, 9168.

151. X. Li, B. Zhang, Z.-G. Xi, S. Luo, J.-P. Cheng, *Adv. Synth. Catal.* **2010**, *352*, 416.

152. M. Ding, F. Zhou, Z.-Q. Qian, J. Zhou, *Org. Biomol. Chem.* **2010**, *8*, 2912.

153. M. Ding, F. Zhou, Y.-L. Liu, C.-H. Wang, X.-L. Zhao, J. Zhou, *Chem. Sci.* **2011**, *2*, 2035.

154. A. Awata, M. Wasai, H. Masu, S. Kado, T. Arai, *Chem. Eur. J.* **2014**, *20*, 2470.

155. X. Chen, W. Zhu, W. Qian, E. Feng, Y. Zhou, J. Wang, H. Jiang, Z.-J. Yao, H. Liu, *Adv. Synth. Catal.* **2012**, *354*, 2151.

156. W. Yang, J. Wang, D.-M. Du, *Tetrahedron: Asymmetry*, **2012**, *23*, 972.

157. C. Reiter, S. Lòpez-Molina, B. Schmid, C. Neiss, A. Görling, S. B. Tsogoeva, *Chem. Cat. Chem.* **2014**, *6*, 1324.

158. S. Shirakawa, L. Wang, R. He, S. Arimitsu, K. Maruoka, *Chem. Asian J.* **2014**, *9*, 1586.

159. L. Zou, X. Bao, Y. Ma, Y. Song, J. Qu, B. Wang, *Chem. Commun.* **2014**, *50*, 5760.

160. M.-X. Zhao, F.-H. Ji, X.-L. Zhao, Z.-Z. Han, M. Shi, *Eur. J. Org. Chem.* **2014**, 644.

161. M. Mechler, R. Peters, *Angew. Chem. Int. Ed.* **2015**, *54*, 10303.

162. C. C. J. Loh, D. Hack, D. Enders, *Chem. Commun.* **2013**, *49*, 10230.

163. L.-H. Zou, A. R. Philipps, G. Raabe, D. Enders, *Chem. Eur. J.* **2015**, *21*, 1004.

164. D.-F. Chen, F. Zhao, Y. Hu, L.-Z. Gong, *Angew. Chem. Int. Ed.* **2014**, *53*, 10763.

165. Z.-Y. Cao, Y.-L. Zhao, J. Zhou, *Chem. Commun.* **2016**, *52*, 2537.

166. M.-X. Zhao, W.-H. Tang, M.-X. Chen, D.-K. Wei, T.-L. Dai, M. Shi, *Eur. J. Org. Chem.* **2011**, 6078.

167. L. Zong, S. Du, K. F. Chin, C. Wang, C.-H. Tan, *Angew. Chem. Int. Ed.* **2015**, *54*, 9390.

168. Q. Zhu, Y. Lu, *Angew. Chem. Int. Ed.* **2010**, *49*, 7753.

169. H. J. Lee, S. H. Kang, D. Y. Kim, *Synlett*, **2011**, 1559.

170. T. Zhang, L. Cheng, S. Hameed, L. Liu, D. Wang, Y.-J. Chen *Chem. Commun.* **2011**, *47*, 6644.

171. M.-X. Zhao, T.-L. Dai, R. Liu, D.-K. Wei, H. Zhou, F.-H. Jia, M. Shi, *Org. Biomol. Chem.* **2012**, *10*, 7970.

172. S.-W. Duan, Y.-Y. Lin, W. Ding, T.-R. Li, D.-Q. Shi, J.-R. Chen, W.-J. Xiao, *Synthesis*, **2013**, *45*, 1647.

173. T. H. Black, S. M. Arrivo, J. S. Schumm, J. M. Knobeloch, *J. Chem. Soc. Chem. Commun.* **1986**, 1524.

174. I. D. Hills, G. C. Fu, *Angew. Chem., Int. Ed.* **2003**, *42*, 3921.

175. (a) S. A. Shaw, P. Aleman and E. Vedeis, *J. Am. Chem. Soc.* **2003**, *125*, 13368. (b) T. A. Duffey, S. A. Shaw, E. Vedejs, *J. Am. Chem. Soc.* **2009**, *131*, 14.

176. E. C. Linton, M. C. Kozlowski, *J. Am. Chem. Soc.* **2008**, *130*, 16162.

177. T. Cao, J. Deitch, E. C. Linton, M. C. Kozlowski, *Angew. Chem. Int. Ed.* **2012**, *51*, 2448.

178. B. M. Trost, M. Osipov, *Angew. Chem. Int. Ed.* **2013**, *52*, 9176.

179. D. Uraguchi, K. Koshimoto, T. Ooi, *J. Am. Chem. Soc.* **2012**, *134*, 6972.

180. M. Kuriyama, S. Tanigawa, Y. Kubo, Y. Demizu, O. Onomura, *Tetrahedron: Asymmetry*, **2010**, *21*, 1370.

181. F. Zhou, C. Tan, J. Tang, Y.-Y. Zhang, W.-M. Gao, H.-H. Wu, Y.-H. Yu, J. Zhou, *J. Am. Chem. Soc.* **2013**, *135*, 10994.

182. W. D. G. Brittain, B. R. Buckley, J. S. Fossey, *Chem. Commun.* 2015, *51*, 17217.

183. G. Zhu, G. Bao, Y. Li, W. Sun, J. Li, L. Hong, R. Wang, *Angew. Chem. Int. Ed.* **2017**, *56*, 5332.

184. E. Sorrentino, S. J. Connon, *Org. Lett.* **2016**, *18*, 5204.

185. Y. Zhi, K. Zhao, A. Wang, U. Englert, G. Raabe, D. Enders, *Adv. Synth. Catal.* **2017**, *359*, 1867.

Chapter 4

Catalytic Asymmetric Synthesis of 3-Substituted-3-Amino-2-Oxindoles

4.1. Introduction

Chiral oxindoles represent an important class of products, which are widely present in nature and exhibit many biological activities [1]. Among these compounds, chiral 3-substituted-3-amino-2-oxindoles bearing a quaternary stereogenic center at the 3-position have been recognized as privileged substructures in new drug discovery. For example, this skeleton is present in many bioactive molecules and drug candidates for the treatment of malaria and stress-related disorders as well as in marine alkaloids, such as the chartellines or in terrestrial alkaloids isolated from the leaves of the Malaysian plant *Psychotria rostrate*. The biological activity of these compounds is greatly affected by the nature of the substituent at the C-3 position, as well as the absolute configuration of the stereogenic center. Consequently, the development of efficient methods to synthesize such products is of great importance and constitutes a current open area of research in asymmetric catalysis [2] with a special mention for organocatalysis [2b, e]. The principal methodology to prepare chiral quaternary 3-amino-2-oxindoles is based on enantioselective catalytic nucleophilic additions to isatin imines, including Mannich reactions, aza-Morita–Baylis–Hillman (MBH) reactions, aza-Henry reactions, additions of heteronucleophiles, Strecker reactions, hydrophosphonylations among others.

Moreover, asymmetric catalytic domino and tandem reactions have recently allowed a wide range of these products to be highly-efficiently achieved. A third route to chiral 3-substituted-3-amino-2-oxindoles is the direct asymmetric amination of 3-substituted oxindoles. Other methodologies, such as enantioselective 1,3-dipolar cycloadditions and substitution reactions of 3-substituted oxindoles containing a C-3 leaving group, have also been successfully developed. Most of these reactions have been promoted with a wide variety of chiral organocatalysts but chiral metal catalysts have also proved to be highly efficient for a range of enantioselective transformations. The goal of this chapter is to provide a comprehensive overview of the major developments in the catalytic asymmetric synthesis of 3-substituted-3-amino-2-oxindoles reported in the last decade. It is divided into four parts, dealing successively with enantioselective nucleophilic additions to isatin imines, asymmetric domino and tandem reactions, asymmetric direct aminations of 3-substituted oxindoles, and miscellaneous enantioselective reactions.

4.2. Nucleophilic Additions to Isatin Imines

Nucleophilic additions to isatin imines represent a direct method to synthesize 3-substituted-3-amino-2-oxindoles, not only because of the easy access to isatin imines, but also by the possibility of using a wide range of nucleophiles, thus increasing the structural diversity of the resulting products. The first asymmetric catalytic versions were reported only in the 2010s despite tremendous achievements in catalytic asymmetric imine addition reactions [3]. Ever since, a number of catalytic asymmetric nucleophilic additions to isatin imines have been developed, including Mannich reactions, aza-MBH reactions, aza-Henry reactions, additions of heteronucleophiles, Strecker reactions, hydrophosphonylation reactions among others.

4.2.1. *Enantioselective Mannich reactions*

The Mannich reaction, a widely applied means of producing synthetically valuable β-amino carbonyl compounds starting from cheap and readily available substrates, involves an aldehyde, an amine, and a ketone reacting

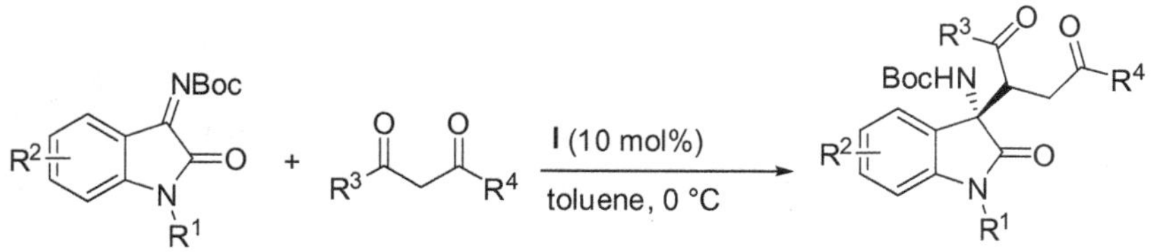

R^1 = H, Me, Bn, Boc
R^2 = H, 5-Me, 5-OMe, 5-Br, 5-OCF$_3$, 6-Cl, 6-Br, 7-F, 7-Cl, 7-Br, 7-Me
R^3 = R^4 = OMe, OEt, OBn, Me, Ph

Scheme 4.1.   Mannich reaction of isatin imines with 1,3-dicarbonyl compounds catalyzed with a quinidine-derived thiourea.

in a three-component, one-pot synthesis [4]. As an alternative, the reaction can be performed as a nucleophilic addition of a C-nucleophile to a preformed imine. A wide variety of organocatalysts [5] have been used to promote asymmetric Mannich reactions. Among them, the quinidine-derived thiourea depicted in Scheme 4.1 was employed by Wang *et al.* to promote the synthesis of chiral 3-amino-2-oxindoles from the Mannich reaction of isatin imines with 1,3-dicarbonyl compounds [6]. Therefore, a range of chiral 3-amino-2-oxindoles was achieved in high yields (85 to >99%) and good to excellent enantioselectivities (75–98% ee) starting from both malonates and diketones.

In 2016, Trivedi *et al.* reinvestigated these reactions using a chiral *Cinchona* alkaloid-derived squaramide [7]. A range of chiral 3-amino-2-oxindoles were achieved under mild reaction conditions in high to quantitative yields (78–99%) and uniformly excellent enantioselectivities (92–99% ee), as shown in Scheme 4.2.

Earlier in 2012, Shibata *et al.* reported the enantioselective decarboxylative Mannich reaction of malonic acid half thioesters with isatin imines performed in the presence of a *N*-heteroarenesulfonyl *Cinchona* alkaloid [8]. The reaction led to the corresponding chiral 3-amino-2-oxindoles in good to high yields (58–92%) and enantioselectivities (75–83% ee), as shown in Scheme 4.3. The utility of this methodology was demonstrated by its application to a total synthesis of the gastrin/cholecystokinin-B-receptor AG-041R.

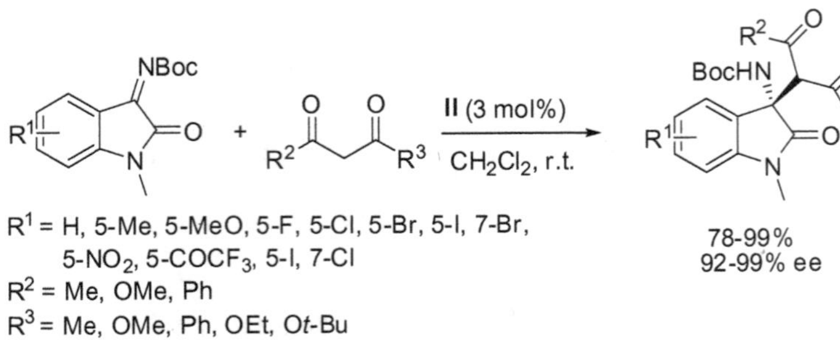
NBoc
II (3 mol%)
CH2Cl2, r.t.
BocHN
R1 = H, 5-Me, 5-MeO, 5-F, 5-Cl, 5-Br, 5-I, 7-Br, 5-NO2, 5-COCF3, 5-I, 7-Cl
R2 = Me, OMe, Ph
R3 = Me, OMe, Ph, OEt, Ot-Bu
78-99%
92-99% ee

Scheme 4.2. Mannich reaction of isatin imines with 1,3-dicarbonyl compounds catalyzed with a *Cinchona* alkaloid-derived squaramide catalyst.

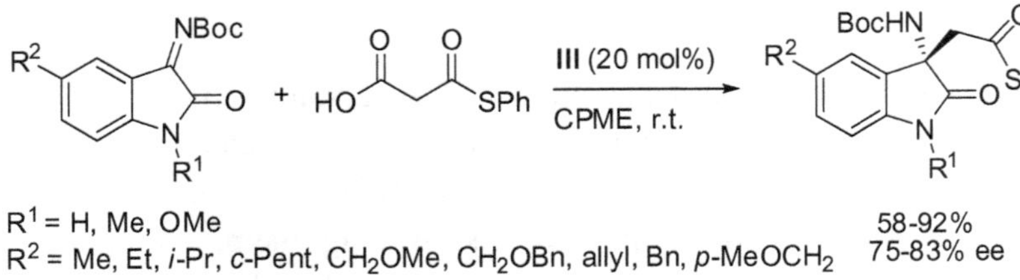
NBoc
HO
SPh
III (20 mol%)
CPME, r.t.
BocHN
R1 = H, Me, OMe
R2 = Me, Et, i-Pr, c-Pent, CH2OMe, CH2OBn, allyl, Bn, p-MeOCH2
58-92%
75-83% ee

Scheme 4.3. Mannich reaction of isatin imines with malonic acid half thioesters catalyzed by an *N*-heteroarenesulfonyl *Cinchona* alkaloid.

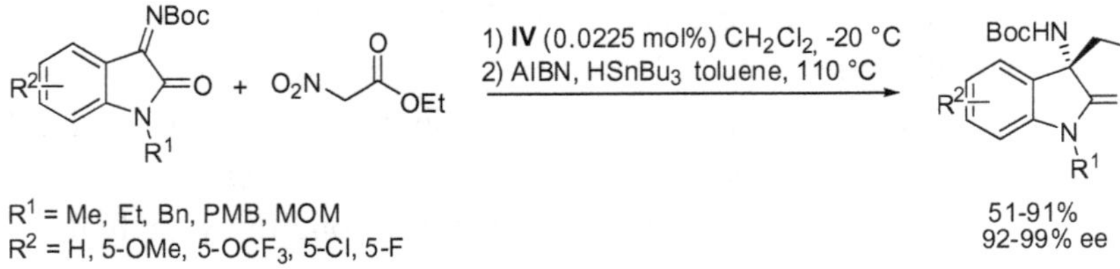
NBoc
O2N
OEt
1) IV (0.0225 mol%) CH2Cl2, -20 °C
2) AIBN, HSnBu3 toluene, 110 °C
BocHN
R1 = Me, Et, Bn, PMB, MOM
R2 = H, 5-OMe, 5-OCF3, 5-Cl, 5-F
51-91%
92-99% ee

Scheme 4.4. Mannich reaction of isatin imines with ethyl nitroacetate catalyzed with a *Cinchona* alkaloid.

More recently, Enders *et al.* demonstrated that a remarkably low catalyst loading (0.0225 mol%) of another *Cinchona* alkaloid promoted the enantioselective Mannich reaction of ethyl nitroacetate with isatin imines, affording after a subsequent denitration the corresponding chiral 3-amino-2-oxindoles in moderate to high yields (51–91%) and uniformly high enantioselectivities (92–99% ee), as shown in Scheme 4.4 [9]. The utility of this methodology was demonstrated by its application in the total synthesis of AG-041 R, and to that of an intermediate in the synthesis of natural product physoveninel.

In 2014, Wu *et al.* reported the enantioselective Mannich reaction of isatin imines with pyrazoleamides catalyzed by a *Cinchona* alkaloid-derived thiourea [10]. In the presence of 10 mol% of this catalyst and molecular sieves as an additive in acetonitrile at 25°C, the reaction afforded the corresponding chiral 3-amino-2-oxindoles in high yields (84–97%), diastereoselectivities (88–98% de), and enantioselectivities (96–99% ee), as shown in Scheme 4.5. The pyrazoleamides could be readily converted into the corresponding chiral β-amino esters through alcoholysis in high yield (89%) and excellent enantioselectivity (99% ee).

Soon after, an enantioselective organocatalyzed vinylogous Mannich reaction of isatin imines with γ-butenolides was reported by Wang *et al.* [11]. It was catalyzed by 10 mol% of a quinidine-derived organocatalyst in dichloromethane at −30°C, providing the corresponding chiral Mannich products in uniformly high yields (84–97%), low to moderate diastereoselectivities (18–58% de), and generally high enantioselectivities (83–94% ee), as summarized in Scheme 4.6.

The chiral amino acid depicted in Scheme 4.7 was employed as an organocatalyst by Peng *et al.* in the enantioselective Mannich reaction of isatin imines with hydroxyacetone [12]. Surprisingly, when the reaction

R^1 = Me, Bn
R^2 = H, 5-F, 5-Cl, 5-Br, 5-Me, 5-OMe, 6-Cl, 6-Br, 7-F, 7-Cl, 7-Br, 7-Me
Ar = Ph, *p*-$F_3CC_6H_4$, *p*-BrC_6H_4, *p*-ClC_6H_4, *p*-FC_6H_4, *p*-PhC_6H_4, *p*-$MeOC_6H_4$, 1-Naph

84-97%
88-98% de
96-99% ee

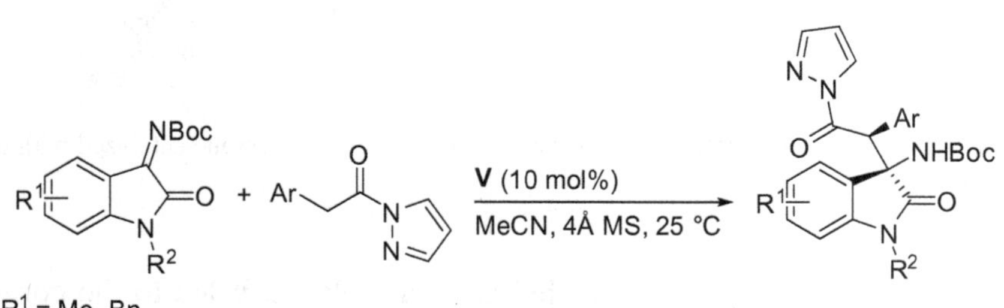

Scheme 4.5. Mannich reaction of isatin imines with pyrazoleamides catalyzed with a *Cinchona* alkaloid-derived thiourea.

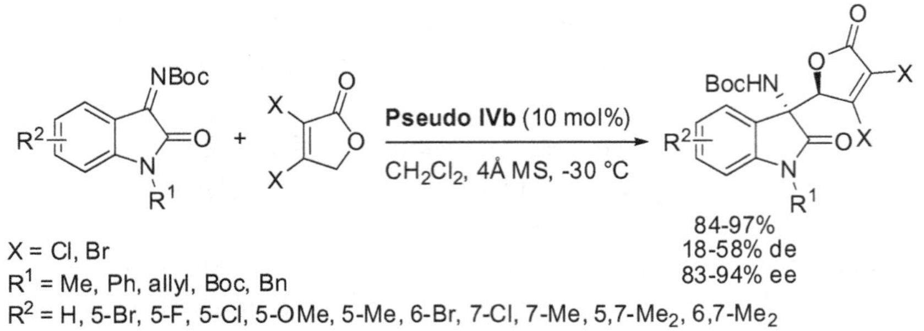

Scheme 4.6.   Vinylogous Mannich reaction of isatin imines with γ-butenolides catalyzed with a quinidine-derived catalyst.

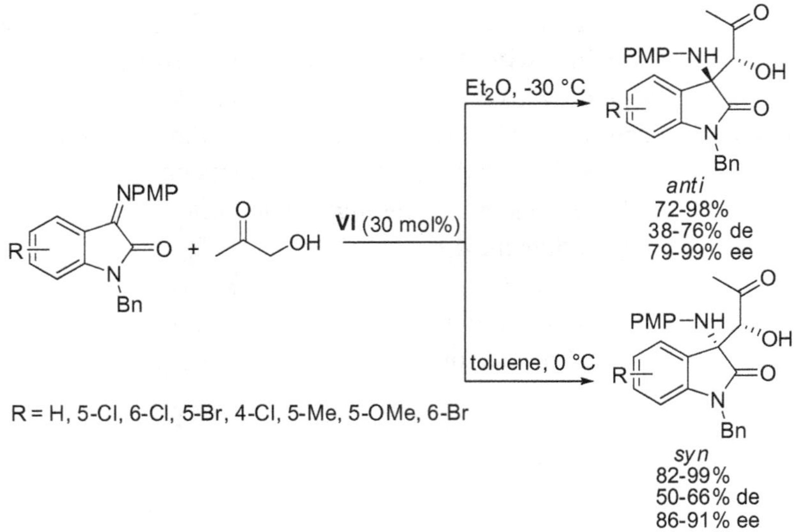

Scheme 4.7.   Mannich reactions of isatin imines with hydroxyacetone catalyzed with an amino acid.

was performed at −30°C in diethyl ether as solvent, it led to the corresponding *anti*-Mannich products as major diastereomers in high yields (72–98%), low to good diastereoselectivities (38–76% de), and good to excellent enantioselectivities (79–99% ee), as shown in Scheme 4.7. On the other hand, when toluene was used as solvent at 0°C, the process afforded the *syn*-Mannich products as major diastereomers in high yields

(82–99%), moderate diastereoselectivities (50–66% de), and high enantioselectivities (86–91% ee). The authors explained these results on the basis of differential *E/Z* ratios of isatin imines in different solvents.

In 2016, Silvani and Lesma described the synthesis of chiral 3-amino-2-oxindole butenolides on the basis of an enantioselective organocatalytic Mannich reaction between isatin-derived benzhydrylketimines and trimethylsiloxyfuran [13]. Using 10 mol% of another type of organocatalysts, such as the chiral phosphoric acid depicted in Scheme 4.8, the process led at –40°C in THF to the corresponding butenolides in moderate to good yields (42–81%), moderate diastereoselectivities (up to 44% de), and moderate to excellent enantioselectivities (up to 96% ee).

In addition to organocatalysts, metal chiral catalysts [14] have been recently applied to promote enantioselective Mannich reactions. As an example, Feng *et al.* reported the enantioselective Mannich reaction of isatin imines with silyl ketene imines catalyzed with a combination of Zn(OTf)$_2$ and a chiral *N,N'*-dioxide ligand [15]. As shown in Scheme 4.9, the process afforded a range of chiral β-amino nitriles exhibiting vicinal tetrasubstituted stereocenters in excellent yields up to 98%, diastereoselectivities up to >90% de, and enantioselectivities up to 99% ee.

4.2.2. *Enantioselective aza-Morita–Baylis–Hillman reactions*

The MBH reaction involves the carbon–carbon coupling of the α-position of an activated alkene with a carbon electrophile containing an

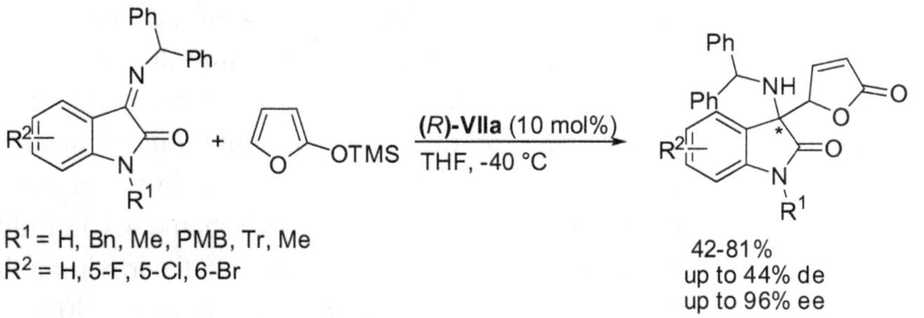

Scheme 4.8.   Mannich reaction of isatin imines with trimethylsiloxyfuran catalyzed with a phosphoric acid.

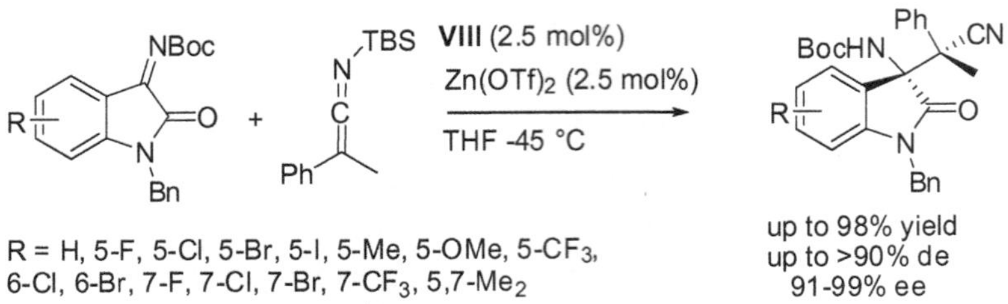

Scheme 4.9. Zinc-catalyzed Mannich reaction of isatin imines with silyl ketene imines.

electron-deficient sp^2 carbon atom, such as an aldehyde (X=O), catalyzed by a tertiary amine or phosphine [16]. This operationally simple and atom-economic process allows the direct preparation of α-methylene-β-hydroxycarbonyl compounds. When activated imines are used instead of aldehydes, the process is called aza-MBH reaction, affording the corresponding α-methylene-β-aminocarbonyl derivatives. Several advantages of the aza-MBH reaction are related to the fact that the products are multifunctional, the catalysts employed are most of the time organic, and the reaction conditions used often mild. Actually, this reaction is one of the best illustrations of organocatalysis [5] for green chemistry when amines or phosphines are employed as catalysts. In 2013, Shi *et al.* reported a highly enantioselective aza-MBH reaction of isatin imines with methyl vinyl ketone [17]. The corresponding chiral highly functionalized 3-amino-2-oxindoles were achieved in both excellent yields (up to 98%) and enantioselectivities (up to 99% ee) by using either a *Cinchona* alkaloid catalyst, such as β-isocupreidine, or a chiral phosphine at 20 mol% of catalyst loading, as shown in Scheme 4.10. In the case of the former catalyst, the reaction was performed in toluene at 0°C while the phosphoric acid-catalyzed reaction occurred in chloroform at room temperature. In both cases, the products exhibited the same configuration. Soon after, these reactions were also investigated by Jugé and Sasai in the presence of a *P*-chirogenic organocatalyst [18]. As shown in Scheme 4.10, in this case the reaction was performed in MTBE as solvent at 5°C, providing the corresponding chiral aza-MBH product in quantitative yield and 90% ee.

In 2014, Wu and Sha employed a chiral squaramide-derived phosphine catalyst to promote the first enantioselective aza-MBH reaction

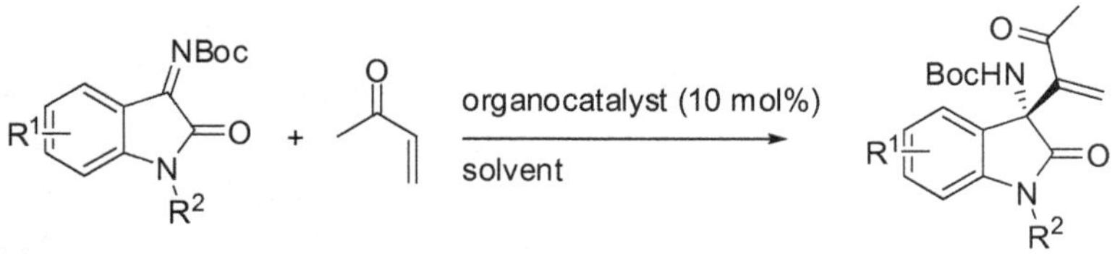

R^1 = H, 5-Me, 5-OMe, 5-F, 5-Cl, 5-Br, 6-Me, 6-Cl, 6-Br, 7-F, 7-Cl, 7-Br, 7-CF$_3$, 5-Cl, 7-Me
R^2 = Bn, Me, allyl

organocatalyst	solvent	yields	ee
β-isocupreidine	toluene at 0 °C	up to 98%	up to 94%
IX	CHCl$_3$ at r.t.	up to 97%	up to >99%
X	MTBE at 5 °C	>99%	90% (R^1 = H, R^2 = Bn)

Scheme 4.10. Aza-MBH reaction of isatin imines with methyl vinyl ketone catalyzed with β-isocupreidine, a chiral phosphine, and a P-chirogenic catalyst.

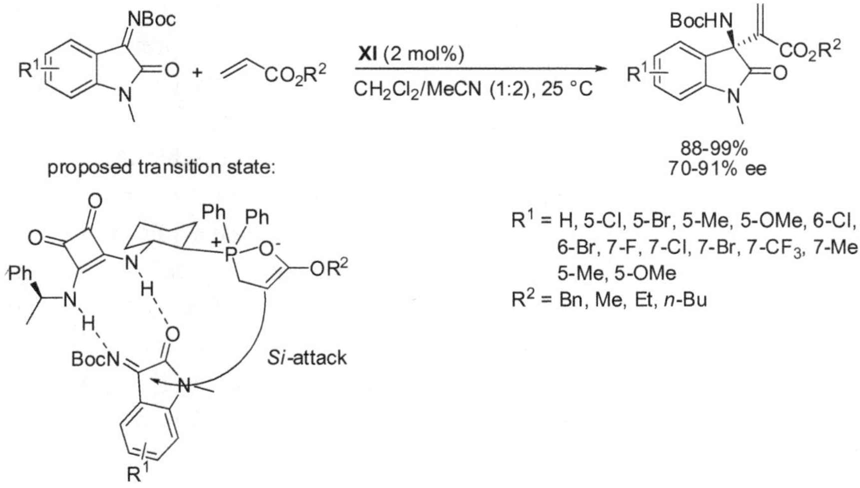

XI (2 mol%)
CH$_2$Cl$_2$/MeCN (1:2), 25 °C

88-99%
70-91% ee

R^1 = H, 5-Cl, 5-Br, 5-Me, 5-OMe, 6-Cl,
6-Br, 7-F, 7-Cl, 7-Br, 7-CF$_3$, 7-Me,
5-Me, 5-OMe
R^2 = Bn, Me, Et, *n*-Bu

proposed transition state:

Scheme 4.11. Aza-MBH reaction of isatin imines with acrylates catalyzed with a squaramide-derived phosphine.

between isatin imines and acrylates (Scheme 4.11) [19]. The reaction was performed with only 2 mol% of this organocatalyst in a 1:2 mixture of dichloromethane and acetonitrile as solvent at 25°C, affording the corresponding chiral 3-amino-2-oxindoles in high yields (88–99%) and moderate to high enantioselectivities (70–91% ee). To explain the results, the authors proposed the transition state depicted in Scheme 4.11 in which the electrophilic squaramide of the catalyst activated the ketimine through hydrogen-bonding interactions. Then, the chiral cyclohexyl scaffold forced the phosphinoyl associated enolate to attack the activated ketimine from the *Si*-face to form the final product exhibiting the (*S*)-configuration.

In 2015, remarkable enantioselectivities of 95–98% ee were also reported by Takizawa *et al.* in enantioselective β-isocupreidine-catalyzed aza-MBH reactions of isatin imines with acrolein [20]. As shown in Scheme 4.12, a range of chiral 3-amino-2-oxindoles was obtained almost enantiopure (95–98% ee) in moderate to good yields (48–83%). The excellent enantioselectivities were obtained uniformly irrespective of the electronic nature of the ketimine moiety using 15 mol% of β-isocupreidine as catalyst at −40°C in a 1:1 mixture of toluene and CPME as solvent. While the use of this organocatalyst led to the formation of products exhibiting the (*S*)-configuration, it was demonstrated that using α-isocupreine at 20 mol% instead of 15 mol% of β-isocupreidine under the same reaction conditions allowed the corresponding (*R*)-configured adducts to be obtained in excellent enantioselectivities (83–96% ee) along with moderate to good yields (37–79%), as shown in Scheme 4.12. To explain these results, a model for the enantioselectivity is proposed in Scheme 4.12. Since proton transfer is a known rate-determining step in aza-MBH reactions [21], the proton shift mediated by the acidic unit on the catalyst could proceed smoothly *via* an intermediate conformation with the least steric hindrance between the quinuclidine moiety of the catalyst and the aromatic ring of the substrate to result in the formation of the (*S*)-product by using β-isocupreidine or the (*R*)-product with α-isocupreine.

In 2015, Chimni *et al.* reported organocatalyzed aza-MBH reactions of isatin imines with maleimides using β-isocupreidine as catalyst [22]. It must be noted that maleimides as MBH donors were challenging in these reactions since they are more usually employed as Michael acceptors. As shown in Scheme 4.13, a wide variety of chiral 3-amino-2-oxindoles was synthesized in moderate to good yields (30–79%) and enantioselectivities (70–99% ee).

Very recently, Khan *et al.* reported the first asymmetric organocatalytic synthesis of chiral 3-amino-2-oxindoles based on enantioselective

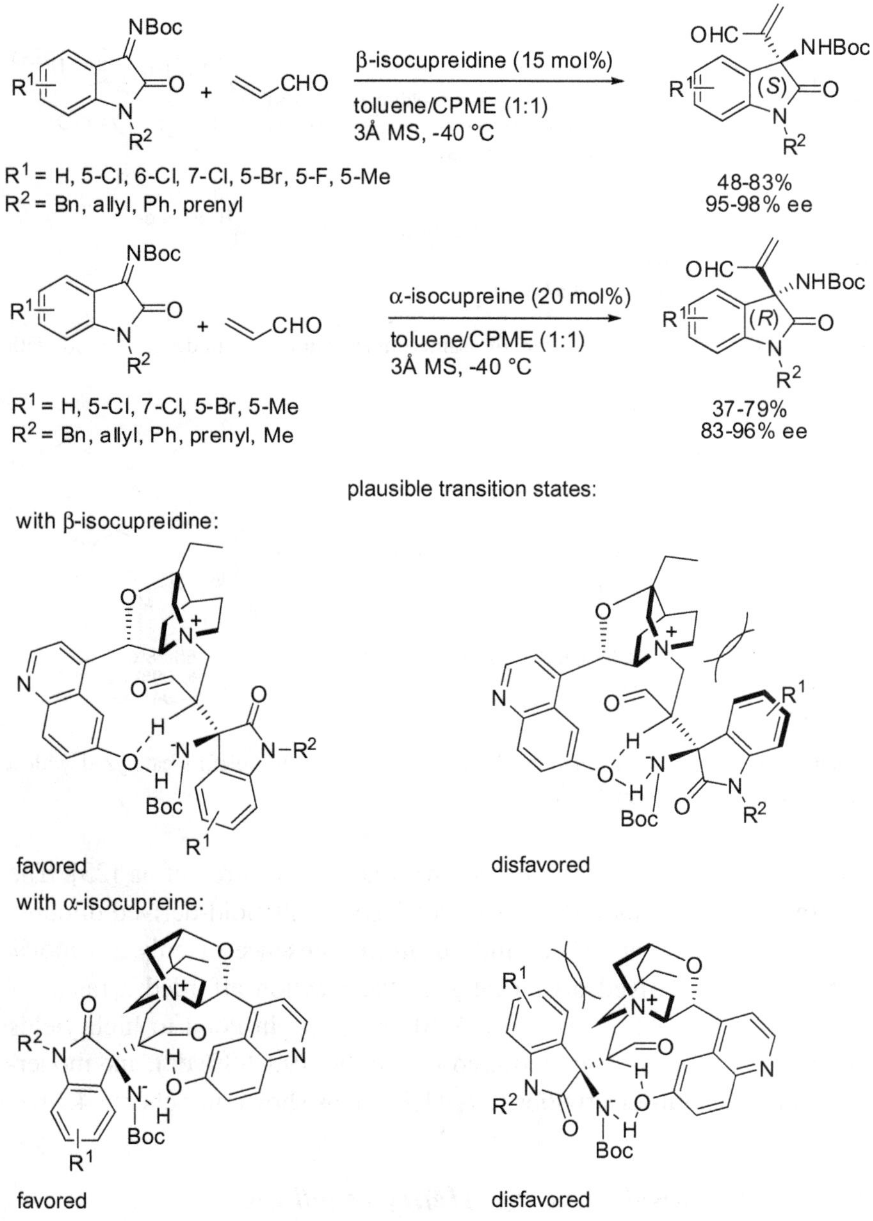

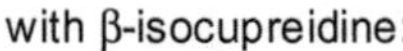

R^1 = H, 5-Cl, 6-Cl, 7-Cl, 5-Br, 5-F, 5-Me
R^2 = Bn, allyl, Ph, prenyl

β-isocupreidine (15 mol%)

toluene/CPME (1:1)
3Å MS, -40 °C

48-83%
95-98% ee

R^1 = H, 5-Cl, 7-Cl, 5-Br, 5-Me
R^2 = Bn, allyl, Ph, prenyl, Me

α-isocupreine (20 mol%)

toluene/CPME (1:1)
3Å MS, -40 °C

37-79%
83-96% ee

plausible transition states:

with β-isocupreidine:

favored

disfavored

with α-isocupreine:

favored

disfavored

Scheme 4.12. Aza-MBH reactions of isatin imines with acrolein catalyzed with β-isocupreidine and α-isocupreine.

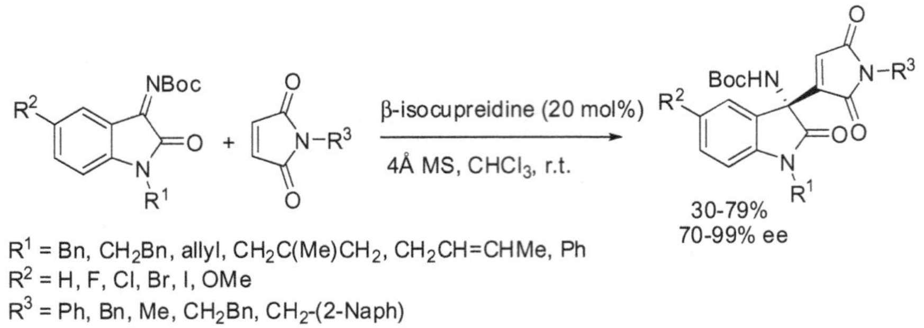

NBoc
β-isocupreidine (20 mol%)
4Å MS, CHCl3, r.t.
BocHN
30-79%
70-99% ee
R1 = Bn, CH2Bn, allyl, CH2C(Me)CH2, CH2CH=CHMe, Ph
R2 = H, F, Cl, Br, I, OMe
R3 = Ph, Bn, Me, CH2Bn, CH2-(2-Naph)

Scheme 4.13. Aza-MBH reaction of isatin imines with maleimides catalyzed with *β*-isocupreidine.

NBoc
Pseudo-I (2.5 mol%)
toluene, -10 °C
BocHN
R1 = H, 5-Me, 5-OMe, 4-Cl, 5-Cl, 6-Cl, 5-Br, 7-F
R2 = Me, Bn, CH2CH(OEt)2
R3 = H, 4-Cl
65-88%
90-98% de
54-94% ee

Scheme 4.14. Aza-MBH reaction of isatin imines with nitroolefins catalyzed with a *Cinchona* alkaloid-derived thiourea.

aza-MBH reaction of isatin imines with activated nitroolefins [23]. The best results were achieved by using a *Cinchona* alkaloid-derived thiourea catalyst in toluene at −10°C. Indeed, in the presence of only 2.5 mol% of this quinine-derived organocatalyst, the reaction afforded a range of chiral densely functionalized aza-MBH products in good to high yields (65–88%), uniformly high diastereoselectivities (90–98% de), and moderate to high enantioselectivities (54–94% ee), as shown in Scheme 4.14.

4.2.3. *Enantioselective aza-Henry reactions*

The catalytic asymmetric Henry reaction, also known as catalytic asymmetric nitroaldol reaction, constitutes a useful synthetic methodology towards chiral *β*-nitro alcohols [24].

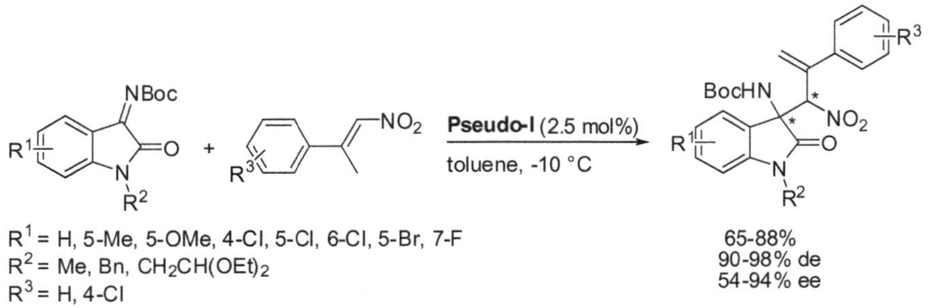

In this reaction, nitromethane (or a related nitroalkane) reacts in the presence of a chiral catalyst with an aldehyde, forming the corresponding optically active β-nitro alcohols. The latter are valuable intermediates in the synthesis of a broad variety of chiral building blocks, such as β-amino alcohols. These reactions and related aza-Henry reactions involving imines have been promoted by both metal catalysts and organocatalysts. While isatins have been widely used as substrates in asymmetric Henry reactions, isatin imines have been less explored as aza-Henry acceptors in enantioselective reactions. Recently, several groups have developed enantioselective aza-Henry reactions of isatin imines catalyzed by either chiral organocatalysts or chiral metal complexes. Among organocatalysts, the *Cinchona* alkaloid catalyst depicted in Scheme 4.15 was employed by Chimni *et al.* at 20 mol% of catalyst loading in THF at room temperature [25]. Therefore, the reaction of variously substituted isatin imines with nitroalkanes led to the corresponding chiral 3-amino-2-oxindoles in moderate to good yields (56–89%) and enantioselectivities (68–80% ee). When nitroalkanes other than nitromethane ($R^3 = H$) were employed as substrates, the corresponding aza-Henry products were obtained as diastereomeric mixtures with low to moderate diastereoselectivities (8–44% de).

Subsequently in 2014, Zhou *et al.* investigated these reactions in the presence of another type of organocatalysts, such as chiral *N,N'*-dioxide catalyst depicted in Scheme 4.16 [26]. The reaction of nitromethane with an isatin imine performed at room temperature in the presence of 20 mol% of this catalyst in THF as solvent afforded the corresponding chiral 3-amino-2-oxindole in 93% yield and 71% ee, as shown in Scheme 4.16.

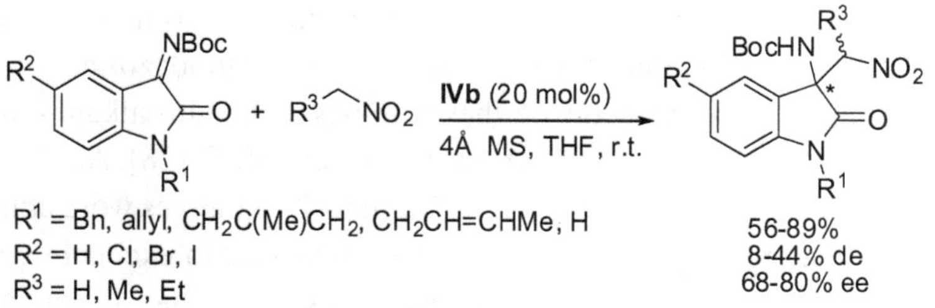

Scheme 4.15.   Aza-Henry reaction of isatin imines with nitroalkanes catalyzed with a *Cinchona* alkaloid.

Scheme 4.16. Aza-Henry reaction of an isatin imine with nitromethane catalyzed with a *N,N′*-dioxide.

Scheme 4.17. Aza-Henry reaction of isatin imines with nitromethane catalyzed with a bifunctional guanidine.

Later in 2015, a chiral bifunctional guanidine was applied as organocatalyst by Feng *et al.* in related reactions (Scheme 4.17) [27]. When using nitromethane as substrate, the reaction performed at −30°C with 10 mol% of this catalyst in toluene provided a range of chiral aza-Henry products in high to quantitative yields (81–99%) and generally high enantioselectivities (85–94% ee).

In addition to chiral organocatalysts, several chiral metal complexes have also been successfully applied as catalysts to promote enantioselective aza-Henry reactions of isatin imines. For example, Arai *et al.* demonstrated that a combination of NiCl$_2$ with a chiral *bis*(imidazoline)pyridine ligand catalyzed the reaction of these imines with nitroalkanes in the presence of a base, such as DIPEA, in toluene at −30°C [28]. As shown in Scheme 4.18, the corresponding chiral 3-amino-2-oxindoles were obtained in low to quantitative yields (19–99%) and moderate to excellent enantioselectivities (57–95% ee). When nitroalkanes other than nitromethane were employed as substrates, the products were obtained as mixtures of diastereomers with a moderate diastereoselectivity (60% de).

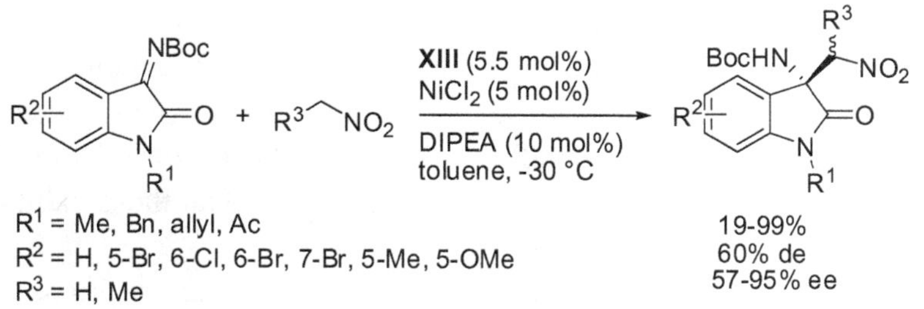

NBoc
XIII (5.5 mol%)
$NiCl_2$ (5 mol%)
DIPEA (10 mol%)
toluene, -30 °C
BocHN
R^3
NO_2
R^2
R^1
R^1 = Me, Bn, allyl, Ac
R^2 = H, 5-Br, 6-Cl, 6-Br, 7-Br, 5-Me, 5-OMe
R^3 = H, Me
19-99%
60% de
57-95% ee

Scheme 4.18. Nickel-catalyzed aza-Henry reaction of isatin imines with nitroalkanes.

NBoc
XIV (10 mol%)
$Cu(BF_4)_2$ (10 mol%)
i-Pr_2NH (13 mol%)
THF, r.t.
BocHN
R^3
NO_2
R^2
R^1
R^1 = Me, Bn, MOM, Boc
R^2 = H, 5-Br, 6-Cl, 6-Br, 7-Br, 5-Me, 5-OMe
R^3 = H, Me
68-99%
20-82% de
81 to >99% ee

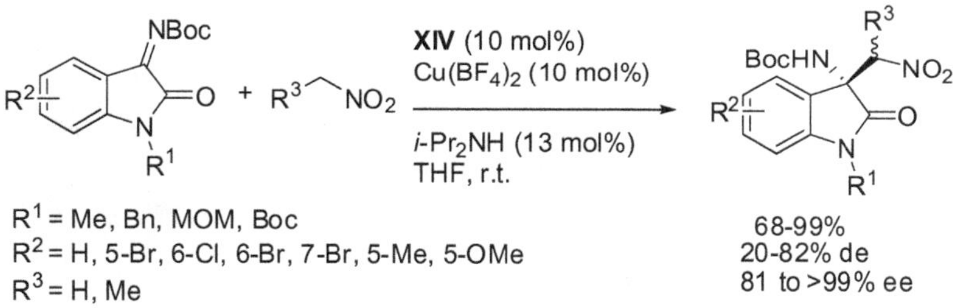

Scheme 4.19. Copper-catalyzed aza-Henry reaction of isatin imines with nitroalkanes.

A comparable study was reported by Pedro *et al.* using a chiral copper complex generated from $Cu(BF_4)_2$ and a chiral *bis*-oxazoline in THF at room temperature in the presence of diisopropylamine as base (Scheme 4.19) [29]. This catalyst efficiently promoted the reaction of a range of isatin imines with nitroalkanes, leading to the corresponding chiral aza-Henry products in good to quantitative yields (68–99%) and high enantioselectivities (81 to >99% ee). When nitroalkanes other than nitromethane were used as substrates, the products were obtained as diastereomeric mixtures with low to high diastereoselectivities (20–82% de).

4.2.4. *Enantioselective additions of alcohols and hydroperoxides*

In 2013, Sha *et al.* reported the first asymmetric organocatalytic addition of alcohols to isatin imines using a quinine-derived thiourea catalyst [30].

Performed at 0°C in a mixture of acetonitrile and ethanol as solvent, it afforded a range of chiral *N,O*-acetals in uniformly excellent yields (92–98%) combined with moderate enantioselectivities (63–78% ee), as shown in Scheme 4.20. Generally, isatin imines bearing a substituent at the six-position provided better enantioselectivities than those exhibiting a substituent at the seven-position.

In 2015, Arai *et al.* applied a chiral metal complex to promote the enantioselective addition of methanol to isatin imines [31]. As shown in Scheme 4.21, the use of a chiral nickel catalyst *in situ* generated from $NiCl_2$ and a chiral *bis*(imidazolidine)pyridine ligand to promote this reaction in toluene at room temperature in the presence of a base, such as

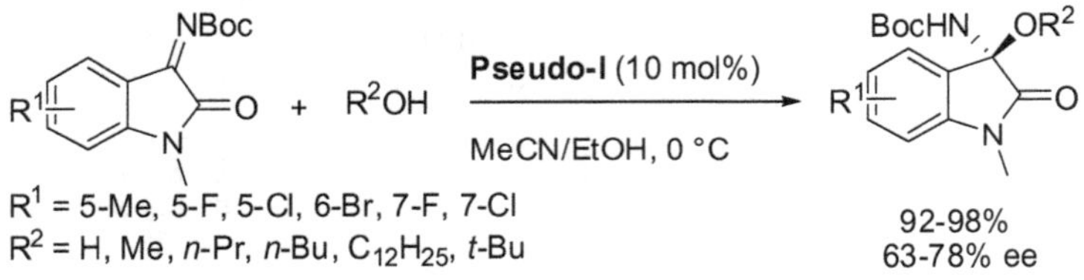

Scheme 4.20. Addition of alcohols to isatin imines catalyzed with a quinine-derived thiourea.

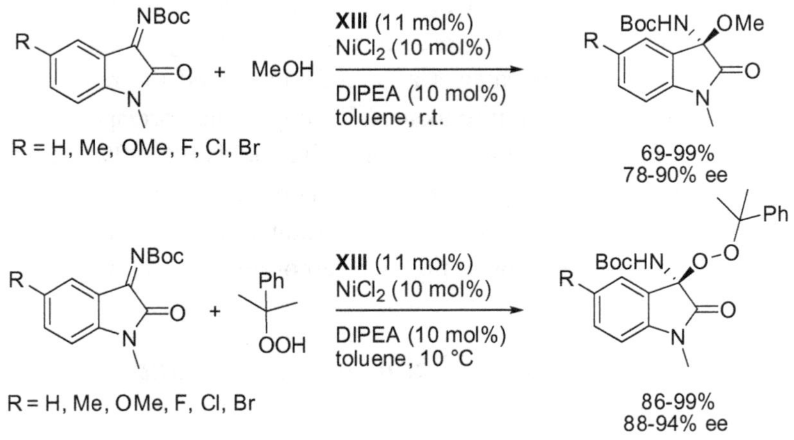

Scheme 4.21. Nickel-catalyzed additions of methanol/cumene hydroperoxide to isatin imines.

DIPEA, led to chiral isatin-derived *N,O*-acetals in good to quantitative yields (69–99%) and high enantioselectivities (78–90% ee). Comparable reaction conditions were applied to the addition of cumene hydroperoxide which provided the corresponding chiral oxindoles in even better yields (86–99%) and enantioselectivities (88–94% ee), as shown in Scheme 4.21.

4.2.5. *Enantioselective Strecker reactions*

The Strecker reaction starting from an aldehyde, ammonia, and a cyanide source is an efficient method for the preparation of α-amino acids and their derivatives. A popular version for asymmetric purposes is based on the use of preformed imines and a subsequent nucleophilic addition of TMSCN in the presence of a chiral catalyst [32]. Intense investigation of the asymmetric Strecker-type reaction has continued over many years, due to the importance of α-amino acid building blocks in medicinal chemistry [33]. Different types of chiral organocatalysts often based on *Cinchona* alkaloids have been found to have catalytic hydrocyanation properties in enantioselective Strecker reactions of isatin imines. Actually, the first asymmetric addition of TMSCN to isatin imines was reported in 2010 by Zhou *et al.* using 10 mol% of a *Cinchona* alkaloid-derived phosphoramide catalyst in DCE as solvent at 0°C [34]. As shown in Scheme 4.22, the corresponding chiral 3-cyano-3-amino-2-oxindole derivatives were obtained in low to good yields (27–72%) and moderate enantioselectivities (39–74% ee).

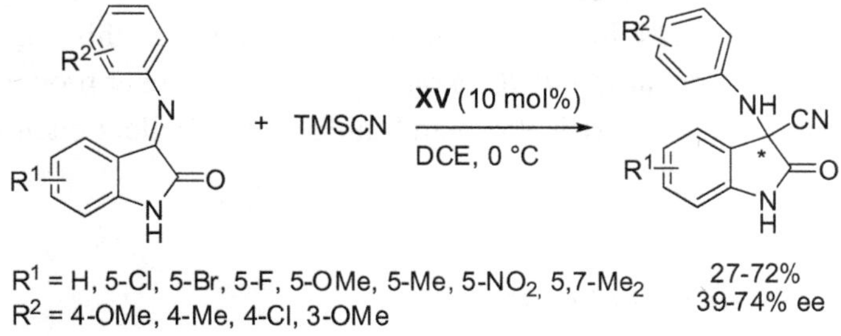

Scheme 4.22.   Strecker reaction of isatin imines catalyzed with a *Cinchona* alkaloid-derived phosphoramide.

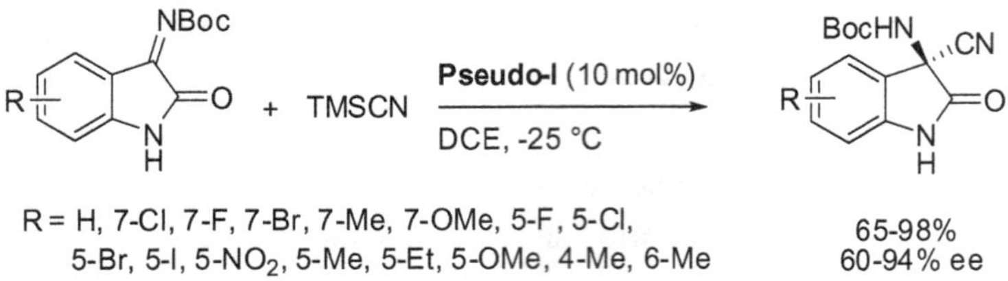

Scheme 4.23. Strecker reaction of isatin imines catalyzed with a quinine-derived thiourea.

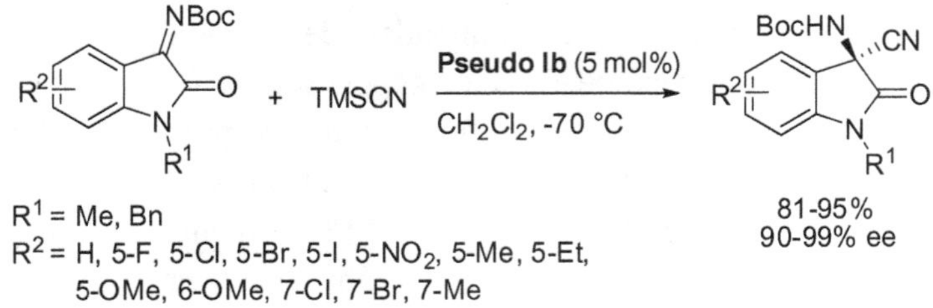

Scheme 4.24. Strecker reaction of isatin imines catalyzed with a *Cinchona* alkaloid-derived thiourea.

An improved version of the asymmetric Strecker reaction of isatin imines was developed by Wang *et al.* using a quinine-derived thiourea catalyst [35]. Performed in the same solvent at −25°C, the reactions afforded comparable products in moderate to quantitative yields (65–98%) and moderate to excellent enantioselectivities (60–94% ee), as shown in Scheme 4.23.

Very good results were also reported by Zhou *et al.* by using another *Cinchona* alkaloid-derived thiourea catalyst in these reactions [36]. As shown in Scheme 4.24, in the presence of only 5 mol% of this organocatalyst in dichloromethane at −70°C, a range of chiral Strecker products were produced in high yields (81–95%) and uniformly excellent enantioselectivities (90–99% ee).

4.2.6. *Miscellaneous reactions*

In 2008, List *et al.* reported the direct synthesis of a chiral aminal from the reaction of isatin with 2-aminobenzamide [37]. The process was catalyzed

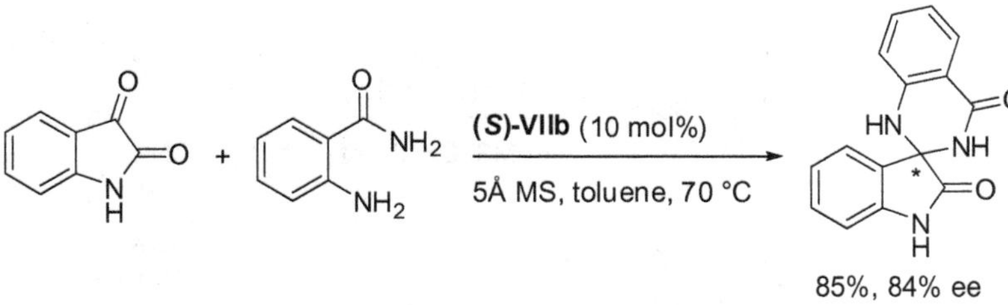

(S)-VIIb (10 mol%)
5Å MS, toluene, 70 °C
85%, 84% ee

Scheme 4.25. Addition of 2-aminobenzamide to isatin catalyzed with a phosphoric acid.

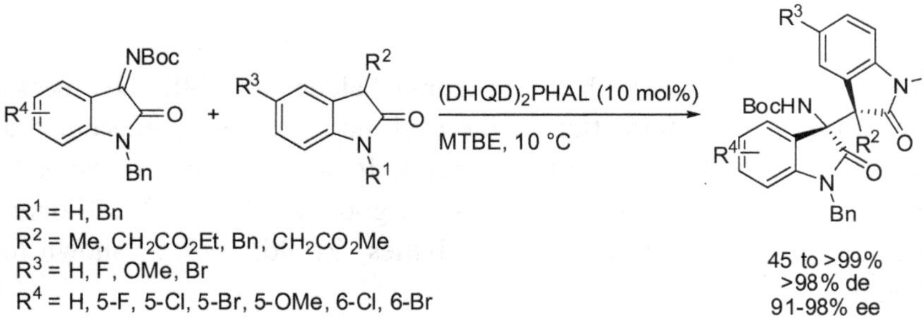

(DHQD)2PHAL (10 mol%)
MTBE, 10 °C
R1 = H, Bn
R2 = Me, CH2CO2Et, Bn, CH2CO2Me
R3 = H, F, OMe, Br
R4 = H, 5-F, 5-Cl, 5-Br, 5-OMe, 6-Cl, 6-Br
45 to >99%
>98% de
91-98% ee

Scheme 4.26. Addition of 3-substituted oxindoles to isatin imines catalyzed with a dimeric *Cinchona* alkaloid.

by 10 mol% of a chiral phosphoric acid in toluene at 70°C, providing the corresponding product in 85% yield and 84% ee (Scheme 4.25). Since this process involved the *in situ* generation of an intermediate isatin imine, it was decided to situate this example in this section. Moreover, it must be noted that it could also be situated in Section 4.3 dealing with domino reactions.

In 2014, Zhu *et al.* reported the organocatalytic addition of oxindoles to isatin imines, providing the corresponding chiral *bis*-oxindoles bearing two vicinal quaternary carbon centers as single diastereomers (>98% de) in moderate to quantitative yields (45 to >99%) and uniformly excellent enantioselectivities (91–98% ee), as shown in Scheme 4.26 [38]. The process was performed at 10°C in MTBE as solvent, in the presence of a *Cinchona* alkaloid catalyst (10 mol%).

In 2014, Chi *et al.* reported the application of *N*-heterocyclic carbene (NHC) catalysts to promote the chemoselective cross-aza-benzoin

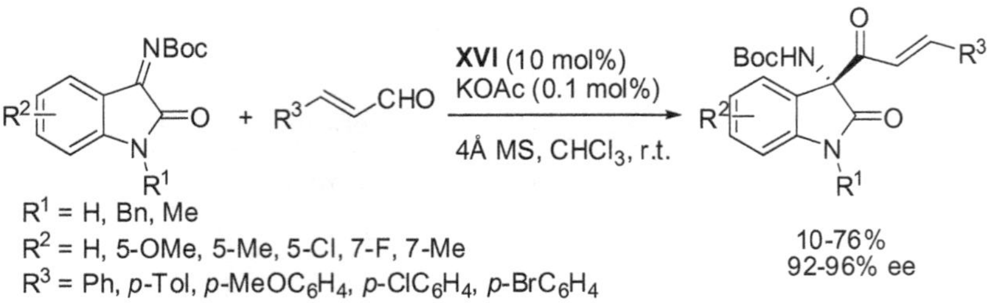

NBoc
R^2
O
N
R^1
+ R^3
CHO
XVI (10 mol%)
KOAc (0.1 mol%)
4Å MS, $CHCl_3$, r.t.
BocHN
O
R^3
R^2
O
N
R^1
R^1 = H, Bn, Me
R^2 = H, 5-OMe, 5-Me, 5-Cl, 7-F, 7-Me
R^3 = Ph, p-Tol, p-$MeOC_6H_4$, p-ClC_6H_4, p-BrC_6H_4
10-76%
92-96% ee

Scheme 4.27. NHC-catalyzed cross-aza-benzoin reaction of isatin imines with α,β-unsaturated aldehydes.

reaction of isatin imines with α,β-unsaturated aldehydes [39]. The process evolved through the formation of the acyl anion of the α,β-unsaturated aldehyde catalyzed by the NHC catalyst. It provided the corresponding chiral 3-amino-2-oxindoles in low to good yields (10–76%) albeit with uniformly excellent enantioselectivities (92–96% ee), as shown in Scheme 4.27.

The Pictet–Spengler reaction is an important acid-catalyzed transformation frequently used in organic synthesis, as well as by various organisms for the synthesis of tetrahydro-β-carbolines or tetrahydroisoquinolines from carbonyl compounds and phenyl ethylamines or tryptamines, respectively [40]. In 2011, Bencivenni *et al.* reported the first asymmetric organocatalytic Pictet–Spengler reaction of isatin derivatives [41]. A chiral phosphoric acid was identified as the most efficient catalyst to promote the condensation/cyclization reaction of isatins with tryptamine to afford at 40°C in DMF as solvent the corresponding chiral spiroindolones in good to excellent yields (74–97%) and enantioselectivities (71–95% ee), as shown in Scheme 4.28. Soon after, Franz *et al.* reported the Pictet–Spengler reaction of isatins with 5-methoxytryptamine catalyzed in dichloromethane as solvent at room temperature by a related chiral phosphoric acid [42]. As shown in Scheme 4.28, the enantiomeric products were achieved in good to quantitative yields (60–99%) and low to high enantioselectivities (16–94% ee). It must be noted that in spite of the fact that these reactions did not start from isatin imines, it was decided to situate them in this section since their mechanism involved the *in situ* generation of isatin imines.

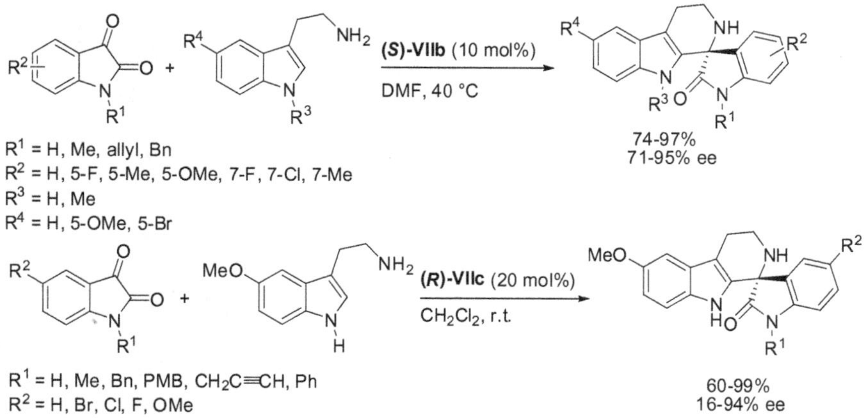

R^1 = H, Me, allyl, Bn
R^2 = H, 5-F, 5-Me, 5-OMe, 7-F, 7-Cl, 7-Me
R^3 = H, Me
R^4 = H, 5-OMe, 5-Br

(*S*)-**VIIb** (10 mol%)
DMF, 40 °C

74-97%
71-95% ee

R^1 = H, Me, Bn, PMB, CH$_2$C≡CH, Ph
R^2 = H, Br, Cl, F, OMe

(*R*)-**VIIc** (20 mol%)
CH$_2$Cl$_2$, r.t.

60-99%
16-94% ee

Scheme 4.28.   Pictet–Spengler reactions of isatin derivatives with tryptamines catalyzed with phosphoric acids.

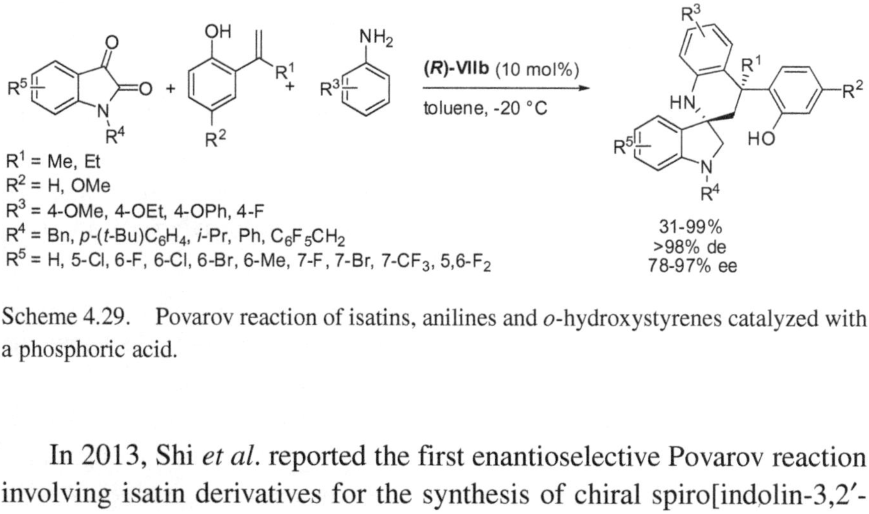

R^1 = Me, Et
R^2 = H, OMe
R^3 = 4-OMe, 4-OEt, 4-OPh, 4-F
R^4 = Bn, *p*-(*t*-Bu)C$_6$H$_4$, *i*-Pr, Ph, C$_6$F$_5$CH$_2$
R^5 = H, 5-Cl, 6-F, 6-Cl, 6-Br, 6-Me, 7-F, 7-Br, 7-CF$_3$, 5,6-F$_2$

(*R*)-**VIIb** (10 mol%)
toluene, -20 °C

31-99%
>98% de
78-97% ee

Scheme 4.29.   Povarov reaction of isatins, anilines and *o*-hydroxystyrenes catalyzed with a phosphoric acid.

In 2013, Shi *et al.* reported the first enantioselective Povarov reaction involving isatin derivatives for the synthesis of chiral spiro[indolin-3,2′-quinolines] bearing a tetrasubstituted stereogenic center [43]. As shown in Scheme 4.29, the Povarov reaction between isatins, anilines, and α-alkyl *o*-hydroxystyrenes was catalyzed by a chiral phosphoric acid, leading to the corresponding chiral 3-amino-2-oxindoles in low to quantitative yields (31–99%), almost complete diastereoselectivity (>98% de), and high enantioselectivities (78–97% ee). It must be noted that this work was situated in this section since the isatin imines were *in situ* generated but

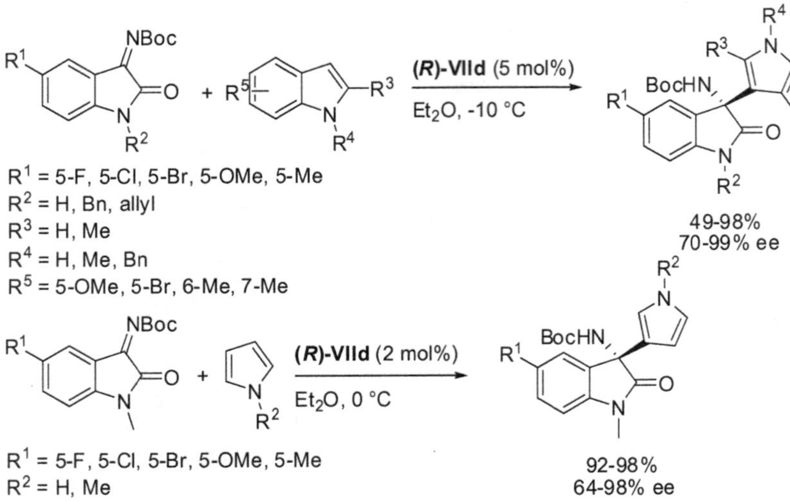

R^1 = 5-F, 5-Cl, 5-Br, 5-OMe, 5-Me
R^2 = H, Bn, allyl
R^3 = H, Me
R^4 = H, Me, Bn
R^5 = 5-OMe, 5-Br, 6-Me, 7-Me

(*R*)-VIId (5 mol%)
Et$_2$O, -10 °C

49-98%
70-99% ee

R^1 = 5-F, 5-Cl, 5-Br, 5-OMe, 5-Me
R^2 = H, Me

(*R*)-VIId (2 mol%)
Et$_2$O, 0 °C

92-98%
64-98% ee

Scheme 4.30. Friedel–Crafts reactions of isatin imines with indoles and pyrroles catalyzed with a phosphoric acid.

these results could also be situated in Section 4.3 dealing with domino reactions.

The Friedel–Crafts reaction is an important reaction for the construction of carbon–carbon and carbon–nitrogen bonds, providing important building blocks for pharmaceuticals [44].

In 2012, Wang *et al.* reported the first asymmetric aza-Friedel–Crafts reaction of isatin imines with either indoles or pyrroles [45]. As shown in Scheme 4.30, 5 mol% of a chiral phosphoric acid was found to efficiently promote at −10°C in diethyl ether the enantioselective aza-Friedel–Crafts reaction of isatin imines with indoles, providing the corresponding chiral 3-indolyl-3-amino-2-oxindoles in moderate to quantitative yields (49–98%) and enantioselectivities (70–99% ee).

It must be noted that these products constitute promising intermediates in the synthesis of drug candidates and alkaloids. Using only 2 mol% of the same catalyst in diethyl ether at 0°C, allowed the enantioselective aza-Friedel–Crafts reaction of isatin imines with pyrroles to afford the corresponding products in uniformly excellent yields (92–98%) and moderate to high enantioselectivities (64–98% ee), as shown in Scheme 4.30.

4.3. Domino and Tandem Reactions

The economic interest in combining asymmetric metal catalysis with the concept of domino reactions is obvious, and has allowed high molecular complexity to be easily reached with remarkable levels of stereocontrol on the basis of simple operational one-pot procedures, and advantages of savings in solvent, time, energy, and costs. In the last decade, an increasing number of powerful enantioselective domino processes catalyzed by various types of chiral organocatalysts [5] as well as chiral metals have been developed [46]. It is important to remember that a domino reaction has been defined by Tietze as a reaction which involves two or more bond-forming transformations, taking place under the same reaction conditions, without adding additional reagents and catalysts, and in which the subsequent reactions result as a consequence of the functionality formed by bond formation or fragmentation in the previous step [47].

4.3.1. *Organocatalyzed reactions*

In 2014, Wu *et al.* disclosed the enantioselective domino Mannich/cyclization reaction of isatin imines with 4-bromo-3-oxobutanoates catalyzed by a *Cinchona* alkaloid-derived squaramide [48]. As shown in Scheme 4.31, the corresponding highly functionalized chiral domino products were obtained in both excellent yields (90–97%) and enantioselectivities (94–98% ee).

Earlier in 2011, Yuan *et al.* reported the synthesis of chiral spirooxindoles based on an enantioselective organocatalytic domino aldol/cyclization

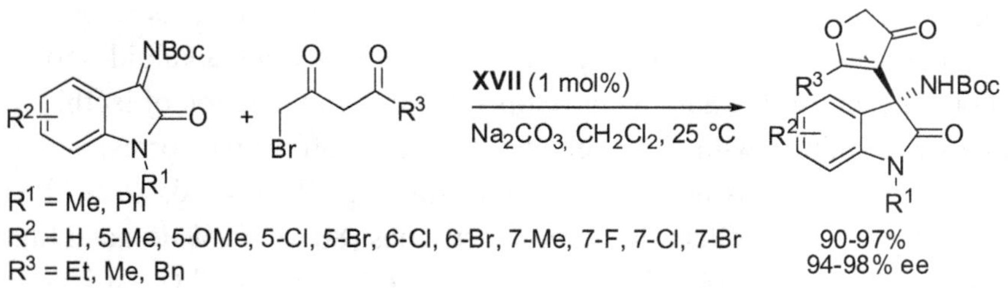

R¹ = Me, Ph
R² = H, 5-Me, 5-OMe, 5-Cl, 5-Br, 6-Cl, 6-Br, 7-Me, 7-F, 7-Cl, 7-Br
R³ = Et, Me, Bn

90-97%
94-98% ee

Scheme 4.31.   Domino Mannich/cyclization reaction of isatin imines with 4-bromo-3-oxobutanoates catalyzed with a *Cinchona* alkaloid-derived squaramide.

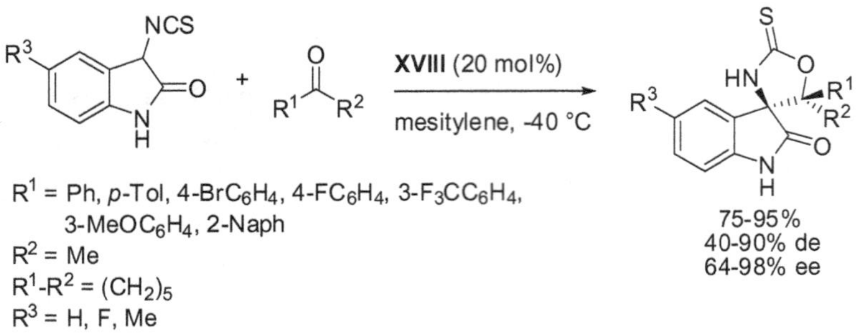

R^1 = Ph, *p*-Tol, 4-BrC$_6$H$_4$, 4-FC$_6$H$_4$, 3-F$_3$CC$_6$H$_4$,
   3-MeOC$_6$H$_4$, 2-Naph
R^2 = Me
R^1-R^2 = (CH$_2$)$_5$
R^3 = H, F, Me

75-95%
40-90% de
64-98% ee

Scheme 4.32. Domino aldol/cyclization reaction of 3-isothiocyanato oxindoles with ketones catalyzed with a thiourea.

reaction of 3-isothiocyanato-2-oxindoles with ketones [49]. The reaction was catalyzed at −40°C by a chiral tertiary amine-thiourea in mesitylene as solvent, leading to the corresponding highly functionalized chiral domino products in good to excellent yields (75–95%), moderate to high diastereoselectivities (40–90% de), and moderate to excellent enantioselectivities (64–98% ee), as shown in Scheme 4.32. The best results were achieved by using acetophenones as substrates.

Several enantioselective organocatalytic domino Michael/cyclization reactions have been recently developed. For example, Wang *et al.* reported a highly diastereo- and enantioselective domino Michael–cyclization reaction of isothiocyanato-2-oxindoles with electron-deficient olefins catalyzed by a *Cinchona* alkaloid-derived thiourea [50]. The process was performed in dichloromethane at room temperature and provided the corresponding chiral functionalized 3,2′-pyrrolidinyl spirooxindole derivatives as almost single diastereomers (>90% de) in high yields (86–99%) and good to excellent enantioselectivities (78–96% ee), as shown in Scheme 4.33 (Eq. (4.1)). Using another chiral *Cinchona* alkaloid-derived thiourea catalyst, the authors developed the domino reaction of isothiocyanato-2-oxindoles with methyleneindolinones to afford the corresponding chiral spirocyclic oxindoles with uniformly excellent yields (92–99%), general good to excellent diastereoselectivities (86 to >99% de), and high enantioselectivities (86–98% ee), as shown in Scheme 4.33 (Eq. (4.2)).

Concurrently with this, Huang and Wang also developed the reactions between 3-isothiocyanato oxindoles with methyleneindolinones in the

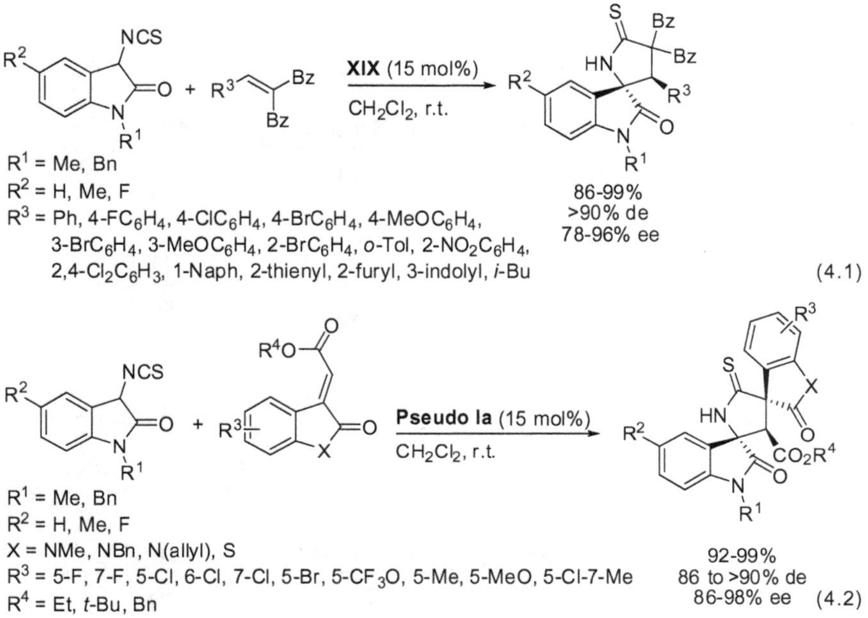

XIX (15 mol%)
CH2Cl2, r.t.
R1 = Me, Bn
R2 = H, Me, F
R3 = Ph, 4-FC6H4, 4-ClC6H4, 4-BrC6H4, 4-MeOC6H4, 3-BrC6H4, 3-MeOC6H4, 2-BrC6H4, o-Tol, 2-NO2C6H4, 2,4-Cl2C6H3, 1-Naph, 2-thienyl, 2-furyl, 3-indolyl, i-Bu
86-99%
>90% de
78-96% ee
(4.1)
Pseudo Ia (15 mol%)
CH2Cl2, r.t.
R1 = Me, Bn
R2 = H, Me, F
X = NMe, NBn, N(allyl), S
R3 = 5-F, 7-F, 5-Cl, 6-Cl, 7-Cl, 5-Br, 5-CF3O, 5-Me, 5-MeO, 5-Cl-7-Me
R4 = Et, t-Bu, Bn
92-99%
86 to >90% de
86-98% ee
(4.2)

Scheme 4.33. Domino Michael–cyclization reactions of isothiocyanato-2-oxindoles with electron-deficient olefins and methyleneindolinones catalyzed with *Cinchona* alkaloid-derived thioureas.

presence of a trifunctional catalyst (Scheme 4.34, Eq. (4.3)) to provide the corresponding chiral spirocyclic oxindoles in remarkable yields (96–99%), diastereoselectivity (>90% de), and enantioselectivities (90–99% ee) [51]. On the other hand, another quinine-derived thiourea (Scheme 4.34, Eq. (4.4)) emerged as optimal catalyst to promote the reaction between methyleneindolinones bearing an ester moiety to give the corresponding domino products in moderate to quantitative yields (29–99%), excellent diastereoselectivity, (>90% de), and high enantioselectivities (85–99% ee).

An asymmetric domino Michael–cyclization reaction between 3-isothiocyanato oxindoles and various aryl-substituted unsaturated pyrazolones was developed by Wang *et al.*, leading to the corresponding chiral spiro[oxindole/thiobutyrolactam/pyrazolone] structures exhibiting three contiguous stereogenic centers [52]. The process was catalyzed by a chiral tertiary amine-thiourea, and afforded these complex highly functionalized

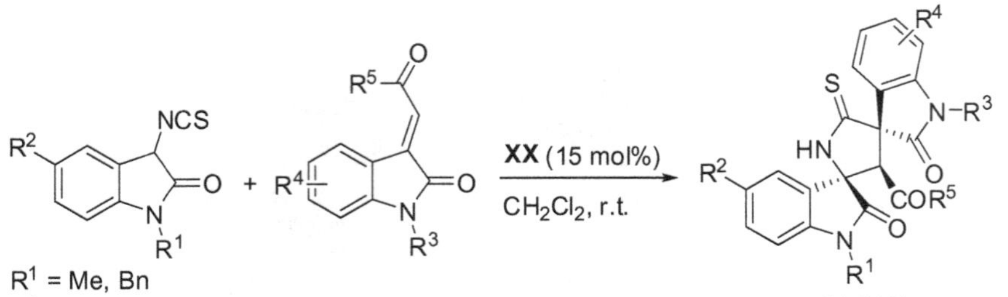

R^1 = Me, Bn
R^2 = H, Cl, Me
R^3 = Me, Ac
R^4 = 5-F, 5-Cl, 5-NO$_2$, 5-CF$_3$O, 5-Me, 5-OMe, 6-Br, 7-F
R^5 = Ph, 4-MeOC$_6$H$_4$, 4-FC$_6$H$_4$, 2-furyl, 2-thienyl

96-99%
>90% de
90-99% ee

(4.3)

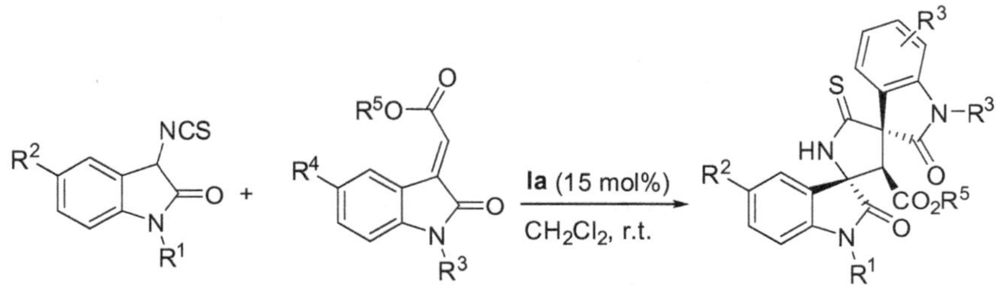

R^1 = Me, Bn
R^2 = H, Cl, MeO
R^3 = Me, Ac, Bn
R^4 = 5-F, 5-Cl, 5-NO$_2$, 5-Me, 5-OMe
R^5 = OEt, t-Bu

29-99%
>90% de
85-99% ee

(4.4)

Scheme 4.34. Domino Michael–cyclization reactions of 3-isothiocyanato oxindoles with methyleneindolinones catalyzed with *Cinchona* alkaloid-derived thioureas.

polycyclic products in good to quantitative yields (72–96%), moderate to excellent diastereoselectivities (66 to >90% de), and high enantioselectivities (80–99% ee), as shown in Scheme 4.35. With alkyl-substituted unsaturated pyrazolones, only low enantioselectivities were observed (11% ee) in spite of high yield (82%) and excellent diastereoselectivity (90% de).

The utility of *N*-heterocyclic carbenes (NHCs) as organocatalysts in domino/tandem reactions has received growing attention in the past few years [53]. In 2012, Jiao *et al.* demonstrated the application of this type of organocatalysts for the addition of α,β-unsaturated aldehydes to *N*-arylisatin

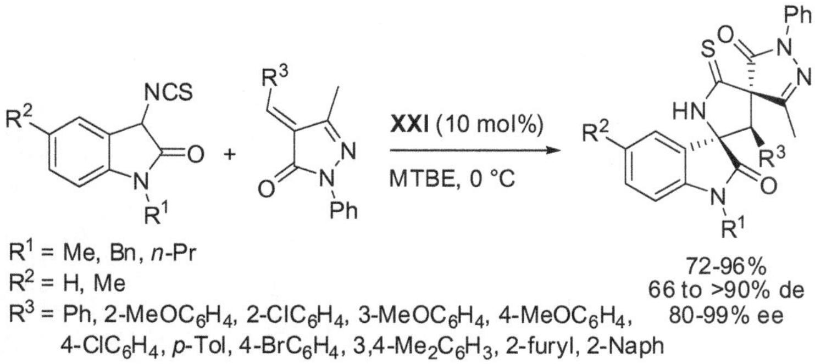

NCS
R2
R1
O
+
R3
N
N
Ph
XXI (10 mol%)
MTBE, 0 °C
Ph
S
HN
R3
R1
R1 = Me, Bn, n-Pr
R2 = H, Me
R3 = Ph, 2-MeOC6H4, 2-ClC6H4, 3-MeOC6H4, 4-MeOC6H4,
4-ClC6H4, p-Tol, 4-BrC6H4, 3,4-Me2C6H3, 2-furyl, 2-Naph
72-96%
66 to >90% de
80-99% ee

Scheme 4.35. Domino Michael–cyclization reaction of 3-isothiocyanato oxindoles with unsaturated pyrazolones catalyzed with a tertiary amine-thiourea.

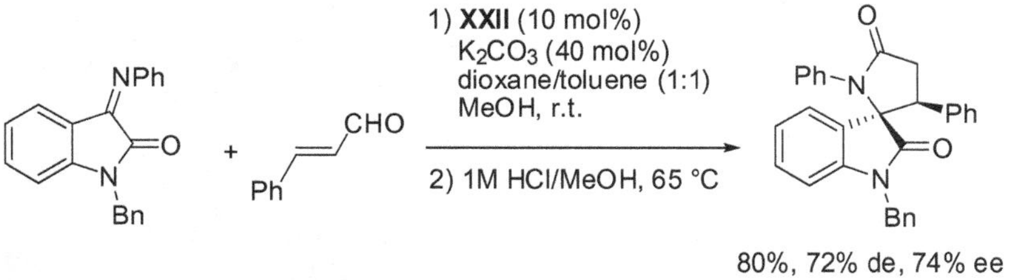

NPh
O
N
Bn
+
CHO
Ph
1) XXII (10 mol%)
K2CO3 (40 mol%)
dioxane/toluene (1:1)
MeOH, r.t.
2) 1M HCl/MeOH, 65 °C
Ph–N
Ph
80%, 72% de, 74% ee

Scheme 4.36. NHC-catalyzed domino addition–cyclization reaction of an *N*-phenyl isatin imine with cinnamaldehyde.

imines to develop a one-pot procedure for the synthesis of chiral spirocyclic γ-lactam oxindoles [54]. As shown in Scheme 4.36, by using 10 mol% of a NHC catalyst, the homoenolate equivalent of cinnamaldehyde was added to an *N*-phenyl isatin imine to give after subsequent acid hydrolysis the corresponding spirocyclic γ-lactam oxindole in 80% yield, 72% de and 74% ee. The mechanism of this domino reaction involved the addition of the NHC catalyst to the α,β-unsaturated aldehyde to give a Breslow intermediate serving as homoenolate equivalent. Then, the latter added to the isatin imine to give an intermediate which then cyclized to afford the final product.

This type of reaction was also investigated by Chi *et al.* by using another NHC catalyst [55]. As shown in Scheme 4.37, the annulation reaction of various α,β-unsaturated aldehydes with a series of *N*-aryl isatin

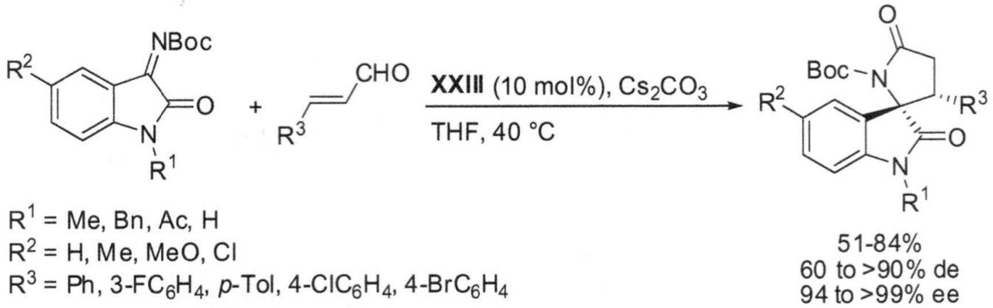

NBoc
CHO
XXIII (10 mol%), Cs2CO3
THF, 40 °C
Boc
R1 = Me, Bn, Ac, H
R2 = H, Me, MeO, Cl
R3 = Ph, 3-FC6H4, p-Tol, 4-ClC6H4, 4-BrC6H4
51-84%
60 to >90% de
94 to >99% ee

Scheme 4.37. NHC-catalyzed domino addition/cyclization reaction of isatin imines with α,β-unsaturated aldehydes.

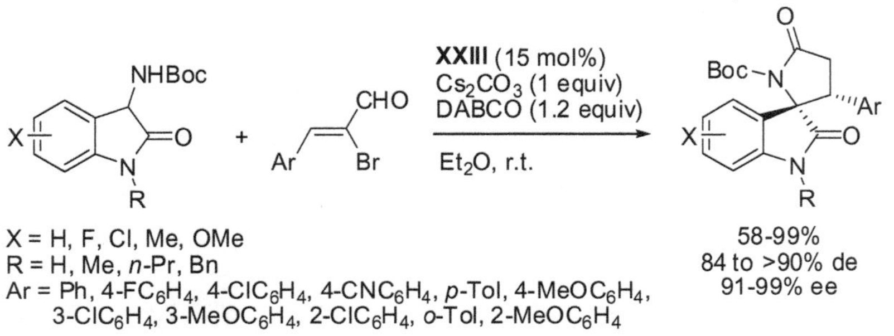

NHBoc
CHO
Ar
Br
XXIII (15 mol%)
Cs2CO3 (1 equiv)
DABCO (1.2 equiv)
Et2O, r.t.
Boc
X = H, F, Cl, Me, OMe
R = H, Me, n-Pr, Bn
Ar = Ph, 4-FC6H4, 4-ClC6H4, 4-CNC6H4, p-Tol, 4-MeOC6H4, 3-ClC6H4, 3-MeOC6H4, 2-ClC6H4, o-Tol, 2-MeOC6H4
58-99%
84 to >90% de
91-99% ee

Scheme 4.38. NHC-catalyzed domino Michael–lactamization reaction of 3-amino-2-oxindoles with α-bromoenals.

imines led to the corresponding chiral spirocyclic-γ-lactams in moderate to good yields (51–84%), good to high diastereoselectivities (60 to >90% de), and uniformly excellent enantioselectivities (94 to >99% ee).

In addition, Ye and Sun recently reported enantioselective NHC-catalyzed formal [3+2] annulations occurring between α-bromoenals and 3-amino-2-oxindoles [56]. As shown in Scheme 4.38, the domino process was catalyzed by 15 mol% of the same NHC catalyst in the presence of two bases, such as Cs_2CO_3 and DABCO at room temperature. The reaction occurred through the Michael addition of the 3-amino-2-oxindole *via* its enolate to an α,β-unsaturated acyl azolium, generated from the α-bromoenal and the catalyst, to give an intermediate which subsequently cyclized through lactamization to give the final chiral spirocyclic oxindolo-γ-lactam. A series of chiral cycloadducts were synthesized in

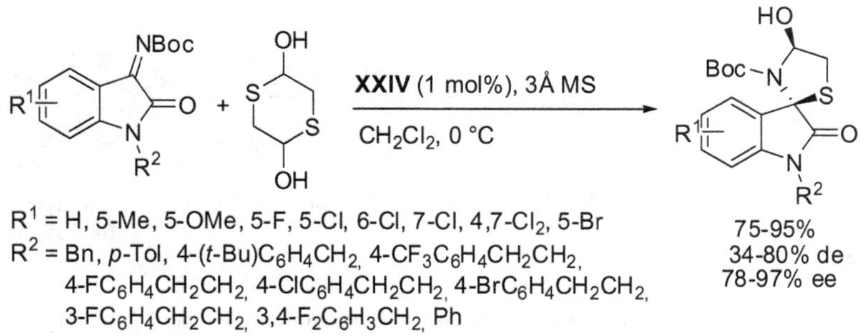

R^1 = H, 5-Me, 5-OMe, 5-F, 5-Cl, 6-Cl, 7-Cl, 4,7-Cl$_2$, 5-Br
R^2 = Bn, *p*-Tol, 4-(*t*-Bu)C$_6$H$_4$CH$_2$, 4-CF$_3$C$_6$H$_4$CH$_2$CH$_2$,
    4-FC$_6$H$_4$CH$_2$CH$_2$, 4-ClC$_6$H$_4$CH$_2$CH$_2$, 4-BrC$_6$H$_4$CH$_2$CH$_2$,
    3-FC$_6$H$_4$CH$_2$CH$_2$, 3,4-F$_2$C$_6$H$_3$CH$_2$, Ph

75-95%
34-80% de
78-97% ee

Scheme 4.39. Domino addition–cyclization reaction of isatin imines with 1,4-dithiane-2,5-diol catalyzed with a squaramide.

good to quantitative yields (58–99%), high to complete diastereoselectivities (84 to >90% de), and excellent enantioselectivities (91–99% ee).

In 2016, Li and Li reported an efficient asymmetric [3+2] annulation reaction between isatin imines and 1,4-dithiane-2,5-diol as equal equivalent of 2-mercaptoacetaldehyde [57]. The domino reaction catalyzed with a chiral tertiary amine-squaramide catalyst began with the addition of 2-mercaptoacetaldehyde to the isatin imine, leading to an aldehyde intermediate which subsequently cyclized into the corresponding chiral spirocyclic 3-aminooxindole. As shown in Scheme 4.39, a range of these products were achieved in high yields (75–95%), moderate to good diastereoselectivities (34–80% de), and high enantioselectivities (78–97% ee).

An asymmetric tandem Mannich–hydroamination reaction was developed by Liu *et al.* [58].

It occurred between isatin imines and a propargylated malononitrile to give the corresponding densely functionalized chiral tricyclic products in good to high yields (56–91%) and enantioselectivities (64–96% ee). As shown in Scheme 4.40, it involved the sequential treatment of the substrates with a chiral *Cinchona* alkaloid catalyst, BF$_3$(Et$_2$O), and XPhosAuNTf$_2$.

4.3.2. *Metal-catalyzed reactions*

In 2013, Matsunaga and Kanai reported the first catalytic asymmetric addition of isothiocyanato oxindoles to aldehydes by using a chiral

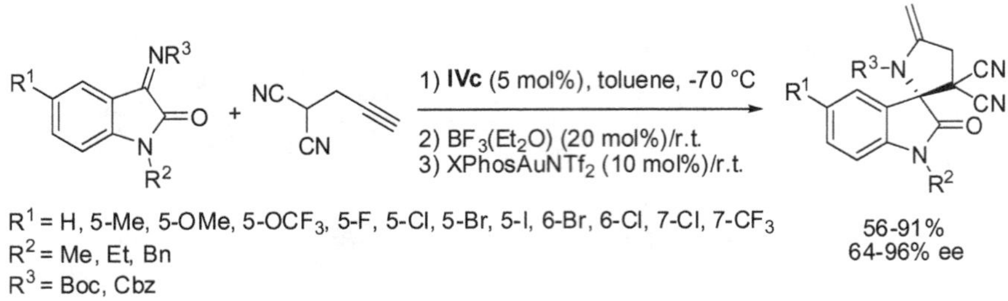

NR3
R1
O
N
R2
+
NC
CN
1) IVc (5 mol%), toluene, -70 °C
2) BF3(Et2O) (20 mol%)/r.t.
3) XPhosAuNTf2 (10 mol%)/r.t.
R3-N
CN
CN
R1
O
N
R2
R1 = H, 5-Me, 5-OMe, 5-OCF3, 5-F, 5-Cl, 5-Br, 5-I, 6-Br, 6-Cl, 7-Cl, 7-CF3
R2 = Me, Et, Bn
R3 = Boc, Cbz
56-91%
64-96% ee

Scheme 4.40. Tandem multicatalyzed Mannich–hydroamination reaction of isatin imines with a propargylated malononitrile.

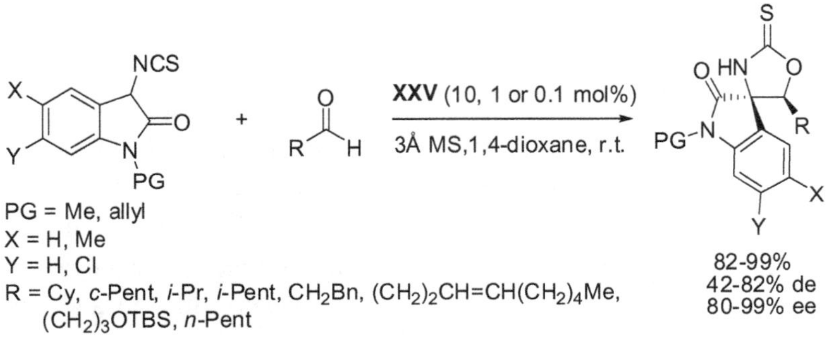

NCS
X
Y
O
N
PG
+
O
R
H
XXV (10, 1 or 0.1 mol%)
3Å MS,1,4-dioxane, r.t.
S
O HN
O
R
PG-N
X
Y
PG = Me, allyl
X = H, Me
Y = H, Cl
R = Cy, c-Pent, i-Pr, i-Pent, CH2Bn, (CH2)2CH=CH(CH2)4Me,
(CH2)3OTBS, n-Pent
82-99%
42-82% de
80-99% ee

Scheme 4.41. Nickel-catalyzed domino aldol-type–cyclization reaction of isothiocyanato oxindoles with aldehydes.

dinuclear nickel Schiff base catalyst [59]. As shown in Scheme 4.41, the reaction of a range of aliphatic aldehydes with isothiocyanato oxindoles afforded the corresponding spirooxindoles in high yields, diastereo-, and enantioselectivities of up to 99%, 82% de and 99% ee, respectively. The process, evolving through a domino aldol-type–cyclization reaction, was generally promoted by 10 mol% of a chiral dinuclear Ni_2-Schiff base at room temperature in 1,4-dioxane as solvent. It was demonstrated that the catalyst loading could be reduced to 0.1 mol% still providing remarkable enantioselectivity of up to 98% ee (PG = Me, X = Y = H, R = *n*-Pent). The authors found that the dinuclear nickel catalyst was much more efficient than the corresponding dinuclear copper and cobalt complexes, which gave low enantioselectivities (2–21% ee). The substrate scope of the domino aldol-type–cyclization reaction showed that α-branched aliphatic aldehydes gave the corresponding chiral tricyclic products in 66–78% de and 80–92% ee, while linear aliphatic aldehydes exhibited slightly higher diastereo- and enantioselectivities than the α-branched ones, providing products in 88–99% ee and up to 82% de. In contrast to aliphatic aldehydes, the system afforded poor results with aromatic aldehydes. For instance, an enantioselectivity of 33% ee was obtained for the reaction of benzaldehyde with unsubstituted N-methyl oxindole (X = Y = H, PG = Me) in combination with a low diastereoselectivity of 10% de.

In 2014, Chen and Xiao reported an unprecedented zinc-catalyzed asymmetric domino Michael–cyclization reaction of 3-nitro-$2H$-chromenes with 3-isothiocyanatooxindoles [60]. This transformation provided an efficient access to various synthetically important polycyclic spirooxindoles in a highly stereoselective manner under mild conditions. Indeed, these complex and densely functionalized chiral products bearing three consecutive stereogenic centers, including 1,3-non-adjacent tetrasubstituted carbon stereocenters, were remarkably achieved as almost single diastereomers (>90% de in all cases) in good to quantitative yields (72–99%) and with general excellent enantioselectivities of 91 to >99% ee (Scheme 4.42). These exceptional results were obtained by using a combination of $Zn(OTf)_2$ with a (S,S)-*bis*-oxazoline ligand bearing a free NH group which could act as a Lewis base through hydrogen-bonding interaction.

In the same area, Xu and Yuan later developed related asymmetric domino Michael–cyclization reactions by using the (R,R)-enantiomer of

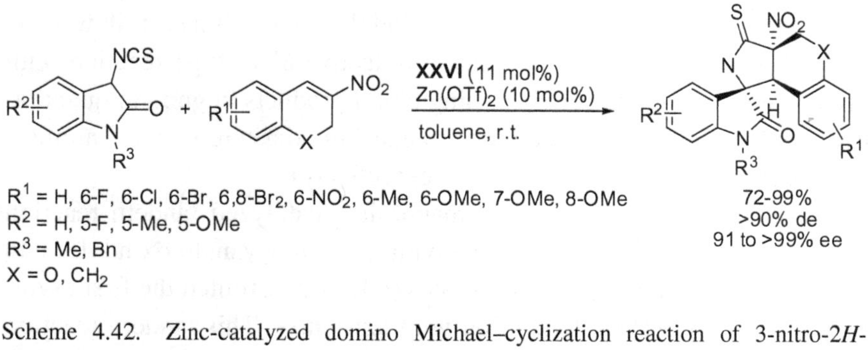

Scheme 4.42.   Zinc-catalyzed domino Michael–cyclization reaction of 3-nitro-$2H$-chromenes with 3-isothiocyanatooxindoles.

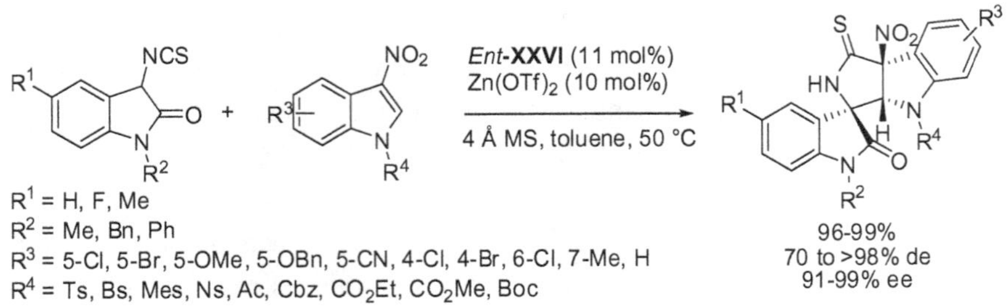

NCS
R1
O
N
R2
+
NO2
R3
N
R4
Ent-XXVI (11 mol%)
Zn(OTf)2 (10 mol%)
4 Å MS, toluene, 50 °C
S
NO2
R3
HN
R1
N
H
R4
N
O
R2
R1 = H, F, Me
R2 = Me, Bn, Ph
R3 = 5-Cl, 5-Br, 5-OMe, 5-OBn, 5-CN, 4-Cl, 4-Br, 6-Cl, 7-Me, H
R4 = Ts, Bs, Mes, Ns, Ac, Cbz, CO2Et, CO2Me, Boc
96-99%
70 to >98% de
91-99% ee

Scheme 4.43. Zinc-catalyzed domino Michael–cyclization reaction of 3-nitroindoles with 3-isothiocyanatooxindoles.

the same ligand [61]. As shown in Scheme 4.43, the reaction occurred between 3-isothiocyanatooxindoles and 3-nitroindoles to afford the corresponding chiral polycyclic spirooxindole derivatives exhibiting three contiguous stereocenters. Performing the reaction with *N*-methyl protected 3-isothiocyanatooxindoles in the presence of the (*R,R*)-*bis*-oxazoline ligand associated to $Zn(OTf)_2$ as precatalyst in toluene at 50°C, the corresponding highly functionalized complex products were impressively obtained in quantitative yields as single stereoisomers (>98% de and >99% ee) in almost all cases of substrates examined (Scheme 4.43).

In 2013, Gong *et al.* reported a synthesis of chiral 3-amino-2-oxindoles based on a multicatalytic asymmetric three-component aldol-type reaction of 3-diazo oxindoles with anilines and glyoxylates [62]. This domino process was catalyzed by a rhodium complex, such as $[Rh_2(OAc)_4]$, combined with a chiral phosphoric acid (Scheme 4.44).

The reaction basically proceeded *via* a rhodium-catalyzed generation of ammonium ylides from 3-diazo oxindoles and anilines, followed by chiral Brønsted acid-catalyzed enantioselective aldol-type reaction with glyoxylates to give the corresponding chiral products in good to quantitative yields (63–99%) and moderate to excellent diastereo- and enantioselectivities (50–94% de, 13–99% ee, respectively).

In 2015, an enantioselective magnesium-catalyzed tandem reaction was developed by Wang *et al.*, involving 3-isothiocyanato oxindoles and *N*-(2-picolinoyl)aziridines as substrates [63]. It constituted the first asymmetric formal [3+3] cycloaddition with aziridines. This tandem reaction was mediated by a magnesium catalyst *in situ* generated from $MgBu_2$ and

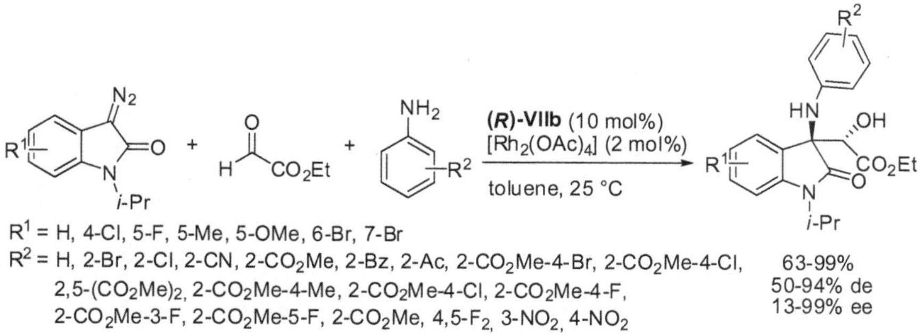

(R)-VIIb (10 mol%)
[Rh2(OAc)4] (2 mol%)
toluene, 25 °C
R1 = H, 4-Cl, 5-F, 5-Me, 5-OMe, 6-Br, 7-Br
R2 = H, 2-Br, 2-Cl, 2-CN, 2-CO2Me, 2-Bz, 2-Ac, 2-CO2Me-4-Br, 2-CO2Me-4-Cl, 2,5-(CO2Me)2, 2-CO2Me-4-Me, 2-CO2Me-4-Cl, 2-CO2Me-4-F, 2-CO2Me-3-F, 2-CO2Me-5-F, 2-CO2Me, 4,5-F2, 3-NO2, 4-NO2
63-99%
50-94% de
13-99% ee

Scheme 4.44. Three-component aldol-type reaction of 3-diazo oxindoles, anilines and glyoxylates catalyzed with a combination of $[Rh_2(OAc)_4]$ and a chiral phosphoric acid.

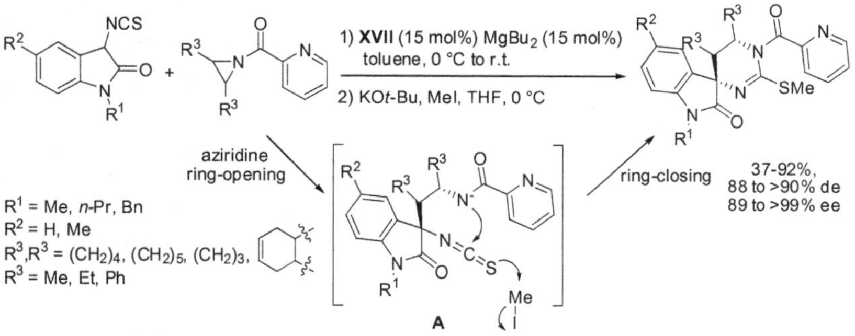

1) XVII (15 mol%) MgBu2 (15 mol%)
toluene, 0 °C to r.t.
2) KOt-Bu, MeI, THF, 0 °C
aziridine ring-opening
ring-closing
R1 = Me, n-Pr, Bn
R2 = H, Me
R3,R3 = (CH2)4, (CH2)5, (CH2)3,
R3 = Me, Et, Ph
A
37-92%,
88 to >90% de
89 to >99% ee

4.45

Scheme 4.45. Magnesium-catalyzed tandem ring-opening–ring-closing reaction of 3-isothiocyanato oxindoles with *N*-(2-picolinoyl)aziridines.

a chiral BINOL-derived fluorinated ligand in toluene at 0°C to room temperature. It began with the ring opening of aziridines with 3-isothiocyanato oxindoles to give intermediates A which subsequently cyclized into final products by treatment with *t*-BuOK/MeI (Scheme 4.45). This novel tandem ring-opening/ring-closing reaction allowed a range of densely functionalized chiral pyrimidine derivatives to be synthesized in moderate to high yields (37–92%) with uniformly high diastereo- and enantioselectivities (88 to >90% de and 89 to >99% ee, respectively).

In 2014, a synthesis of chiral 3-amino-2-oxindoles was based by Zhou *et al.* on a rare example of one-pot process involving asymmetric triple relay catalysis [64]. The sequence began with the hydrogenation of a

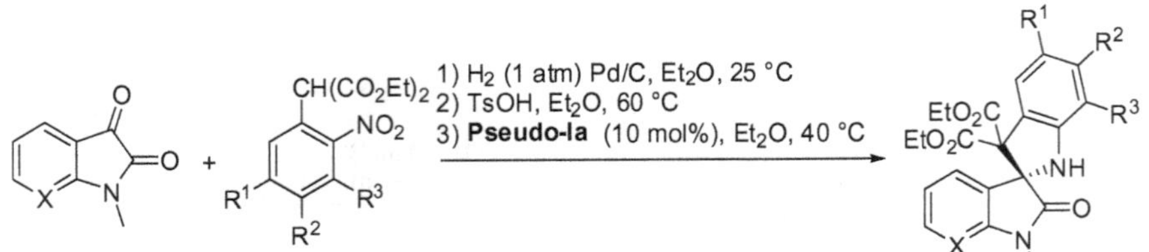

Scheme 4.46. Multicatalyzed tandem hydrogenation–imine formation/[1,5]electrocyclic reaction of malonate-nitroarenes with isatin derivatives.

malonate–nitroarene derivative into the corresponding malonate–aniline derivative. Then, this compound reacted with an isatin derivative in the presence of TsOH to give the corresponding ketimine. When the latter was submitted to a chiral *Cinchona* alkaloid-derived thiourea catalyst, it underwent a [1,5] electrocyclic reaction to afford the corresponding chiral oxindole-based spirocyclic indoline in moderate to high yields (45–85%) and high enantioselectivities (80–97% ee), as shown in Scheme 4.46.

4.4. Direct Aminations of 3-Substituted Oxindoles

4.4.1. *Aminations*

The direct amination of 3-substituted oxindoles using azodicarboxylates is a powerful strategy for enantioselective carbon–nitrogen bond formations [65]. Therefore, the deprotonative activation of 3-substituted oxindoles for direct amination provides another method to prepare chiral quaternary 3-amino-2-oxindoles. In 2009, three independent reports were published simultaneously on the enantioselective amination of oxindoles with diazo compounds catalyzed by dimeric *Cinchona* alkaloids (Scheme 4.47). Therefore, Chen *et al.* reported the use of *bis-Cinchona* alkaloid [(DHQD)$_2$PHAL] to catalyze the enantioselective amination of *N*-unprotected 2-oxindoles with di-*iso*-propyl azodicarboxylate [66]. As shown in Scheme 4.47 (Eq. (4.5)), the reaction led to the corresponding chiral 3-amino-2-oxindoles in moderate to quantitative yields (54 to >99%)

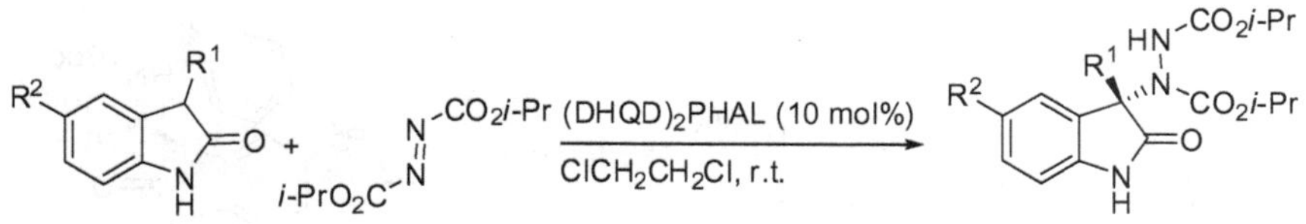

R^1 = Me, Bn, 4-ClC$_6$H$_4$CH$_2$, 4-BrC$_6$H$_4$CH$_2$, 4-MeOC$_6$H$_4$CH$_2$, *p*-TolCH$_2$,
2-BrC$_6$H$_4$CH$_2$, 2-CF$_3$C$_6$H$_4$CH$_2$, 2-MeOC$_6$H$_4$CH$_2$, *o*-TolCH$_2$, 2,6-Me$_2$C$_6$H$_3$CH$_2$,
3,4-OCH$_2$OC$_6$H$_3$CH$_2$, 2-thienylCH$_2$, Ph
R^2 = H, Cl, Br

54 to >99%
63–97% ee

(4.5)

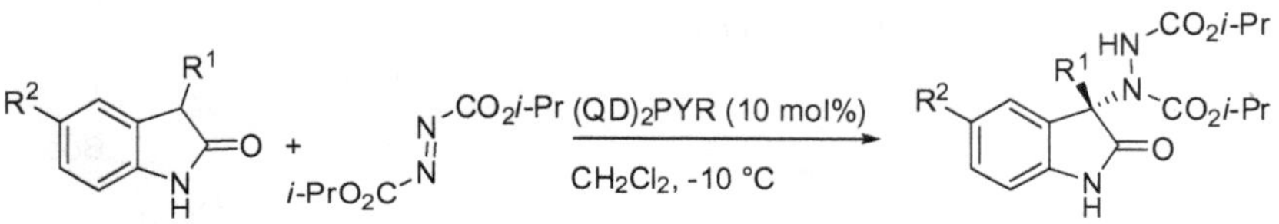

R^1 = *i*-Pr, Bn, 4-ClC$_6$H$_4$CH$_2$, 4-BrC$_6$H$_4$CH$_2$, 3-BrC$_6$H$_4$CH$_2$, 2-FC$_6$H$_4$CH$_2$,
2-NaphthCH$_2$, 2-thienylCH$_2$
R^2 = H, Cl, Br, F, OMe

90–98%
82–94% ee

(4.6)

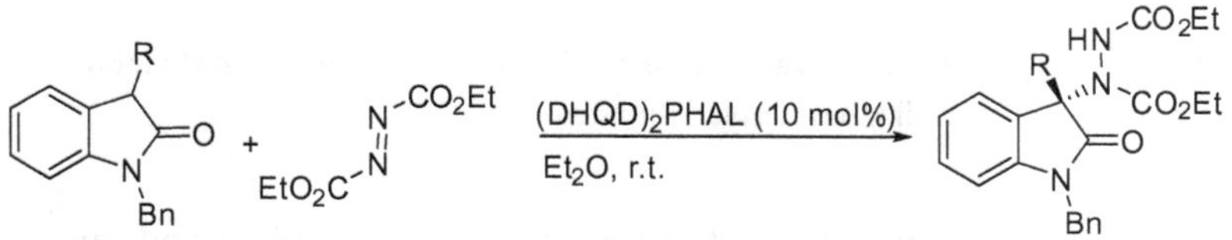

R = Me, allyl, Bn, *p*-BrC$_6$H$_4$CH$_2$, 2-methylallyl, cinnamyl, homoallyl,
CH$_2$C≡CTMS, (CH$_2$)$_3$-C$_6$H$_2$

87–98%
76–99% ee

(4.7)

Scheme 4.47.   Aminations of 3-alkyl-2-oxindoles with azodicarboxylates catalyzed with *Cinchona* alkaloids.

and good to excellent enantioselectivities (63–97% ee). Concurrently, Zhou *et al.* investigated the same reactions in the presence of chiral *bis*-quinidine (QD)$_2$PYR employed at the same catalyst loading of 10 mol% [67]. In this case, the products were obtained in uniformly very high yields (90–98%) and enantioselectivities (82–94% ee), as shown in Scheme 4.47 (Eq. (4.6)). Moreover, Barbas *et al.* also found [(DHQD)$_2$PHAL] to be an efficient organocatalyst for the highly enantioselective amination of *N*-protected oxindoles with diethyl azo-dicarboxylate [68]. As shown in Scheme 4.47 (Eq. (4.7)), the reaction performed at room temperature

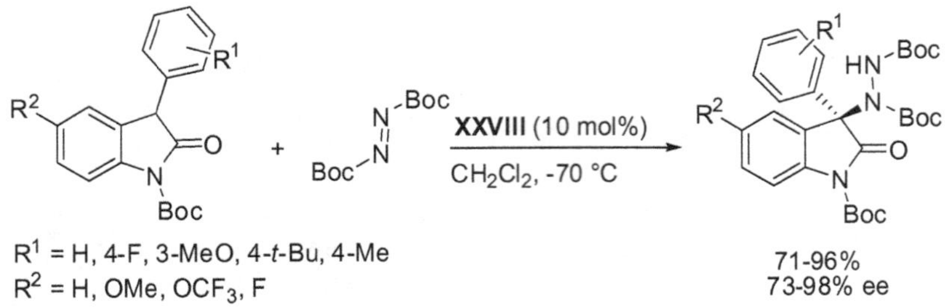

R^1 = H, 4-F, 3-MeO, 4-*t*-Bu, 4-Me
R^2 = H, OMe, OCF$_3$, F

71-96%
73-98% ee

Scheme 4.48. Amination of 3-aryl-2-oxindoles with di-*tert*-butyl azodicarboxylate catalyzed with a dimeric *Cinchona* alkaloid.

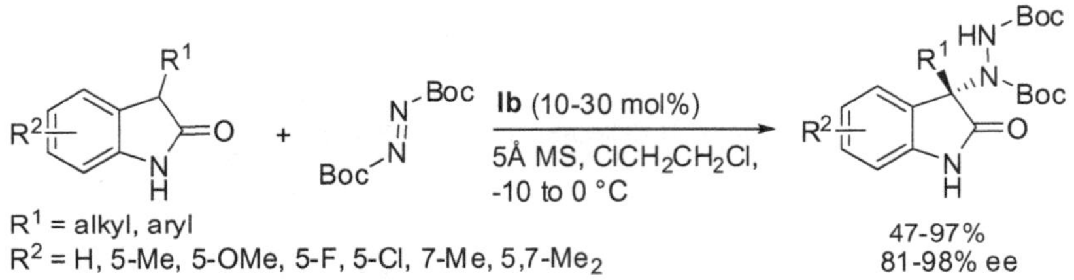

R^1 = alkyl, aryl
R^2 = H, 5-Me, 5-OMe, 5-F, 5-Cl, 7-Me, 5,7-Me$_2$

47-97%
81-98% ee

Scheme 4.49. Amination of 3-aryl/alkyl-2-oxindoles with di-*tert*-butyl azodicarboxylate catalyzed with a *Cinchona* alkaloid-derived thiourea.

yielded the corresponding products in high yields (87–98%) and enanti-oselectivities (76–99% ee). The limitation of these three aminations was the use of only 3-alkyl-2-oxindoles as substrates.

Later in 2010, Barbas *et al.* disclosed an enantioselective amination of 3-aryl-2-oxindoles using di-*tert*-butyl azadicarboxylate (Scheme 4.48) [69]. The reaction promoted by 10 mol% of a dimeric quinidine catalyst at −70°C led to chiral 3-aryl-3-amino-2-oxindole derivatives in good to high yields (71–96%) and enantioselectivities (73–98% ee).

In 2011, Zhou *et al.* showed that a quinine-derived bifunctional thio-urea was capable of catalyzing the direct amination of unprotected 3-aryl- as well as 3-alkyl-substituted oxindoles with di-*tert*-butyl azodicarboxylate (Scheme 4.49) [70]. Indeed, using 10–30 mol% of this organocatalyst, the corresponding chiral 3-amino-2-oxindoles were obtained in moderate to quantitative yields (47–97%) and high enantioselectivities (81–98% ee).

Later in 2013, the same authors reported the first organocata-lytic asymmetric synthesis of 3,3-disubstituted oxindoles featuring two

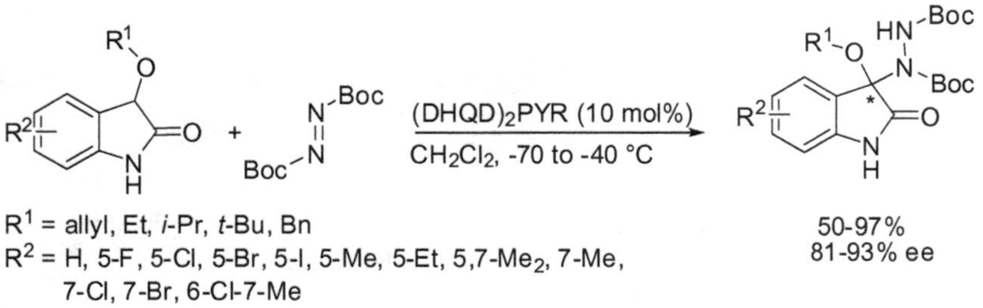

R¹ = allyl, Et, *i*-Pr, *t*-Bu, Bn
R² = H, 5-F, 5-Cl, 5-Br, 5-I, 5-Me, 5-Et, 5,7-Me₂, 7-Me,
    7-Cl, 7-Br, 6-Cl-7-Me

50-97%
81-93% ee

Scheme 4.50. Amination of 3-alkoxy-2-oxindoles with di-*tert*-butyl azodicarboxylate catalyzed with a dimeric *Cinchona* alkaloid.

heteroatoms at the C-3 position *via* amination of 3-alkoxy-2-oxindoles with di-*tert*-butyl azodicarboxylate (Scheme 4.50) [71]. A dimeric *Cinchona* alkaloid was identified as the optimal catalyst for the amination of 3-alkoxy-2-oxindoles with di-*tert*-butyl azodicarboxylate, leading to the corresponding chiral products in moderate to quantitative yields (50–97%) and high enantioselectivities (81–93% ee), as shown in Scheme 4.50.

Earlier in 2010, Feng *et al.* reported the catalysis of the direct amination of oxindoles with azodicarboxylates by a scandium catalyst *in situ* generated from Sc(OTf)$_3$ and a chiral *N,N′*-dioxide ligand (Scheme 4.51) [72]. The reactions performed in dichloromethane at −20°C in the presence of 7.5 mol% of this chiral ligand combined with 5 mol% of Sc(OTf)$_3$ led to the formation of chiral 3-amino-2-oxindoles in good to quantitative yields (70–98%) and high enantioselectivities (83–99% ee). Moreover, the authors demonstrated that the reactions with diethyl azodicarboxylate could be performed with a lower catalyst loading at 30°C. Indeed, these reactions catalyzed by only 0.75 mol% of the chiral ligand combined with 0.5 mol% of Sc(OTf)$_2$ allowed the products to be obtained in high yields (80–95%) and good to high enantioselectivities (74–96% ee), as shown in Scheme 4.51.

4.4.2. *Hydroxyaminations*

The use of nitrosoarenes allows hydroxyamination to be achieved. In 2011, an enantioselective scandium-catalyzed hydroxyamination of

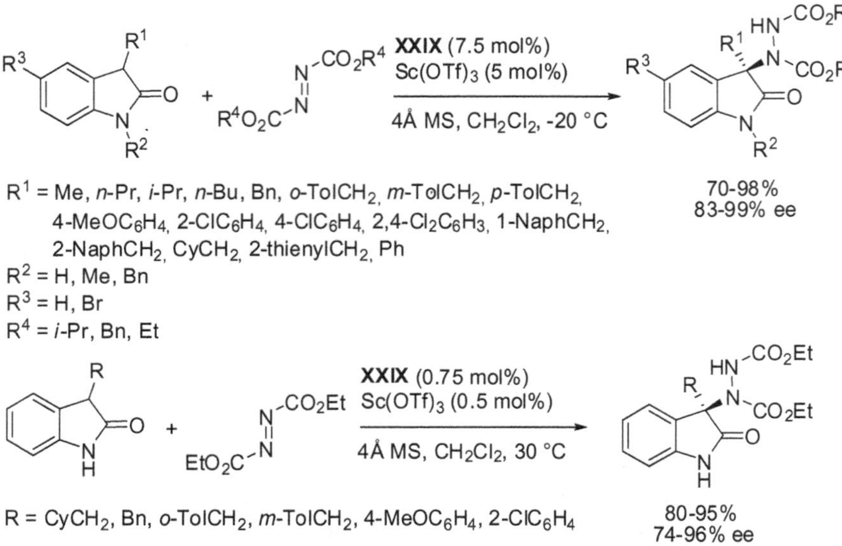

XXIX (7.5 mol%)
Sc(OTf)3 (5 mol%)
4Å MS, CH2Cl2, -20 °C
R1 = Me, n-Pr, i-Pr, n-Bu, Bn, o-TolCH2, m-TolCH2, p-TolCH2, 4-MeOC6H4, 2-ClC6H4, 4-ClC6H4, 2,4-Cl2C6H3, 1-NaphCH2, 2-NaphCH2, CyCH2, 2-thienylCH2, Ph
R2 = H, Me, Bn
R3 = H, Br
R4 = i-Pr, Bn, Et
70-98%
83-99% ee
XXIX (0.75 mol%)
Sc(OTf)3 (0.5 mol%)
4Å MS, CH2Cl2, 30 °C
R = CyCH2, Bn, o-TolCH2, m-TolCH2, 4-MeOC6H4, 2-ClC6H4
80-95%
74-96% ee

Scheme 4.51. Scandium-catalyzed aminations of oxindoles with azodicarboxylates.

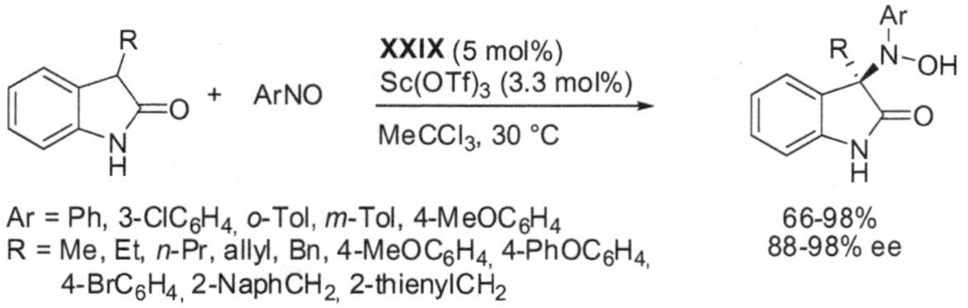

ArNO
XXIX (5 mol%)
Sc(OTf)3 (3.3 mol%)
MeCCl3, 30 °C
Ar = Ph, 3-ClC6H4, o-Tol, m-Tol, 4-MeOC6H4
R = Me, Et, n-Pr, allyl, Bn, 4-MeOC6H4, 4-PhOC6H4, 4-BrC6H4, 2-NaphCH2, 2-thienylCH2
66-98%
88-98% ee

Scheme 4.52. Scandium-catalyzed hydroxyamination of oxindoles with nitrosoarenes.

oxindoles with nitrosoarenes was reported by Feng *et al.* [73]. This process was promoted by a scandium catalyst *in situ* generated from $Sc(OTf)_3$ and a chiral *N,N′*-dioxide ligand. While other rare earth metal sources, such as $La(OTf)_3$, $Lu(OTf)_3$, and $Sm(OTf)_3$, only resulted in trace amounts of products, the use of $Sc(OTf)_3$ allowed a range of *N*-nitroso aldol products to be achieved in good to quantitative yields (66–98%) and high enantioselectivities (88–98% ee) (Scheme 4.52).

Remarkably, the reactions were complete in only 20–60 min. Earlier in 2010, Chen *et al.* reported an organocatalytic enantioselective version of this process [74]. Indeed, the reaction of oxindoles with nitrosobenzene was catalyzed with 10 mol% of cupreine, leading to the corresponding chiral 3-amino-2-oxindoles in good to quantitative yields (70 to >99%) and moderate enantioselectivities (59–72% ee), as shown in Scheme 4.53.

Later, Wang [75] and Rios [76] independently reported similar organocatalytic transformations catalyzed by chiral tertiary amine-thiourea bifunctional organocatalysts. Wang *et al.* employed 20 mol% of a chiral tertiary amine-thiourea (Scheme 4.54, (Eq. (4.8)) to promote the reaction of oxindoles with nitrosobenzene in toluene at 25°C, which afforded the corresponding chiral 3-amino-2-oxindoles in moderate to

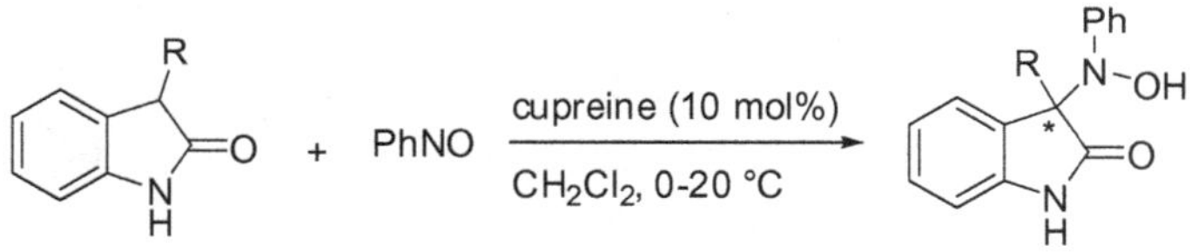

R = Me, Bn, *p*-TolCH$_2$, 4-MeOC$_6$H$_4$CH$_2$, p-CNC$_6$H$_4$CH$_2$, 70 to >99%
3-ClC$_6$H$_4$CH$_2$, 3,4-OCH$_2$O-C$_6$H$_3$CH$_2$ 59-72% ee

Scheme 4.53. Hydroxyamination of oxindoles with nitrosobenzene catalyzed with cupreine.

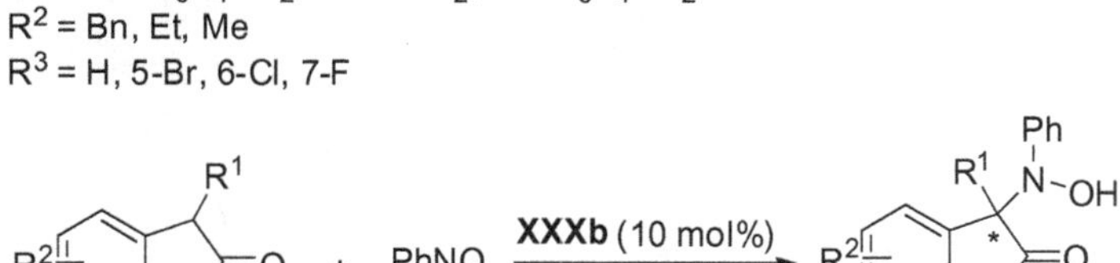

R^1 = Bn, 4-FC$_6$H$_4$CH$_2$, 4-MeOC$_6$H$_4$CH$_2$, 4-ClC$_6$H$_4$CH$_2$,   51-91%
   4-BrC$_6$H$_4$CH$_2$, *m*-TolCH$_2$, 3-ClC$_6$H$_4$CH$_2$ 75-90% ee
R^2 = Bn, Et, Me
R^3 = H, 5-Br, 6-Cl, 7-F

(4.8)

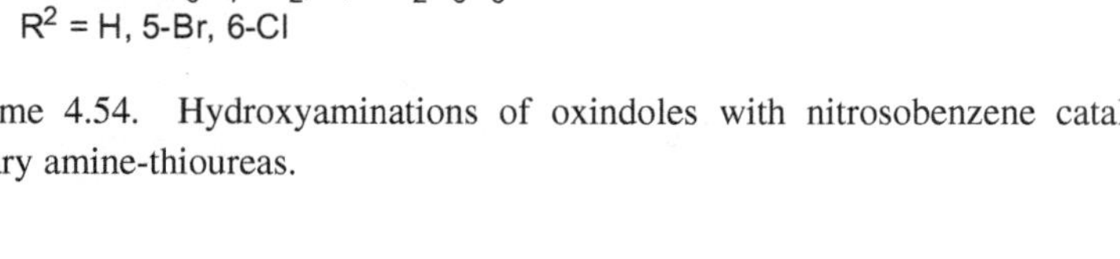

R^1 = Me, Et, CH$_2$CH(Me)$_2$, (CH$_2$)$_2$CH(Me)$_2$, (CH$_2$)$_3$Ph,   54-97%
   4-BrC$_6$H$_4$CH$_2$, 2,6-Cl$_2$C$_6$H$_3$ 23-72% ee
R^2 = H, 5-Br, 6-Cl

(4.9)

Scheme 4.54. Hydroxyaminations of oxindoles with nitrosobenzene catalyzed with tertiary amine-thioureas.

high yields (51–91%) and good to high enantioselectivities (75–90% ee). On the other hand, Rios *et al.* catalyzed these reactions at room temperature with 10 mol% of another tertiary amine-thiourea in MTBE as solvent (Scheme 4.54, Eq. (4.9)). Comparable yields (54–97%) were achieved in combination with lower enantioselectivities (23–72% ee).

4.5. Miscellaneous Reactions

In 2008, Kündig *et al.* reported an asymmetric palladium-catalyzed intramolecular α-arylation of amide enolates containing amino substituents [77]. As shown in Scheme 4.55, the reaction was promoted by a palladium catalyst *in situ* generated from Pd(dba)$_2$ and a chiral NHC ligand in toluene at 50°C or 80°C, leading to the corresponding chiral 3-amino-2-oxindoles in uniformly high yields (88–96%) and good to high enantioselectivities (75–90% ee).

The 1,3-dipolar cycloaddition [78] is a classic reaction in organic chemistry consisting of the reaction of a dipolarophile with a 1,3-dipolar compound that allows the production of various five-membered heterocycles [3a,79].

A wide variety of chiral catalysts has been applied to promote asymmetric versions of this reaction. In particular, 1,3-dipolar cycloadditions of nitrile imines with alkenes represent an attractive strategy to generate pyrazolines, which constitute motifs in a number of bioactive compounds. In 2013, Stanley *et al.* reported a rare example of enantioselective 1,3-dipolar cycloadditions of nitrile imines with a variety of *N*-benzoyl

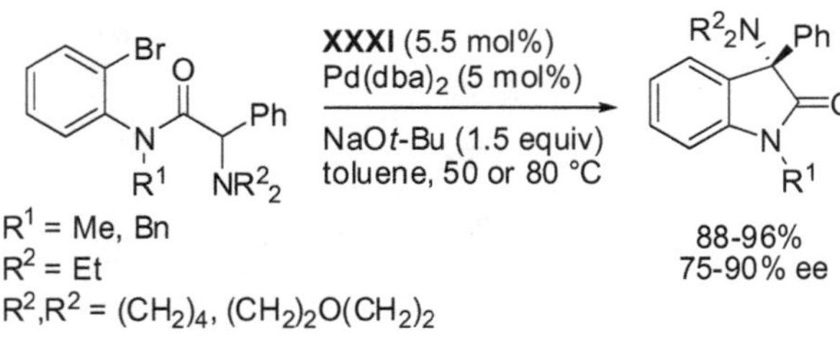

Scheme 4.55. Palladium-catalyzed intramolecular α-arylation of amide enolates containing amino substituents.

R = 4-MeOC$_6$H$_4$, 4-ClC$_6$H$_4$, 4-BrC$_6$H$_4$, 4-F$_3$CC$_6$H$_4$, 4-MeOC$_6$H$_4$, 3,4,5-(MeO)$_3$C$_6$H$_2$, 2-FC$_6$H$_4$, 2-BrC$_6$H$_4$, *t*-Bu
Ar^1 = Ph, 4-CF$_3$C$_6$H$_4$, 3-BrC$_6$H$_4$, 2-FC$_6$H$_4$, 2-BrC$_6$H$_4$
Ar^2 = 4-BrC$_6$H$_4$, 4-FC$_6$H$_4$, 4-CF$_3$C$_6$H$_4$

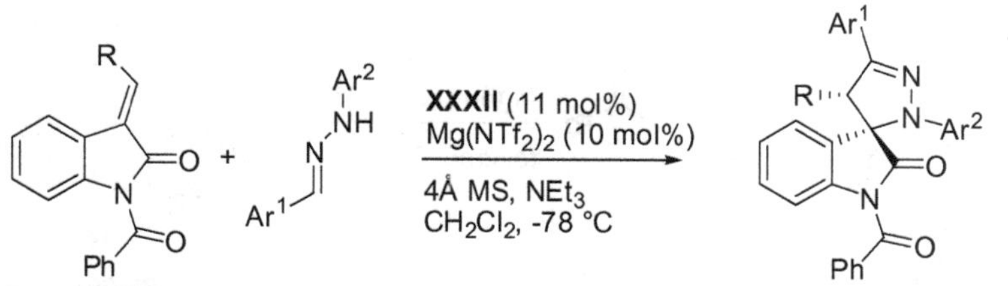

Scheme 4.56. Magnesium-catalyzed 1,3-dipolar cycloaddition of *N*-benzoyl methyleneoxindoles with nitrile imines.

methyleneoxindoles [80]. As shown in Scheme 4.56, the reaction was performed at −78°C in dichloromethane in the presence of a combination of Mg(NTf$_2$)$_2$ and a chiral *bis*(oxazoline) ligand to afford a range of chiral spiro[pyrazolin-3,3′-oxindoles] in moderate to high yields (43–91%), good to high diastereoselectivities (80 to >90% de), and moderate to excellent enantioselectivities (61–99% ee).

In 2015, the same authors reported the first enantioselective 1,3-dipolar cycloaddition between 3-isothiocyanato oxindoles and alkynyl ketones [81]. The process was catalyzed by a combination of MgBu$_2$ and a chiral oxazoline ligand in toluene at 0°C, providing the corresponding chiral spirooxindoles in high yields (83–99%) and good to high enantioselectivities (72–94% ee), as shown in Scheme 4.57. The scope of the reaction was wide since a range of variously substituted aromatic; heteroaromatic as well as aliphatic ketones were compatible with the catalyst system, providing comparable excellent results in reactions with differently alkyl-substituted 3-isothiocyanato oxindoles.

In 2014, Ye *et al.* developed an asymmetric synthesis of spirocyclic oxindolo-β-lactams based on an enantioselective Staudinger-type reaction between isatin imines and ketenes [82]. As shown in Scheme 4.58, the process was catalyzed by 20 mol% of a chiral bifunctional NHC catalyst in THF at −10°C, providing the corresponding chiral tricyclic 3-amino-2-oxindoles in moderate to high yields (70–92%) with high diastereo- and enantioselectivities of up to 90% de and 99% ee, respectively.

R^1 = H, 6-Me
R^2 = Me, *n*-Pr, Bn
R^3 = Ph, *p*-Tol, *n*-Pent
R^4 = Ph, *p*-Tol, 4-ClC$_6$H$_4$, 4-BrC$_6$H$_4$, 4-MeOC$_6$H$_4$, 3-ClC$_6$H$_4$, *m*-Tol, 2-FC$_6$H$_4$, 2-thienyl, 1-thienyl, Et, *c*-Pr

83-99%
72-94% ee

Scheme 4.57. Magnesium-catalyzed 1,3-dipolar cycloaddition of 3-isothiocyanato oxindoles with alkynyl ketones.

R^1 = Me, Bn, Cbz
R^2 = H, F, Me, Cl
R^3 = Ph, *p*-Tol, 4-MeOC$_6$H$_4$, 4-ClC$_6$H$_4$, 4-BrC$_6$H$_4$
Ar = 4-ClC$_6$H$_4$, 4-BrC$_6$H$_4$, *p*-Tol, 4-MeOC$_6$H$_4$

70-92%
up to 90% de
up to 99% ee

Scheme 4.58. NHC-catalyzed Staudinger-type reaction of isatin imines with ketenes.

In 2014, Shi *et al.* reported the asymmetric substitution reaction between isatin-derived hydrazones and *O*-Boc-protected aza-MBH adducts [83]. As shown in Scheme 4.59, the process was catalyzed at 10°C by 10 mol% of (DHQD)$_2$AQN in MTBE, providing the corresponding chiral azo compounds incorporating an oxindole scaffold in moderate to high yields (50–91%) and with good to high enantioselectivities (68–93% ee).

In 2016, Cui and Chen developed organocatalytic asymmetric Mannich-type reactions of 3-aminooxindoles with *in situ* generated *N*-Boc-protected aldimines [84]. The reactions were promoted by 10 mol% of a chiral bifunctional thiourea catalyst at 25°C. As shown in Scheme 4.60, a series of chiral vicinal oxindole-diamines were smoothly obtained in good to quantitative yields (72–99%), moderate to good diastereoselectivities (46–88% de), and high enantioselectivities (77–96% ee).

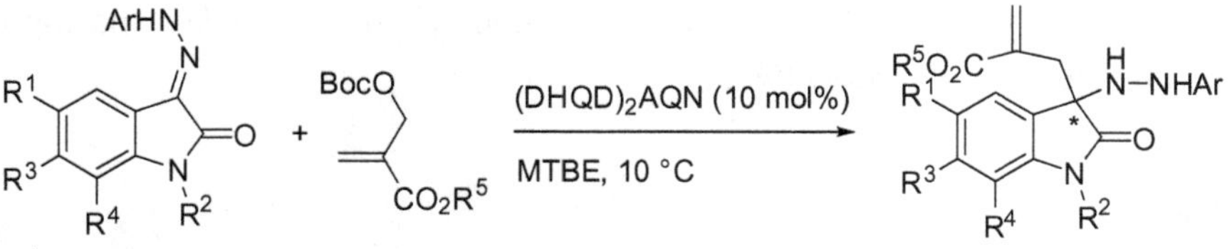

R^1 = H, Bn
R^2 = Me, CH$_2$CO$_2$Et, Bn, CH$_2$CO$_2$Me
R^3 = H, F, OMe, Br
R^4 = H, 5-F, 5-Cl, 5-Br, 5-OMe, 6-Cl, 6-Br
R^5 = Et, Bn, neopentyl

50-91%
68-93% ee

Scheme 4.59. (DHQD)$_2$AQN-catalyzed substitution of isatin-derived hydrazones with *O*-protected aza-MBH adducts.

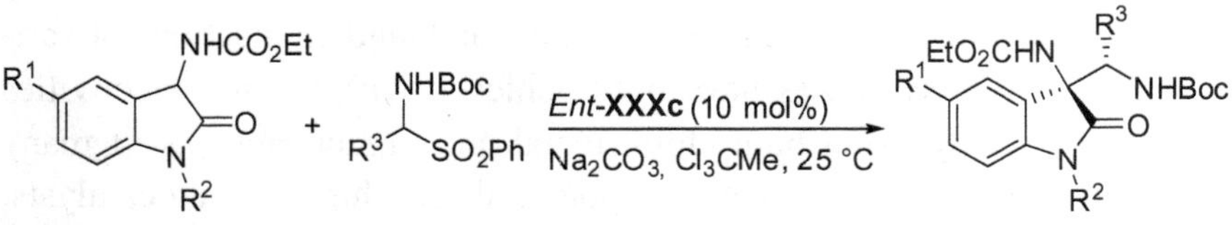

R^1 = H, Me, F
R^2 = Me, Et, Bn
R^3 = Ph, 2-MeOC$_6$H$_4$, *m*-Tol, 3-MeOC$_6$H$_4$, 3,4-(MeO)$_2$C$_6$H$_3$, 3-NO$_2$C$_6$H$_4$, 3-ClC$_6$H$_4$, 3-BrC$_6$H$_4$, 4-MeOC$_6$H$_4$, *p*-Tol, 4-FC$_6$H$_4$, 4-ClC$_6$H$_4$, 4-BrC$_6$H$_4$, 2-furyl, 2-thienyl, 1-Naph

72-99%
46-88% de
77-96% ee

Scheme 4.60. Mannich-type reaction of 3-amino-2-oxindoles with *in situ* generated *N*-Boc-protected aldimines catalyzed with a bifunctional thiourea.

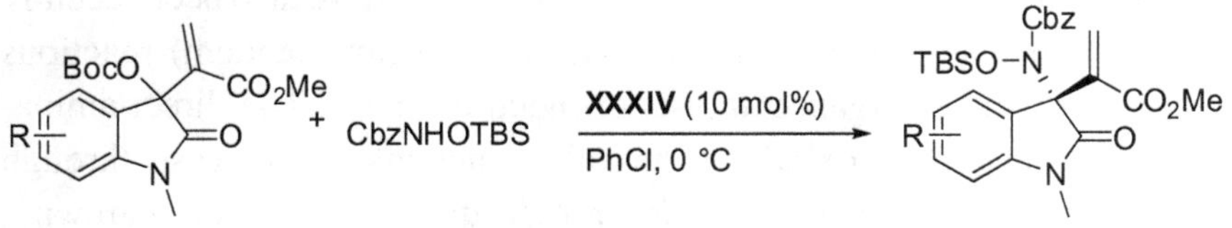

R = 5-Me, 5-OMe, 5-F, 5-Cl, 5-Br, 5-I, 5-CF$_3$O, 7-F

71-97%
85-94% ee

Scheme 4.61. Allylic amination of MBH carbonates of isatins catalyzed with a *Cinchona* alkaloid.

In 2012, Chen *et al.* developed a highly enantioselective allylic amination of MBH carbonates of isatins with *N*-silyloxycarbamates catalyzed by a modified *β*-isocupreidine catalyst [85]. As shown in Scheme 4.61, the corresponding chiral 3-amino-2-oxindoles were obtained in chlorobenzene at 0°C in good to excellent yields (71–97%) and high enantioselectivities (85–94% ee).

4.6. Conclusions

Chiral 3-amino-2-oxindoles are important and ubiquitous motifs in many natural products and biologically active compounds with a range of significant pharmaceutical properties. This chapter demonstrates the impressive amount of enantioselective catalytic methodologies to prepare these complex chiral products that have been employed in the last decade, spanning from enantioselective nucleophilic additions of many types to isatin imines and direct aminations of 3-substituted oxindoles, to more recently developed enantioselective catalytic domino and tandem reactions of various types. These syntheses have been achieved with enantioselectivities often excellent by using both chiral metal and organocatalysts of many types. In particular, the catalytic potential of chiral organocatalysts, including *Cinchona* alkaloids, amino acids, phosphoric acids, phosphines, thioureas, squaramides, *N,N'*-dioxides, phosphoramides, *N*-heterocyclic carbenes, guanidines among others, has been successfully exploited in the last decade. For example, remarkable results have been reported in enantioselective nucleophilic additions to isatin imines, such as Mannich reactions, aza-MBH reactions, aza-Henry reactions, additions of various heteronucleophiles, Strecker reactions, hydrophosphonylations, and miscellaneous additions. Excellent enantioselectivities have also been recently described in many types of fascinating domino (and tandem) reactions catalyzed by either organocatalysts or metal catalysts. The direct amination of 3-substituted oxindoles has also encountered success through catalysis with *Cinchona* alkaloids or *N,N'*-dioxides. The ever-growing need for achieving such biologically important products will prompt organic chemists to develop other enantioselective transformations especially domino reactions. Indeed, designing novel catalytic asymmetric transformations by different combination of substrates and catalysts will increase the library of potentially bioactive 3-amino-2-oxindoles. In this context, a great challenge will be to design and develop efficient (organo)-catalysts providing high stereoselectivity at low catalyst loading.

Appendix A. Catalysts Cited in This Chapter

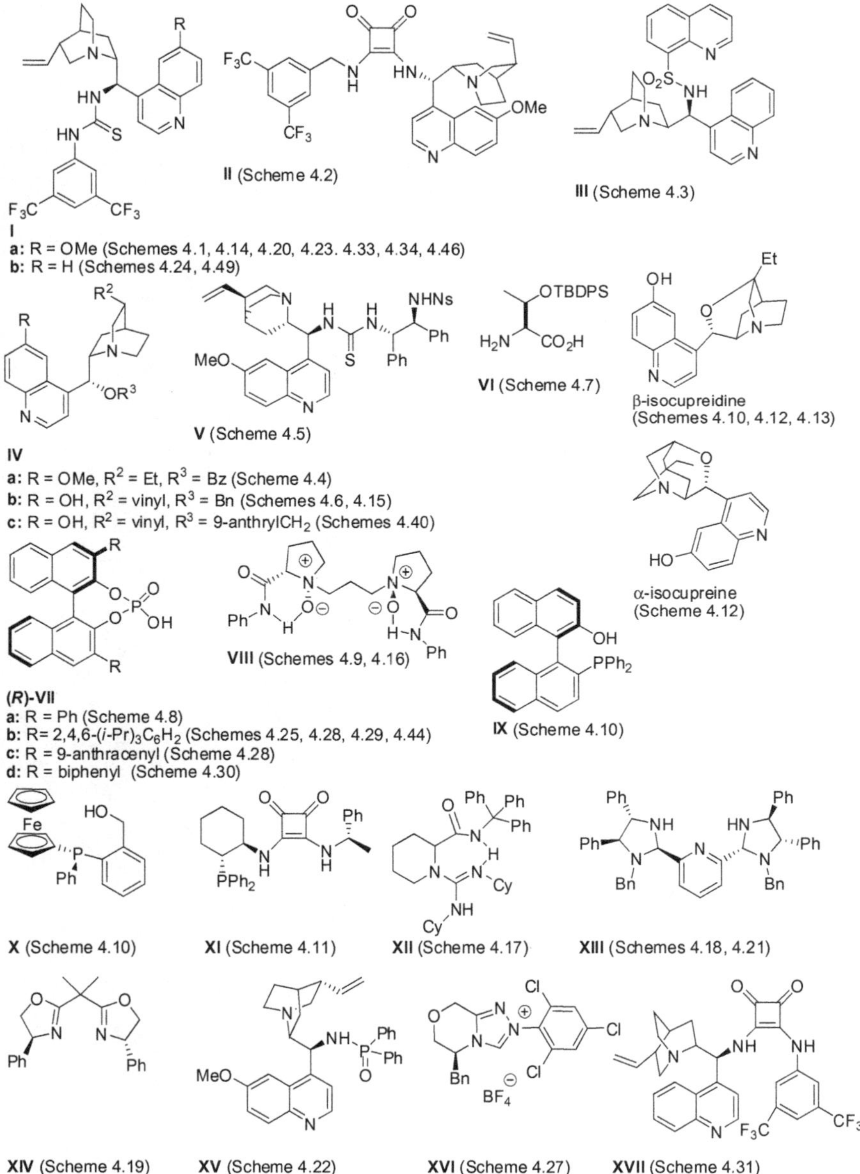

XVIII (Scheme 4.32)

XIX (Scheme 4.33)

XX (Scheme 4.34)

XXI (Scheme 4.35)

XXII (Schemes 4.36, 4.58)

XXIII (Schemes 4.37, 4.38)

XXIV (Scheme 4.39)

XXV (Scheme 4.41)

XXVI (Schemes 4.42, 4.43)

XXVII (Scheme 4.45)

XXVIII (Scheme 4.48)

XXIX (Schemes 4.51, 4.52)

XXX
a: $R_2 = (CH_2)_5$ (Scheme 4.54)
b: R = Me (Scheme 4.54)
c: $R_2 = (CH_2)_4$ (Scheme 4.60)

XXXI (Scheme 4.55)

XXXII (Scheme 4.56)

XXXIII (Scheme 4.57)

XXXIV (Scheme 4.61)

Appendix B. List of Abbreviation

AIBN	Azo-*bis*-isobutyronitrile
Bn	Benzyl
Boc	*tert*-Butoxycarbonyl
Bs	Benzensulfonyl
Bz	Benzoyl
Cbz	Benzyloxycarbonyl
CPME	Cyclopentyl methyl ether
Cy	Cyclohexyl
DABCO	1,4-Diazabicyclo[2.2.2]octane
DCE	Dichloroethane
$(DHQD)_2AQN$	Hydroquinidine anthraquinone-1,4-diyl diether
$(DHQD)_2PHAL$	Hydroquinidine phthalazine-1,4-diyl diether
$(DHQD)_2PYR$	Hydroquinidine 2,5-diphenylpyrimidine-4,6-diyl diether
DIPEA	Diisopropylethylamine
Mes	2,4,6-Trimethylphenyl
MOM	Methoxymethyl
MS	Molecular sieves
MTBE	Methyl *tert*-butyl ether
Naph	Naphthyl
Ns	4-Nitrobenzensulfonyl (nosyl)
Pent	Pentyl
PG	Protecting group
PMB	4-Methoxybenzyl
PMP	4-Methoxyphenyl
$(QD)_2PYR$	Quinidine 2,5-diphenylpyrimidine-4,6-diyl diether
TBDPS	*tert*-Butyldiphenylsilyl
TMS	Trimethylsilyl
Tol	Methylphenyl (tolyl)
Ts	4-Methylbenzensulfonyl (tosyl)
XPhos	2-Dicyclohexylphosphino-2′,4′,6′-triisopropylbiphenyl

References

1. (a) B. S. Jensen, *CNS Drug Rev.* **2002**, *8*, 353; (b) C. Marti, E. M. Carreira, *Eur. J. Org. Chem.* **2003**, 2209; (c) H. Lin, S. J. Danishefsky, *Angew. Chem. Int. Ed.* **2003**, *42*, 36; (d) C. V. Galliford, K. Scheidt, *Angew. Chem. Int. Ed.* **2007**, *46*, 8748; (e) L. Hog, R. Wang, *Adv. Synth. Catal.* **2013**, *355*, 1023; (f) B. M. Trost, D. A. Bringley, T. Zhang, N. Cramer, *J. Am. Chem. Soc.* **2013**, *135*, 16720; (g) B. Yu, D.-Q. Yu, H.-M. Liu, *Eur. J. Med. Chem.* **2015**, *97*, 673.

2. (a) F. Zhou, Y.-L. Liu, J. Zhou, *Adv. Synth. Catal.* **2010**, *352*, 1381; (b) R. Dalpozzo, G. Bartoli, G. Bencivenni, *Chem. Soc. Rev.* **2012**, *41*, 7247; (c) A. Kumar, S. S. Chimni, *RSC Adv.* **2012**, *2*, 9748; (d) S. Mohammadi, R. Heiran, R. P. Herrera, E. Marqués-Lopez, *Chem. Cat. Chem.* **2013**, *5*, 2131; (e) P. Chauhan, S. S. Chimni, *Tetrahedron: Asymmetry*, **2013**, *24*, 343; (f) Y.-L. Liu, F. Zhu, C. Wang, J. Zhou, *Chin. J. Org. Chem.* **2013**, *33*, 1595; (g) J.-S. Yu, F. Zhou, Y.-L. Liu, J. Zhou, *Synlett.* **2015**, *26*, 2491; (h) J. Kaur, S. S. Chimni, S. Mahajan, A. Kumar, *RSC Adv.* **2015**, *5*, 52481; (i) M. G. Ziarani, R. Moradi, N. Lashgari, *Tetrahedron: Asymmetry*, **2015**, *26*, 517; (j) B. Yu, H. Xing, D.-Q. Yu, H.-M. Liu, *Beilstein J. Org. Chem.* **2016**, *12*, 1000; (k) R. Dalpozzo, *Adv. Synth. Catal.* **2017**, *359*, 1772.

3. (a) A. Noble, J. C. Anderson, *Chem. Rev.* **2013**, *113*, 2887; (b) S. Kobayashi, Y. Mori, Y. Fossey, M. M. Salter, *Chem. Rev.* **2011**, *111*, 2626; (c) E. Marqués-Lopez, P. Merino, T. Tejero, R. P. Herrera, *Eur. J. Org. Chem.* **2009**, *15*, 2401.

4. For reviews on asymmetric Mannich reactions, see: (a) A. Cordova, *Acc. Chem. Res.* **2004**, *37*, 102; (b) M. M. B. Marques, *Angew. Chem. Int. Ed.* **2006**, *45*, 348; (c) N. R. Candeas, F. P. Montalbano, M. S. D. Cal, P. M. P. Gois, *Chem. Rev.* **2010**, *110*, 6169.

5. For reviews on asymmetric organocatalysis, see: (a) P. I. Dalko, L. Moisan, *Angew. Chem. Int. Ed.* **2001**, *40*, 3726; (b) P. I. Dalko, L. Moisan, *Angew. Chem. Int. Ed.* **2004**, *43*, 5138; (c) A. Berkessel, H. Gröger, In *Asymmetric Organocatalysis–From Biomimetic Concepts to Powerful Methods for Asymmetric Synthesis*, Wiley-VCH: Weinheim, **2005**; (d) J. Seayad, B. List, *Org. Biomol. Chem.* **2005**, *3*, 719; (e) M. S. Taylor, E. N. Jacobsen, *Angew. Chem. Int. Ed.* **2006**, *45*, 1520; (f) P. I. Dalko, In *Enantioselective Organocatalysis*, Wiley-VCH: Weinheim, **2007**; (g) P. I. Dalko, *Chimia* **2007**, *61*, 213; (h) H. Pellissier, *Tetrahedron*, **2007**, *63*, 9267; (i) A. G. Doyle, E. N. Jacobsen, *Chem. Rev.* **2007**, *107*, 5713; (j) M. G. Gaunt, C. C. C. Johansson, A. McNally, N. C. Vo, *Drug Discovery Today* **2007**, *2*, 8;

(k) D. W. C. MacMillan, *Nature*, **2008**, *455,* 304; (l) X. Yu, W. Wang, *Chem. Asian. J.* **2008**, *3,* 516; (m) A. Dondoni, A. Massi, *Angew. Chem. Int. Ed.* **2008**, *47,* 4638; (n) P. Melchiorre, M. Marigo, A. Carlone, G. Bartoli, *Angew. Chem. Int. Ed.* **2008**, *47,* 6138; (o) F. Peng, Z. Shao, *J. Mol. Catal. A* **2008**, *285,* 1; (p) C. Palomo, M. Oiarbide, R. Lopez, *Chem. Soc. Rev.* **2009**, *38,* 632; (q) L.-W. Xu, J. Luo, Y. Lu, *Chem. Commun.* **2009**, 1807; (r) M. Bella, T. Gasperi, *Synthesis*, **2009**, *10,* 1583; (s) H. Pellissier, In *Recent Developments in Asymmetric Organocatalysis,* Royal Society of Chemistry: Cambridge, **2010**.

6. W. Yan, D. Wang, J. Feng, P. Li, D. Zhao, R. Wang, *Org. Lett.* **2012**, *14,* 2512.

7. K. S. Rao, P. Ramesh, L. R. Chowhan, R. Trivedi, *RSC Adv.* **2016**, *6,* 84242.

8. N. Hara, S. Nakamura, M. Sano, R. Tamura, Y. Funahashi, N. Shibata, *Chem. Eur. J.* **2012**, *18,* 9276.

9. K. Zhao, T. Shu, J. Jia, G. Raabe, D. Enders, *Chem. Eur. J.* **2015**, *21,* 3933.

10. T. Z. Li, X. B. Wang, F. Sha, X. Y. Wu, *J. Org. Chem.* **2014**, *79,* 4332.

11. Y. Guo, Y. Zhang, L. Qi, F. Tiana, L. Wang, *RSC Adv.* **2014**, *4,* 27286.

12. Q. X. Guo, Y. W. Liu, X. C. Li, L. Z. Zhong, Y. G. Peng, *J. Org. Chem.* **2012**, *77,* 3589.

13. G. Rainlodi, A. Sacchetti, A. Silvani, G. Lesma, *Org. Biomol. Chem.* **2016**, *14,* 7768.

14. (a) R. Noyori, In *Asymmetric Catalysts in Organic Synthesis,* Wiley, New-York, **1994**; (b) *Transition Metals for Organic Synthesis,* M. Beller, C. Bolm, Eds., Wiley-VCH, Weinheim, **1998**, Vols I and II; (c) *Comprehensive Asymmetric Catalysis*, E. N. Jacobsen, A. Pfaltz, H. Yamamoto, Eds., Springer, Berlin, **1999**; (d) *Catalytic Asymmetric Synthesis*, 2nd ed., I. Ojima, Ed., Wiley-VCH, New-York, **2000**; (e) G. Poli, G. Giambastiani, A. Heumann, *Tetrahedron*, **2000**, *56,* 5959; (f) E. Negishi, In *Handbook of Organopalladium Chemistry for Organic Synthesis,* John Wiley & Sons, Inc., Hoboken NJ, **2002**, *2,* 1689; (g) A. de Meijere, P. von Zezschwitz, H. Nüske, B. Stulgies, *J. Organomet. Chem.* **2002**, *653,* 129; (h) *Transition Metals for Organic Synthesis,* 2nd ed., M. Beller, C. Bolm, Eds., Wiley-VCH, Weinheim, **2004**; (i) L. F. Tietze, I. Hiriyakkanavar, H. P. Bell, *Chem. Rev.* **2004**, *104,* 3453; (j) D. J. Ramon, M. Yus, *Chem. Rev.* **2006**, *106,* 2126.

15. J. Zhao, B. Fang, W. Luo, X. Hao, X. Liu, L. Lin, X. Feng, *Angew. Chem. Int. Ed.* **2015**, *54,* 241.

16. For reviews on (aza)-Morita–Baylis–Hillman reactions, see: (a) S. E. Drewes, G. H. P. Roos, *Tetrahedron,* **1988**, *44,* 4653; (b) D. Basavaiah, P. D. Rao, R. S. Hyma, *Tetrahedron,* **1996**, *52,* 8001; (c) P. Langer, *Angew. Chem. Int. Ed.* **2000**, *39,* 3049; (d) D. Basavaiah, A. J. Rao, T. Satyanarayana, *Chem.*

Rev. **2003**, *103*, 811; (e) D. Basavaiah, V. Rao, R. J. Reddy, *Chem. Soc. Rev.* **2007**, *36*, 1581; (f) G. Masson, C. Housseman, J. Zhu, *Angew. Chem. Int. Ed.* **2007**, *46*, 4614; (g) Y.-L. Shi, M. Shi, *Eur. J. Org. Chem.* **2007**, 2905; (h) V. Singh, S. Batra, *Tetrahedron*, **2008**, *64*, 4511; (i) V. Carrasco-Sanchez, M. J. Simirgiotis, L. S. Santos, *Molecules*, **2009**, *14*, 3989; (j) V. Declerck, J. Martinez, F. Lamaty, *Chem. Rev.* **2009**, *109*, 1; (k) G.-N. Ma, J.-J. Jiang, M. Shi, Y. Wie, *Chem. Commun.* **2009**, 5496; (l) Y. Wie, M. Shi, *Chinese Sci. Bull.* **2010**, *55*, 1699; (m) J. Mansilla, J. M. Saa, *Molecules*, **2010**, *15*, 709; (n) D. Basavaiah, B. S. Reddy, S. S. Badsara, *Chem. Rev.* **2010**, *110*, 5447; (o) Y. Wie, M. Shi, *Acc. Chem. Res.* **2010**, *43*, 1005; (p) J. W. Cran, M. E. Krafft, K. A. Seibert, T. F. N. Haxell, J. A. Wright, C. Hirosawa, K. A. Abboud, *Tetrahedron*, **2011**, *67*, 9922; (q) D. Basavaiah, G. Veeraraghavaiah, *Chem. Soc. Rev.* **2012**, *41*, 68; (r) C. G. Lima-Junior, M. L. A. A. Vasconcellos, *Bioorg. Med. Chem.* **2012**, *20*, 3954; (s) T.-Y. Liu, M. Xie, Y.-C. Chen, *Chem. Soc. Rev.* **2012**, *41*, 4101; (t) Y. Wei, M. Shi, *Chem. Rev.* **2013**, *113*, 6659; (u) F.-L. Hu, M. Shi, *Org. Chem. Front.* **2014**, *1*, 587; (v) S. Bhowmik, S. Batra, *Curr. Org. Chem.* **2014**, *18*, 3078; (w) P. Xie, Y. Huang, *Org. Biomol. Chem.* **2015**, *13*, 8578.

17. F. L. Hu, Y. Wei, M. Shi, S. Pindi, G. Li, *Org. Biomol. Chem.* **2013**, *11*, 1921.

18. S. Takizawa, E. Rémond, F. A. Arteaga, Y. Yoshida, V. Sridharan, J. Bayardon, S. Jugé, H. Sasai, *Chem. Commun.* **2013**, *49*, 8392.

19. X. Zhao, T. Z. Li, J. Y. Qian, F. Sha, X. Y. Wu, *Org. Biomol. Chem.* **2014**, *12*, 8072.

20. Y. Yoshida, M. Sako, K. Kishi, H. Sasai, S. Hatakeyama, S. Takizawa, *Org. Biomol. Chem.* **2015**, *13*, 9022.

21. (a) R. E. Plata, D. A. Singleton, *J. Am. Chem. Soc.*, **2015**, *137*, 3811; (b) P. Verma, P. Verma, R. B. Sunoj, *Org. Biomol. Chem.* **2014**, *12*, 2176; (c) C. Lindner, Y. Liu, B. Karaghiosoff, B. Maryasin, H. Zipse, *Chem. Eur. J.* **2013**, *19*, 6429.

22. A. Kumar, V. Sharma, J. Kaur, N. Kumar, S. S. Chimni, *Org. Biomol. Chem.* **2015**, *13*, 5629.

23. M. K. Choudhary, T. Menapara, R. Tak, R. I. Kureshy, N.-u. H. Khan, *Chemistry, Select*, **2017**, *2*, 2224.

24. For reviews on catalytic asymmetric Henry reactions, see: (a) C. Palomo, M. Oiarbide, A. Laso, *Eur. J. Org. Chem.* **2007**, 2561; (b) J. Boruwa, N. Gogoi, P. P. Saikia, N. C. Barua, *Tetrahedron: Asymmetry*, **2006**, *17*, 3315.

25. A. Kumar, J. Kaur, S. S. Chimni, *RSC Adv.* **2014**, *4*, 24816.

26. Y. H. Wang, Y. L. Liu, Z. Y. Cao, J. Zhou, *Asian J. Org. Chem.* **2014**, *3*, 429.

27. B. Fang, X. Liu, J. Zhao, Y. Tang, L. Lin, X. Feng, *J. Org. Chem.* **2015**, *80*, 3332.

28. T. Arai, E. Matsumara, H. Masu, *Org. Lett.* **2014**, *16*, 2768.

29. M. Holmquist, G. Blay, J. R. Pedro, *Chem. Commun.* **2014**, *50*, 9309.

30. T. Z. Li, X. B. Wang, F. Sha, X. Y. Wu, *Tetrahedron*, **2013**, *69*, 7314.

31. T. Arai, K. Tsuchiya, E. Matsumara, *Org. Lett.* **2015**, *17*, 2416.

32. For reviews on Strecker reactions, see: (a) H. Gröger, *Chem. Rev.* **2003**, *103*, 2795; (b) D. Enders, J. P. Shilvock, *Chem. Soc. Rev.* **2000**, *29*, 359.

33. (a) T. Vilaivan, W. Bhanthumnavin, Y. Sritana-Anant, *Curr. Org. Chem.* **2005**, *9*, 1315; (b) Y. Ohfune, T. Shinada, *Eur. J. Org. Chem.* **2005**, 5127.

34. Y. L. Liu, F. Zhou, J. J. Cao, C. B. Ji, M. Ding, J. Zhou, *Org. Biomol. Chem.* **2010**, *8*, 3847.

35. D. Wang, J. Liang, J. Feng, K. Wang, Q. Sun, L. Zhao, D. Li, W. Yan, R. Wang, *Adv. Synth. Catal.* **2013**, *355*, 548.

36. Y. L. Liu, J. Zhou, *Chem. Commun.* **2013**, *49*, 4421.

37. X. Cheng, S. Vellatath, R. Goddard, B. List, *J. Am. Chem. Soc.* **2008**, *130*, 15786.

38. Z. Tang, Y. Shi, H. Mao, X. Zhu, W. Li, Y. Cheng, W. H. Zheng, C. Zhu, *Org. Biomol. Chem.* **2014**, *12*, 6085.

39. J. Xu, C. Mou, T. Zhu, B. A. Song, Y. R. Chi, *Org. Lett.* **2014**, *16*, 3272.

40. M. Chrzanowska, M. D. Rozwadowska, *Chem. Rev.* **2004**, *104*, 3341.

41. S. Duce, F. Pesciaioli, L. Gramigna, L. Bernardi, A. Mazzanti, A. Ricci, G. Bartoli, G. Bencivenni, *Adv. Synth. Catal.* **2011**, *353*, 860.

42. J. J. Badillo, A. Silva-Garcia, B. H. Shupe, J. C. Fettinger, A. K. Franz, *Tetrahedron Lett.* **2011**, *52*, 5550.

43. F. Shi, G.-J. Xing, R.-Y. Zhu, W. Tan, S. Tu, *Org. Lett.* **2013**, *15*, 128.

44. For reviews on asymmetric organocatalytic Friedel–Crafts reactions, see: (a) S. L. You, Q. Cai, M. Zeng, *Chem. Soc. Rev.* **2009** *38*, 2190; (b) V. Terrasson, R. M. de Figueiredo, J. M. Campagne, *Eur. J. Org.* **2010**, 2635; (c) P. Chauhan, S. S. Chimni, *RSC Adv.* **2012**, *2*, 6117.

45. J. Feng, W. Yan, D. Wang, P. Li, Q. Sun, R. Wang, *Chem. Commun.* **2012**, *48*, 8003.

46. (a) H. Clavier, H. Pellissier, *Adv. Synth. Catal.* **2012**, *354*, 3347; (b) H. Pellissier, *Adv. Synth. Catal.* **2016**, *358*, 2194; (c) H. Pellissier, *Curr. Org. Chem.* **2016**, *20*, 234.

47. (a) L. F. Tietze, U. Beifuss, *Angew. Chem. Int. Ed. Engl.* **1993**, *32*, 131; (b) L. F. Tietze, *Chem. Rev.* **1996**, *96*, 115; (c) L. F. Tietze, G. Brasche, K. Gericke, *Domino Reactions in Organic Synthesis*, Wiley-VCH, Weinheim,

2006; (d) L. F. Tietze, *Domino Reactions — Concepts for Efficient Organic Synthesis*, Wiley-VCH, Weinheim, **2014**.

48. X. B. Wang, T. Z. Li, F. Sha, X. Y. Wu, *Eur. J. Org. Chem.* **2014**, 739.

49. W.-B. Chen, Z.-J. Wu, J. Hu, L.-F. Cun, X.-M. Zhang, W.-C. Yuan, *Org. Lett.* **2011**, *13*, 2472.

50. Y.-M. Cao, F.-F. Shen, F.-T. Zhang, R. Wang, *Chem. Eur. J.* **2013**, *19*, 1184.

51. H. Wu, L.-L. Zhang, Z.-Q. Tian, Y.-D. Huang, Y.-M. Wang, *Chem. Eur. J.* **2013**, *19*, 1747.

52. Q. Chen, J. Liang, S. Wang, D. Wang, R. Wang, *Chem. Commun.* **2013**, *49*, 1657.

53. (a) K. Zeitler, *Angew. Chem. Int. Ed.* **2005**, 44, 7506; (b) N. Marion, S. Diez-Gonzalez, S. P. Nolan, *Angew. Chem. Int. Ed.* **2007**, *46*, 2988; (c) D. Enders, O. Niemeier, A. Henseler, *Chem. Rev.* **2007**, *107*, 5606; (d) S. E. Denmark, G. L. Beutner, *Angew. Chem. Int. Ed.* **2008**, *47*, 1560; (e) E. M. Phillips, A. Chan, K. A. Scheidt, *Aldrichimica Acta*, **2009**, *42*, 55; (f) C. Grondal, M. Jeanty, D. Enders, *Nature Chem.* **2010**, *2*, 167; (g) A. Grossmann, D. Enders, *Angew. Chem. Int. Ed.* **2011**, *50*, 2; (h) A. Grossman, D. Enders, *Angew. Chem. Int. Ed.* **2012**, *51*, 314.

54. B. Zhang, P. Feng, L. H. Sun, Y. Cui, S. Ye, N. Jiao, *Chem. Eur. J.* **2012**, *18*, 9198.

55. H. Lv, B. Tiwari, J. Mo, C. Xing, Y. R. Chi, *Org. Lett.* **2012**, *14*, 5412.

56. K.-Q. Chen, Y. Li, C.-L. Zhang, D.-Q. Sun, S. Ye, *Org. Biomol. Chem.* **2016**, *14*, 2007.

57. B.-X. Feng, J.-D. Yang, J. Li, X. Li, *Tetrahedron Lett.* **2016**, *57*, 3457.

58. X. J. Chen, X. Chen, X. Ji, H. L. Jiang, Z. J. Yao, H. Liu, *Org. Lett.* **2013**, *15*, 1846.

59. S. Kato, M. Kanai, S. Matsunaga, *Chem. Asian J.* **2013**, *8*, 1768.

60. F. Tan, L.-Q. Lu, Q.-Q. Yang, W. Guo, Q. Bian, J.-R. Chen, W.-J. Xiao, *Chem. Eur. J.* **2014**, *20*, 3415.

61. J.-Q. Zhao, Z.-J. Wu, M.-Q. Zhou, X.-Y. Xu, X.-M. Zhang, W.-C. Yuan, *Org. Lett.* **2015**, *17*, 5020.

62. L. Ren, X.-L. Lian, L.-Z. Gong, *Chem. Eur. J.* **2013**, *19*, 3315.

63. L. Wang, D. Yang, D. Li, R. Wang, *Org. Lett.* **2015**, *17*, 3004.

64. X.-P. Yin, X.-P. Zeng, Y.-L. Liu, F.-M. Liao, J.-S. Yu, F. Zhou, J. Zhou, *Angew. Chem. Int. Ed.* **2014**, *53*, 13740.

65. (a) F. Zhou, F.-M. Liao, J.-S. Yu, J. Zhou, *Synthesis*, **2014**, *46*, 2983; (b) A. Russo, C. D. Fusco, A. Lattanzi, *RSC Adv.* **2012**, *2*, 385; (c) A. M. R. Smith, K. K. Hii, *Chem. Rev.* **2011**, *111*, 1637.

66. L. Cheng, L. Liu, D. Wang, Y.-J. Chen, *Org. Lett.* **2009**, *11*, 3874.

67. Z.-Q. Qian, F. Zhou, T.-P. Du, B.-L. Wang, M. Ding, X.-L. Zhao, J. Zhou, *Chem. Commun.* **2009**, 6753.

68. T. Bui, M. Borregan, C. F. Barbas, *J. Org. Chem.* **2009**, *74*, 8935.

69. T. Bui, G. Hernandez-Torres, C. Milite, C. F. Barbas, *Org. Lett.* **2010**, *12*, 5696.

70. F. Zhou, M. Ding, Y.-L. Liu, C.-H. Wang, C.-B. Ji, Y.-Y. Zhang, J. Zhou, *Adv. Synth. Catal.* **2011**, *353*, 2945.

71. X.-P. F. Zhou, C. Zeng, X.-L. Wang, J. Zhao, J. Zhou, *Chem. Commun.* **2013**, *49*, 2022.

72. Z. Yang, Z. Wang, S. Bai, K. Shen, D. Chen, X. Liu, L. Lin, X. Feng, *Chem. Eur. J.* **2010**, *16*, 6632.

73. K. Shen, X. Liu, G. Wang, L. Lin, X. Feng, *Angew. Chem. Int. Ed.* **2011**, *50*, 4684.

74. T. Zhang, L. Cheng, L. Liu, D. Wang, Y.-J. Chen, *Tetrahedron: Asymmetry*, **2010**, *21*, 2800.

75. L.-N. Jia, J. Huang, L. Peng, L.-L. Wang, J.-F. Bai, F. Tian, G.-Y. He, X.-Y. Xu, L.-X. Wang, *Org. Biomol. Chem.* **2012**, *10*, 236.

76. X. Companyo, G. Valero, O. Pineda, T. Calvet, M. Font-Bardia, A. Moyano, R. Rios, *Org. Biomol. Chem.* **2012**, *10*, 431.

77. Y.-X. Jia, J. M. Hillgren, E. L. Watson, S. P. Marsden, E. P. Kündig, *Chem. Commun.* **2008**, 4040.

78. R. Huisgen, *Angew. Chem.* **1963**, *75*, 604.

79. (a) K. V. Gothelf, K. A. Jørgensen, *Chem. Rev.* **1998**, *98*, 863; (b) G. Pandey, P. Banerjee, S. R. Gadre, *Chem. Rev.* **2006**, *106*, 4484; (c) H. Pellissier, *Tetrahedron*, **2007**, *63*, 3235; (d) C. Najera, J. M. Sansano, *Org. Biomol. Chem.* **2009**, *7*, 4567; (e) M. Kissane, A. R. Maguire, *Chem. Soc. Rev.*, **2010**, *39*, 845; (f) J. Adrio, J. C. Carretero, *Chem. Commun.* **2011**, *47*, 6784; (g) A. Moyano, R. Rios, *Chem. Rev.* **2011**, *111*, 4703; (h) Y. Xing, N. X. Wang, *Coord., Chem. Rev.* **2012**, *256*, 938; (i) A. D. Lim, J. A. Codelli, S. E. Reisman, *Chem. Sci.* **2013**, *4*, 650; (j) C. Najera, J. M. Sansano, *J. Organomet. Chem.* **2014**, *771*, 78; (k) E. E. Maroto, M. Izquierdo, S. Reboredo, J. Marco-Martinez, S. Filippone, N. Martin, *Acc. Chem. Res.* **2014**, *47*, 2660; (l) J. Adrio, J. C. Carretero, *Chem. Commun.* 2014, **50**, 12434; (m) T. Hashimoto, K. Maruoka, *Chem. Rev.* **2015**, *115*, 5366.

80. A. L. Gerten, M. C. Slade, K. M. Pugh, L. M. Stanley, *Org. Biomol. Chem.* **2013**, *11*, 7834.

81. L. Wang, D. Yang, D. Li, X. Liu, Q. Zhao, R. Zhu, B. Zhang, R. Wang, *Org. Lett.* **2015**, *17*, 4260.

82. H. M. Zhang, Z. H. Gao, S. Ye, *Org. Lett.* **2014**, *16*, 3079.

83. H.-B. Yang, Y.-Z. Zhao, R. Sang, M. Shi, *J. Org. Chem.* **2014**, *79*, 3519.

84. J. Shan, B. Cui, Y. Wang, C. Yang, X. Zhou, W. Han, Y. Chen, *J. Org. Chem.* **2016**, *81*, 5270.

85. H. Zhang, S.-J. Zhang, Q.-Q. Zhou, L. Dong, Y.-C. Chen, *Beilstein J. Org. Chem.* **2012**, *8*, 1241.

Chapter 5

Enantioselective Synthesis of 3-Substituted-3-Hydroxyoxindoles

5.1. Introduction

The 3-substituted-3-hydroxyoxindoles are the core structure present in many natural products and medicinally relevant molecules and are key intermediates for drug development programmes (for reviews on oxindole alkaloids, see Ref. [1]), for example (*R*)-4, 6-dibromo-3-hydroxy-3-(2-oxopropyl)indolin-2-one also known as Convolutamydine A, which possesses a 2-oxopropyl group at C-3, inhibit the differentiation of pro-myelocytic leukemia cells HL-60 [2]. Paratunamide D is a novel 3-hydroxyoxindole alkaloid containing a secologanin unit and shows moderate cytotoxicity against human epidermoid carcinoma KB cells [3]. Maremycins also show cytotoxicity against mouse fibroblasts L-929 and human leukemia K562, while celogentin K is a potential inhibitor of tubulin polymerization [4]. The other examples are CPC-1 [5], diazonamide A [6], arundaphine [7], TMC-95A-D [8], etc. which show broad spectrum of interesting biological activities such as potent anti-cancer, antioxidant, antimalarial, anti-HIV and neuroprotective properties. The structure–activity correlation shows that the biological activity of 3-hydroxyoxindole are sensitive to the nature of the substituent at the C-3 position, as well as the absolute configuration of the stereogenic center. In this context, the development of efficient methods for the construction of the multifunctionalized framework containing oxindole moiety has significant synthetic and biological importance, and is the current area of research in asymmetric catalysis. In the last few years, variety of organocatalytic methods has been developed including additions to isatin such as Friedel–Crafts reaction, Mannich reaction, Henry reaction, Strecker reaction, Morita–Baylis–Hillman (MBH) reaction, intramolecular arylation, anilide cyclizations, allylation, crotylation and prenylation of isatins, hydroxylation reaction, etc. In line with the organocatalytic methods, the metal catalyzed nucleophilic addition to the isatins has also been developed for the synthesis of 3-substituted-3-hydroxyoxindoles. The number of publications related to the synthesis of 3-substituted-3-hydroxyoxindoles has exponentially increased over last 10–12 years. This chapter will feature the strategies for enantioselective synthesis of 3-hydroxyoxindole using both metal catalyst as well as organocatalyst. This chapter is organized based on the named reactions of isatin derivatives with different nucleophiles to afford 3-hydroxyoxindole derivatives.

5.2. Asymmetric Aldol Reaction of Isatins

The aldol reaction is a versatile method for enantioselective carbon–carbon bond formation which has been developed widely in the last few decades [9]. The asymmetric aldol reaction can be classified as "Indirect aldol reaction" and "Direct aldol reaction", where the former reaction is very often carried out using preformed enolate while the latter one uses the direct form of ketone in non-activated form. In 2000, the first report on the direct aldol reaction of aldehyde derivatives with acetone catalyzed by proline provided aldol addition product in 81 to >99% ee and 12–93% yield [10]. This was further succeeded by the development of new catalysts based on proline especially prolinamides [11]. Although, the first aldol reaction of isatin with acetone for the synthesis of tetrasubstituted-3-hydroxyoxindole was disclosed in 1933 [12], but its asymmetric version was developed in 2005 by Tomasini and co-workers [13]. They performed organocatalytic asymmetric aldol reaction of 4,6-dibromoisatin with acetone catalyzed by H-D-Pro-(R)-β^3-Phg-OBn **I** to obtain 3-hydroxyoxindoles in quantitative yield with 68% ee, which could be enantioenriched

Table 5.1. Organocatalytic enantioselective aldol reaction of 4,6-dibromoisatin with acetone.

$(R)/(S)$-Convolutamydine A

Sr. no.	Catalyst/reaction conditions	Yield (%)	ee (%)
1.	**H-D-Pro-(R)-β_3Phg-OBn I** (10 mol%) −15°C	82	97 (R)
2.	**D-Leucinol II** (20 mol%) DCM, rt, 36 h	98	94 (R)
3.	**N-(2-thiophenesulfonyl) prolinamide III** (5 mol%), H$_2$O, 0°C	99	95 (R)
4.	**L-leucinol amino alcohol IV** (10 mol%) Water, DCM, rt	95	80 (S)
5.	**Primary–tertiary 1–2 diamine V** (20 mol%), TfOH, 1,4-dioxane (0.5 mL), water, 5°C	95	96 (S)

to 97% ee after recrystallization (Table 5.1). The use of primary amino alcohols L-Leucinol **II** for aldol reaction of 4,6-dibromoisatin with acetone has been developed to provide product in 94% ee [14]. To gain insight into the mechanism, various experiments were designed that shows the importance of the hydroxyl group of the catalyst. The hydrogen bonding between the keto group of isatin and hydroxyl group of catalyst is a prerequisite for a highly enantioselective reaction. The reaction of 4,6-dibromoisatin with acetone catalyzed by N-(2-thiophenesulfonyl) prolinamide **III** was reported to afford (S)-convolutamydine A in 95% ee [15]. Further, the application of *in situ* generated organocatalyst for the aldol reaction of 4,6-dibromoisatin with acetone has been explored by using D-leucinol amino alcohol **IV**, which was generated *in situ* from the corresponding amino acid L-Leu through the BH$_3$-mediated reduction. The anti-leukaemia agent (R)-convolutamydine A was synthesized in 95% yield and with 80% ee under mild conditions (Table 5.1) [16]. The same reaction could be catalyzed by using chiral primary–tertiary 1–2 diamine **V** as a catalyst to obtain (S) isomers of convolutamydine in 95% yield with 96% ee (Table 5.1).

The asymmetric organocatalytic reaction of acetone with isatins catalyzed by dipeptide catalyst H-D-Pro-(R)-β^3-Phg-OBn **I** [17] or by thiopeptide **VI** [18] provides chiral 3-substituted-3-hydroxy-2-oxindoles in 92–99% yield with 72–77% ee and 54–68% yield with 56–68% enantiomeric excess, respectively (Table 5.2). The use of chiral primary–tertiary 1,2-diamine **V** as a catalyst furnished aldol products in 6–95% yield and 31 to >98% ee [19]. Recently, same reaction of isatins with acetone was reported by Guan. The 1,2-diaminocyclohexane (DACH)-derived chiral catalyst **VII** catalyzed the reaction to afford the aldol adduct in 60–98% yield with 88–98% ee (Table 5.2) [20]. Singh *et al.* used prolinamide ((2S,4R)-4-hydroxy-N-(S)-1-phenylethyl) pyrrolidine-2-carboxamide **VIII** as organocatalysts for enantioselective aldol reaction of isatins with acetone [21]. The product was obtained in 79–99% yield with 24–74% ee. N-aryl-(S)-prolinamide **IX** were also used as organocatalysts for the same

Table 5.2.   Organocatalytic asymmetric reaction of acetone with isatin derivatives.

R^1= H, Me, Et, Bn, allyl
R^2= H, 5-F, 5-Cl, 5-Br, 5-I, 5-MeO, 5-Me, 5-NO$_2$, 4-Br

Sr. no.	Catalyst/reaction conditions	Yield (%)	ee (%)
1.	**H-D-Pro-(R)-β$_3$Phg-OBn I** (10 mol%) −15°C	92–99	72–77(R)
2.	**Primary–tertiary 1–2 diamine V** (20 mol%) 1,4-dioxane (0.5 mL), water, 5°C	6–95	31 to >98 (S)
3.	**Thio-peptide VI** (10 mol%), ball milling	54–68	56–86 (S)
4.	**1,2-diaminocyclohexane (DACH)-derived chiral catalyst VII** (20 mol%), Water, DCM, 10°C	60–98	88–98 (R)
5.	**(2S, 4R)-4-hydroxy-N-(S)-1-phenylethyl) pyrrolidine-2-carboxamide VIII** (10 mol%) THF, −35°C	72–99	24–74(S)
6.	**N-Aryl-(S)-prolinamide IX** (20 mol%) Ethanol, −35°C	76–99	42–87(S)

reaction to provide adduct in 72–99% yield and 42–87% ee. The substitution of halogen group at C-5 position of isatin led to product with lower ee while the nitro group at C-5 of isatin provide product in 87% ee. The electron-releasing groups were found to lower the yield and ee of product (Table 5.2).

The chiral amino oxazolines **X** were synthesized and used as organo-catalyst for intermolecular asymmetric aldol reaction of acetone with isatin to afford the product in 31–46% yield with 23–61% ee (Table 5.3) [22]. The chitosan as a heterogeneous organocatalyst **XI** has been exploited for the aldol reaction of acetone with isatins to furnish aldol adducts in 95–97% yield with 5–25% enantiomeric excess (Table 5.3) [23]. The reaction of isatin derivatives with acetone catalyzed by *N*-(2-thiophenesulfonyl) prolinamide **III** was reported to provide in 59–99% yield with 77–97% ee. The hydrogen bonding between the amidic proton

Table 5.3. Chiral organocatalytic reaction of acetone with isatin derivatives.

R^1 = H, NO_2, F, Br
R^2 = H, methyl, benzyl, allyl

Sr. no.	Catalyst/reaction conditions	Yield (%)	ee (%)
1.	**Chiral amino oxazoline X** (20 mol%) THF (0.5 mL), 5–8°C, 72 h	31–46	23–61
2.	**Chitosan organocatalyst XI** (20 mol%) THF, 5–8°C, 72 h	95–97	5–25
3.	*N*-**(2-thiophenesulfonyl) prolinamide III** (5 mol%) TFA (7.5 mol%), water, 0°C	59–99	77–97
4.	**Primary–tertiary diamine organocatalyst XII (1 mol%)** DNP (1 mol%), water, rt	61–90	55–80
5.	**Prolinamide XIII** (5 mol%) TFA, −0°C	91	63

and sulfur atom of thiophene in the organocatalyst plays an important role in orienting the reactants for high enantioselectivity. The preferential formation of the *R* enantiomer has been attributed to *anti-trans*-transition state, since *syn-trans*-transition state leads to the formation of the *S* enantiomer, which is destabilized by a steric repulsion between 4-bromo and 2-thienyl groups (Table 5.3). The same reaction was reported using solvent-free protocol and prolinamides **XIII** as organocatalysts to provide the product in up to 91% yield with 63% enantiomeric excess (Table 5.3) [24]. The primary–tertiary diamine-based organocatalysts **XII** was developed and used for the asymmetric aldol reaction of isatin derivative with acetone to furnish product in up to 90% yield with 80% ee.

The enantioselective addition of acetone to isatin derivatives catalyzed by *N*-3-pyridyl prolinamide catalyst **XIV–XXI** was also reported (Scheme 5.1) [25]. A number of *N*-pyridylprolinamide and *N*-quinolinylprolinamide organocatalysts were screened for this reaction and prolinamide **XIV** in combination with AcOH furnished *S*-isomer of the product with 72% enantiomeric excess (Scheme 5.1). While in the case of amino pyridine (**XIV–XVI**) and aminoquinoline (**XVII–XVIII**)-based catalysts, the relative position of the pyridine nitrogen atom and the amino group have a major impact on the enantiomeric excess of the reaction. The reaction was performed with prolinamide catalyst in order to analyze the effect of the pyridine nitrogen, which provided the product with 65% ee but the reaction was slow. Further, to evaluate the role of the amidic proton, the same reaction was performed with organocatalyst **XX**, which provided trace amounts of product. This highlighted the importance of the amidic proton for obtaining the product in high enantiomeric excess.

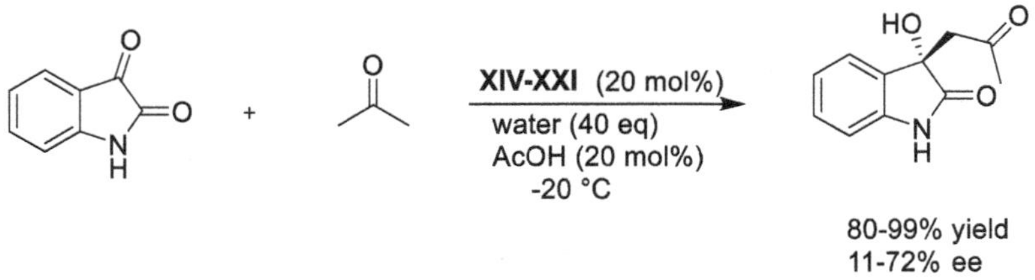

XIV-XXI (20 mol%)

water (40 eq)
AcOH (20 mol%)
-20 °C

80-99% yield
11-72% ee

Scheme 5.1.   *N*-aryl and *N*-heteroaryl pyrrolidine amide organocatalytic aldol reaction of isatins with acetone.

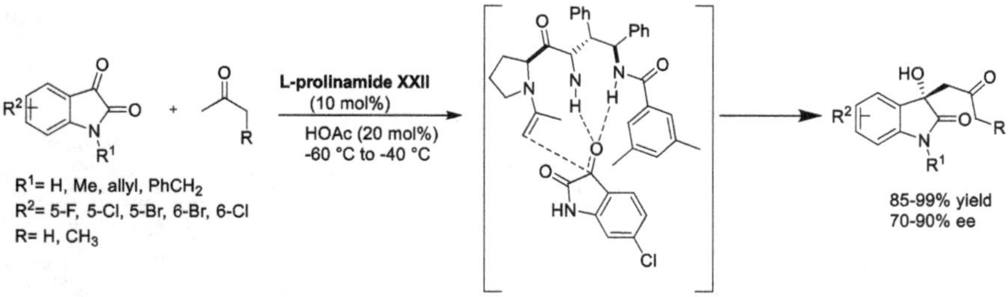

L-prolinamide XXII
(10 mol%)

HOAc (20 mol%)
-60 °C to -40 °C

R^1= H, Me, allyl, $PhCH_2$
R^2= 5-F, 5-Cl, 5-Br, 6-Br, 6-Cl
R= H, CH_3

85-99% yield
70-90% ee

Scheme 5.2. Proline catalyzed aldol reaction of isatin with acetone and 2-butanone.

L-Proline derived bifunctional organocatalysts **XXII** in combination with acetic acid as an additive was used for organocatalytic aldol reaction of isatin with ketones (acetone and 2-butanone) to afford the aldol adduct in 85–99% yield and 70–99% ee (Scheme 5.2). This methodology has been further extended for the synthesis of (*S*)-convolutamydine A in 45% yield and 87% enantiomeric excess. The enantiofacial discrimination in the transition state was governed by the catalyst which controls the enamine geometry and double hydrogen bonds with isatin, which fixes its orientation in the transition state [26].

Several asymmetric 3-substituted-3-hydroxyoxindoles were synthesized from aldol reaction of isatin derivatives with unactivated ketones catalyzed by quinidine thiourea **XXIII** to provide in 70–99% yield with 73–97% ee (Table 5.4) [27]. The product was further transformed into 3-hydroxyindolin-2-one which is the lead compound for the treatment of Ewing's sarcoma discovered by Toretsky and co-workers [28]. In the proposed transition state, the quinuclidine nitrogen of catalyst deprotonates acetone and after deprotonation, the enolate associates with the quinuclidine nitrogen through ionic interactions. Simultaneously, the thiourea moiety of quinuclidine thiourea catalyst forms hydrogen bonds with isatin carbonyl group thus activating it. The enolate attacks the *Si*-face of isatin to afford (*R*)-configured product. The Cinchonine-based urea **XXIV** as a catalyst was also used for aldol reaction of unactivated ketones with isatin derivatives. The aldol adduct was isolated in 85–98% yield with 78–91% ee (Table 5.4). The application of urea catalyst has been further extended to acetone as well as acetaldehyde [29]. The product was further transformed into ester that could be used as an intermediate

Table 5.4. Chiral thiourea/urea catalyzed reaction of isatin derivatives with unactivated ketones.

R^1= H
R^2= H, 4-Cl, 4-Br, 5-F
R= H, Me, Ph, 1-Np, 2-Np, Ar-CH=CH$_2$

m-CPBA, TFA, DCM, rt, 12 h

87% yield
89% ee
R = PMP

Sr. no.	Catalyst/reaction conditions	Yield (%)	ee (%)
1.	**Quinidine thiourea XXIII** (10 mol%) THF, 5°C	70–99	73–97
2.	**Cinchonine urea XXIV** (20 mol%) dioxane, 5°C, 72 h	85–98	78–91
3.	**[2.2] paracyclophane-based thiourea XXV** (20 mol%) THF, water	59–92	45–88 (*S*)

for convolutamydine syntheses. Song *et al.* developed a novel [2.2] paracyclophane-based thiourea catalyst **XXV** that catalyzed the asymmetric aldol reaction of isatins with enolizable ketones in the presence of H$_2$O. The aldol adducts were obtained in up to 92% yield and up to 88% ee [30]. The protecting group on the isatin was found to have crucial effect on the enantiomeric excess while the electronic properties of substituents on the isatin aromatic ring had little effect on the reactivity of reaction (Table 5.4).

The enantioselective aldol reaction of ketone with isatin derivatives catalyzed by carbohydrate-derived amino alcohol **XXVI/XXVII** provided aldol adducts in 80–99% yield with 55–75% ee (Table 5.5) [31]. Ren *et al.* reported asymmetric aldol reaction of isatins with ketones catalyzed by quinidine catalyst bearing 2-aminopyrimidin-4(1*H*)-one **XXVIII** to afford product in 45–92% yield with 60–94% ee (Table 5.5) [32]. The different isatin derivatives were well tolerated under the optimized conditions. The acetophenone bearing electron-withdrawing groups (EWGs) demonstrated high reactivity in comparison to those bearing electron-donating group. In 2015, Kumar *et al.* disclosed asymmetric aldol reaction of isatin with acetone using phthalimide prolinamide organocatalysts **XXIX** under

Table 5.5.   Carbohydrate-derived amino alcohol/phthalimide prolinamide catalyzed enantioselective aldol reaction of ketone with isatin derivatives.

R^1 = H, Cl, Br, Me
R^2 = H, Me, Bn
R^3 = Me, Ph

Sr. no.	Catalyst/reaction conditions	Yield (%)	ee (%)
1.	**Carbohydrate-derived amino alcohol XXVI/XXVII** (10 mol%) DCM, 0°C	80–99	55–75
2.	**Quinidine-2-aminopyrimidin-4(1H)-one derivative XXVIII** (20 mol%), THF, rt	45–92	60–94 (*R*)
3.	**Phthalimide prolinamide organocatalyst XXIX** (20 mol%) Neat, –20°C	69–78	51–71

solvent and additive free conditions [33]. The aldol adduct was isolated in 69–78% yield with 51–71% ee (Table 5.5). In the proposed transition state, initially the catalyst activates the ketone to form enamine and amidic NH activates the isatin *via* hydrogen bonding resulting in the formation of compact transition state which facilitates the attack of enamine on the *Re*-face of isatin to afford (*S*)-configured product.

The (*S*)-pyrrolidine tetrazole organocatalyst **XXX** has been developed to catalyze the aldol reaction of α-branched aldehyde with isatins to afford aldol product with two contiguous stereocenters in 46–92% yield and 49–93% ee [34]. The catalyst **XXX** proved to be highly efficient for both α-branched and linear aldehydes (Scheme 5.3).

The asymmetric aldol reaction of isatin with acetaldehyde derivatives catalyzed by 4-hydroxy-diarylprolinol **XXXI** and chloroacetic acid as an additive have been performed to obtain aldol adduct in 55–86% yield and 81–86% ee. The synthesized aldol adduct have been further transformed to *ent*-convolutamydine and CPC-1 (Scheme 5.4) [35]. A similar transformation with wider substrate scope was reported using the same catalyst without acid additive. The aldol adduct were obtained in 65–95%

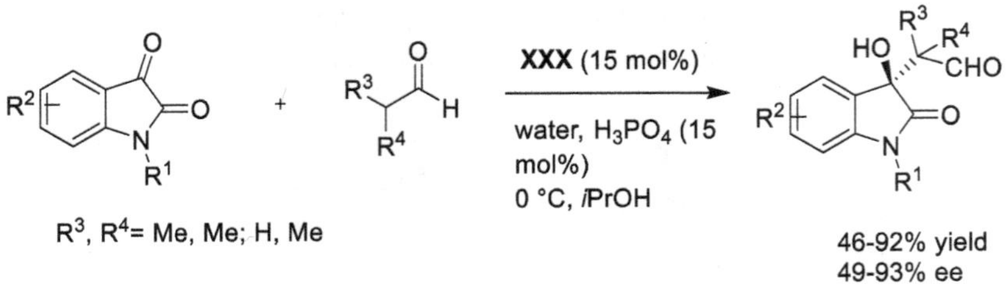

R³, R⁴= Me, Me; H, Me

46-92% yield
49-93% ee

Scheme 5.3.   Organocatalytic aldol reaction of linear aldehydes with isatins.

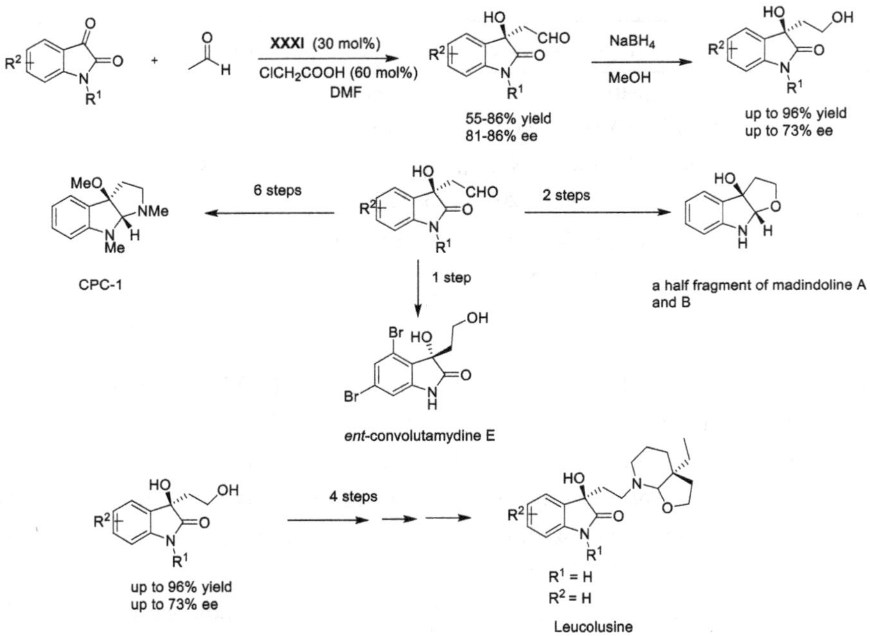

55-86% yield
81-86% ee

up to 96% yield
up to 73% ee

CPC-1

6 steps

2 steps

HO

a half fragment of madindoline A
and B

1 step

ent-convolutamydine E

up to 96% yield
up to 73% ee

4 steps

$R^1 = H$
$R^2 = H$

Leucolusine

Scheme 5.4.   Asymmetric aldol reaction of isatin with acetaldehyde derivatives catalyzed by 4-hydroxy-diarylprolinol.

yield and 77–98% enantiomeric excess after reduction with sodium borohydride. This approach was used for the first enantioselective synthesis of optically active (–)-donaxaridine and (*R*)-chimonamidine [36]. The 9-amino-(9-deoxy)-*epi*-quinine (9-NH$_2$-*epi*QN) catalyzed aldol addition of acetaldehyde to isatins has also been disclosed to provide aldol adducts in 72–96% yield and up to 93% ee [37]. Same catalyst was used for total synthesis of Leucolusine *via* reaction of isatins with acetaldehyde to afford product in up to 96% yield with up to 78% ee (Scheme 5.4) [38].

Yuan and co-workers have developed the reaction of isatin derivatives with acetaldehyde catalyzed by chiral diarylprolinol **XXXIII** as a catalyst followed by reduction with sodium borohydride in the absence of acid additive to afford the aldol adduct in 65–95% ee with 77–98% yield (Table 5.6).

This approach was further used for the enantioselective synthesis of (*R*)-convolutamydine B, (-)-Donaxaridine and (*R*)-chimonamidine in excellent ee. The same reaction was reported by Cheng *et al.* to afford product in 90–97% yield with 69–82% ee catalyzed by L-*tert*-leucine derivative in conjunction with $(H_4SiW_{12}O_{40})_{0.25}$ catalyst **XXXII** (Table 5.6) [39].

The combination of quinine derived amine **XXXIV** and benzoic acid as cocatalyst catalyzed the aldol addition reaction of acetaldehyde with isatin derivatives followed by reduction using sodium borohydride to provide the aldol adduct in 90–97% yield with 30–93% ee (Scheme 5.5) [40]. In the proposed transition state, the quinine amine form enamine

Table 5.6. Organocatalytic asymmetric reaction of isatin derivatives with acetaldehyde catalyzed by chiral diarylprolinol.

R^1 = H, methyl, allyl, benzyl
R^2 = H, 4-Cl, 5-Cl, 6-Cl, 4-Br, 5-Br, 6-Br

R^1= H (-)-Donaxaridine
86% yield, 99% ee
R^1= Me (*R*)-Chimonamidine
72% yield, 96% ee

Sr. no.	Catalyst/reaction conditions	Yield (%)	ee (%)
1.	**Diarylprolinol XXXIII** (20 mol%) DME, −10°C then NaBH$_4$ (5 eq.), MeOH, 0°C	77–98	65–95
2.	**L-tert-leucine derivative in conjunction with $(H_4SiW_{12}O_{40})_{0.25}$ catalyst XXXII** (20 mol%), neat then NaBH$_4$, MeOH	90–97	69–82

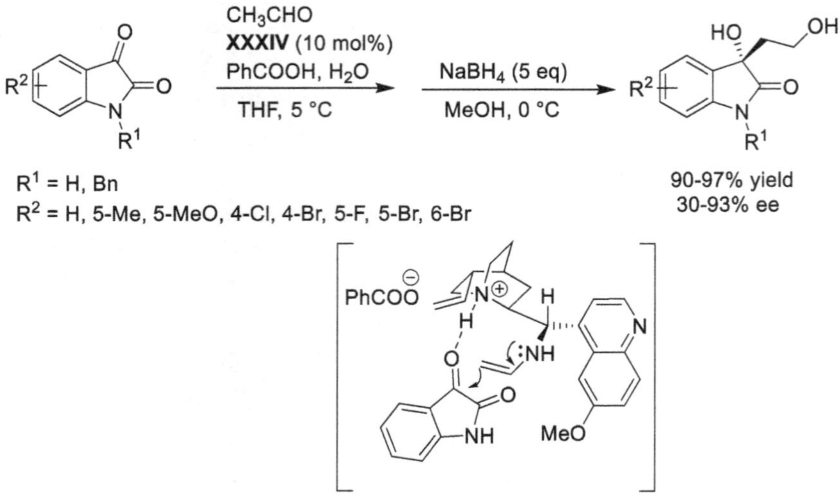

Scheme 5.5. Quinine derived amine catalyzed aldol addition reaction of acetaldehyde with isatin derivatives.

intermediate with acetaldehyde. The benzoic acid accelerates the formation of enamine through acid catalysis and simultaneously, it also protonates the quinuclidine nitrogen to form ammonium salt which then forms hydrogen bond with isatin carbonyl groups. The enamine attacks at the *Re*-face of isatin to provide (*S*)-configured product.

The enantioselective aldol reaction of cyclohexanone with isatins catalyzed by **XXXV** afforded chiral 3-cycloalkane-3-hydroxyoxindoles in quantitative yield with 99:1 d.r. (*syn/anti*) with 99% enantiomeric excess (*syn* isomer) (Scheme 5.6) [41]. The authors found that the prolinamide catalyst failed to catalyze this reaction, in contrast to their successful use in the aldol reaction of cyclohexanone with aldehydes [42], therefore highlighting the role of primary amine organocatalysts in the catalysis of sterically demanding molecules.

1,2-Diaminocyclohexane-hexanedioic acid **XXXVI** has also been used as a catalyst to catalyze the aldol reaction of cyclic ketones with isatin derivatives in MeOH-H$_2$O to afford the corresponding adduct in 71–90% yield, up to 99:1 *anti/syn* d.r. and >99% ee (Scheme 5.6) [43].

Recently, same research group reported enantioselective aldol addition reaction of cyclohexanone to isatins catalyzed by chiral phosphoramide

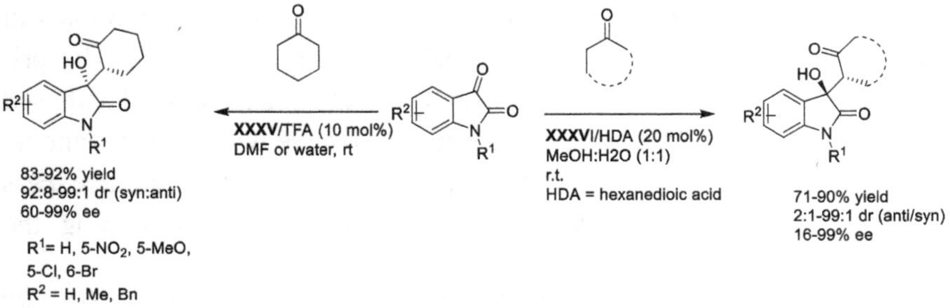

83-92% yield
92:8-99:1 dr (syn:anti)
60-99% ee

R^1= H, 5-NO$_2$, 5-MeO,
5-Cl, 6-Br
R^2 = H, Me, Bn

XXXV/TFA (10 mol%)
DMF or water, rt

XXXVI/HDA (20 mol%)
MeOH:H2O (1:1)
r.t.
HDA = hexanedioic acid

71-90% yield
2:1-99:1 dr (anti/syn)
16-99% ee

Scheme 5.6. 1,2-Diaminocyclohexane-hexanedioic acid catalyzed aldol reaction of cycloketones with isatin derivatives.

XXXVII to furnish adduct in 49–94% yield with 76:24 to 90:10 d.r. and 75:25 to 93:7 e.r. (Table 5.7) [44]. The 1,2-diaminocyclohexane (DACH) derived organocatalysts **XXXVIII** that contains an additional phenolic hydroxyl group have been used to catalyze the reaction of isatin with cyclohexanone. The different aldol adduct was isolated in 81–98% yield with 60–98% ee (Table 5.7). In the proposed transition state, the protonated amino group of catalyst activates the aldehydes *via* hydrogen bonding and primary amine of catalyst forms enamine with ketone. The non-covalent π–π interactions between the aldehydes and aromatic rings of catalyst provide further stability to transition state. Same reaction was reported by Qi and co-workers using solvent-free protocol and prolinamides **XXXIX** as organocatalysts to obtain the desired product in 22–98% yield, 53 to >99% enantiomeric excess and >99:1 d.r. (Table 5.7) [45]. In the proposed transition state, the pyrrolidine moiety of the catalyst reacts with cyclohexanone to form enamine and simultaneously, aldehydes get activated by hydrogen bonding with N–H bonding of two amides of the catalyst. The enamine attacked at *Re*-face of carbonyl group to afford (2*S*, 1′*R*)-configured product (Table 5.7). Ishimaru and co-workers reported *N*-aryl-L-valinamide **XL** catalyzed enantioselective aldol reaction of carbonyl compounds with isatins under mild conditions using PTSA.H$_2$O as additive to afford aldol adduct in 85–93% yield, 83 to >99% ee and 89:11–71:29 d.r. (Table 5.7) [46]. The axially unfixed biaryl-based water-compatible bifunctional organocatalyst **XLI** was developed and employed for the direct aldol reactions of isatin derivatives with enolisable ketones in water to access structurally diverse 3-alkyl-3-hydroxy-2-oxindoles in 70 to >99% yield, >99:1 d.r. and 67–98% ee (Table 5.7) [47]. The primary–tertiary diamine-based organocatalysts **XII** was developed and used for the asymmetric aldol reaction of isatin derivative with cyclohexanone. The product was obtained in 90–94% yield, 70–95% enantiomeric excess with 73:27–96:4 d.r. [48]. It was found that at high catalyst loading, the

Table 5.7. Organocatalytic asymmetric aldol reaction of isatin derivatives with cyclohexanone.

R^1 = H, NO_2, F, Br
R^2 = H, methyl, benzyl

Sr. no.	Catalyst/reaction conditions	Yield (%)	ee (%)	d.r. (%)
1.	**Chiral phosphoramide XXXVII** (10 mol%) BzOH (10 mol%), 3°C, 48 h	49–94	75:25–93:7	76:24–90:10
2.	**1,2-diaminocyclohexane (DACH) derived organocatalyst XXXVIII** (10 mol%), TFA (20 mol%), EtOH, rt, 24 h	81–98	60–98	—
3.	**Prolinamide catalyst XXXIX** (5 mol%), TFA (7.5 mol%), 0°C	22–98	53 to >99	>99:1
4.	**N-aryl-L-valinamide XL** (25 mol%), water (20 mol%)	85–93	83 to >99	89:11–71:29
5.	**Axially unfixed biaryl-based water-compatible organocatalyst XLI** (15 mol%), p-TsOH.H_2O (10 mol%), water, rt	70 to >99	67–98	>99:1
6.	**Primary–tertiary diamine-based organocatalyst XII** (1 mol%), DNP (1 mol%), water, rt	90–94	70–95	Up to 96:4
7.	**Amino amide organocatalyst XLII** (15 mol%), THF, rt	54–97	12–98	83:17–99:1
8.	**Chiral 2-*aza*norbornane-based organocatalyst XLIII** (20 mol%), EtOH, −35°C	79–99	24–74	—

retro-aldol reaction occurs that changes the reaction course from kinetic control to thermodynamic control. While at low catalyst loading, the retro-aldol is very slow and the shifting of the reaction from kinetic control to thermodynamic control was not possible (Table 5.7).

The asymmetric aldol reaction of isatins with ketones catalyzed by amino amide organocatalysts **XLII** affords product in 54–97% yield, 12–98% ee and 83:17–99:1 d.r. (Table 5.7) [49]. Chiral 2-azanorbornane-based organocatalysts **XLIII** was developed and its catalytic activity was explored for an enantioselective aldol reaction of isatins with ketones. The products were obtained in 79–99% yield, 24–74% ee (Table 5.7) [50].

An amino acid sulphonamides **XLIV** catalyzed aldol reaction of isatins with cyclohexanone derivatives have been developed under neat conditions. The 3-hydroxyoxindoles were obtained in 32–96% yield, 13–91% enantiomeric excess and 62:38–98:2 d.r. The various derivatives of ketones such as acetone, cyclopentanone and cycloheptanone were well tolerated (Scheme 5.7) [51].

Luo and co-workers have demonstrated the application of phenylalanine lithium salt **XLV** as an effective novel catalyst for enantioselective aldol reaction of isatin derivatives with ketones. The 3-hydroxyoxindoles were obtained in 73–97% yield with 66–90% ee. These catalysts can be easily prepared from inexpensive natural chiral amino acids (Scheme 5.8) [52].

The synthesis of 3-hydroxyoxindoles *via* aldol reaction of ketones with isatins have been reported to be catalyzed by chitosan aerogel in water. This environment friendly approach provides the product in 91–96% yield, 66–94% ee and up to 96:4 d.r. (Scheme 5.9) [53].

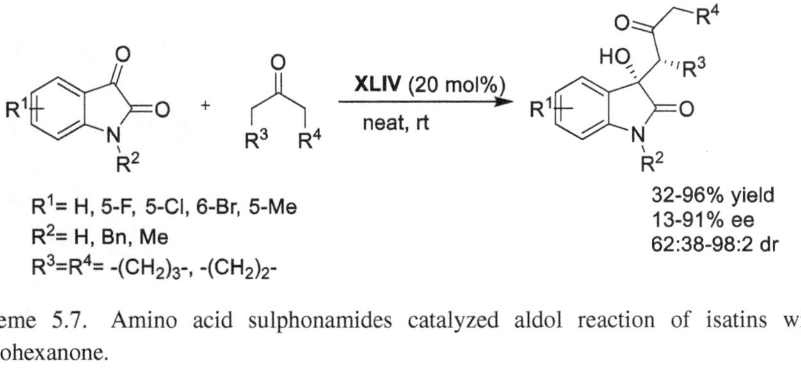

R^1= H, 5-F, 5-Cl, 6-Br, 5-Me
R^2= H, Bn, Me
R^3=R^4= -(CH$_2$)$_3$-, -(CH$_2$)$_2$-

32-96% yield
13-91% ee
62:38-98:2 dr

Scheme 5.7. Amino acid sulphonamides catalyzed aldol reaction of isatins with cyclohexanone.

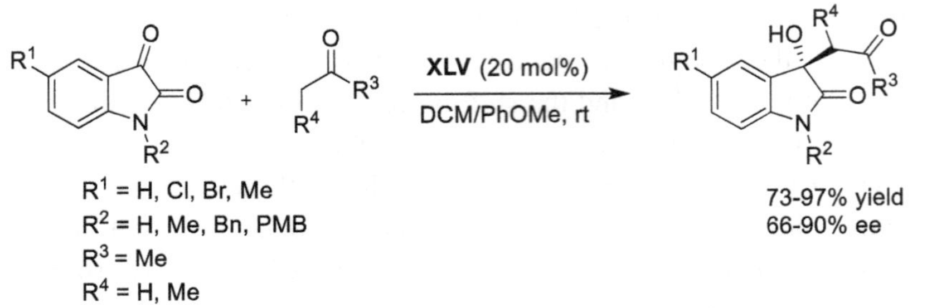

R¹ = H, Cl, Br, Me
R² = H, Me, Bn, PMB
R³ = Me
R⁴ = H, Me

XLV (20 mol%)
DCM/PhOMe, rt

73-97% yield
66-90% ee

Scheme 5.8. Enantioselective aldol reaction of isatin derivatives with ketones catalyzed by phenylalanine lithium salt.

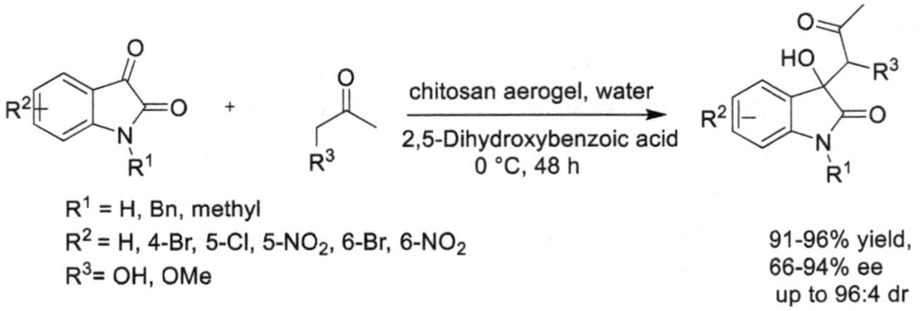

R¹ = H, Bn, methyl
R² = H, 4-Br, 5-Cl, 5-NO_2, 6-Br, 6-NO_2
R³ = OH, OMe

chitosan aerogel, water
2,5-Dihydroxybenzoic acid
0 °C, 48 h

91-96% yield,
66-94% ee
up to 96:4 dr

Scheme 5.9. Enantioselective aldol reaction of ketones with isatins catalyzed by chitosan aerogel in water.

The aldol reaction of aliphatic ketones or aldehydes with isatins using flexible *N*-(2,6-difluorophenyl)-L-valinamide **XLVI** as catalyst and Malonic acid as additive provides adduct in 65–82% yield, 33–94% ee [54]. The transition state was proposed which shows that malonic acid plays an important role for acceleration of reaction as well as for high stereoselectivity. This was further confirmed by DFT calculations (Scheme 5.10).

N-Prolinylanthranilamide-based pseudopeptide organocatalysts **XLVII** have been used to catalyze the reaction of 2,2-dimethyl-1,3-dioxane-5-one with isatin derivatives. The chiral product was obtained in 65–75% yield, 75–90% ee and up to 22:1 d.r. [55]. The secondary amide group of catalyst and tetrazole unit were found to be the key factors for good stereoselectivity. This approach was further used for the synthesis of natural TMC-95A (Scheme 5.11).

XLVI (25 mol%)
malonic acid (1.0 eq)
dry EtOH (1.0 mL), rt

R^1= Me, -(CH$_2$)$_5$-
R^2= H, methoxy, methyl
R= H, 5-methoxy, 5-methyl

65-82% yield,
33-94% ee
up to 84:16 dr

Scheme 5.10. Organocatalytic aldol reaction of aliphatic ketones or aldehydes with aromatic aldehydes or isatins.

XLVII (20 mol%)
*i*PrOH/H$_3$PO$_4$/water
4 °C, 6 days

R = H, 7-Cl, 7-Br, 7-I, 5-Br, 4-Br

65-75% yield,
75-90% ee
up to 22:1 dr

Intermediate for natural TMC-95A

Scheme 5.11. Organocatalytic asymmetric reaction of 2,2-dimethyl-1,3-dioxane-5-one with isatin derivatives.

The *Cinchona* amine catalyst in combination with TCA as an additive was found to be an efficient catalyst for the reaction of pyruvic aldehydes dimethyl acetal with isatin derivatives. Both the enantiomers of the product were obtained using pseudo-enantiomeric organocatalysts. The reaction was catalyzed using 9-NH$_2$-*epi*-CD and 9-NH$_2$-*epi*-CN **XLVIII/ XLIX** to afford the product in 87–96% yield and 89–97% enantiomeric excess [56]. The transition state was proposed in which initially the primary amine group of catalyst generates enamine with pyruvic aldehydes dimethyl acetal and simultaneously the carbonyl group of isatin gets activated *via* hydrogen bonding with protonated tertiary amine (Scheme 5.12).

Melchiorre reported an asymmetric vinylogous aldol reaction of α-branched enals with isatin derivatives. The diarylprolinol **L/LI** as a catalyst were used to catalyze the reaction isatin derivative with enals to provide the aldol adduct in 28–92% yield with 85–95% ee [57]. The use of aryl substituent at the α-branched enal position with isatins provide the spirooxindoles lactol as a product in 99% ee (Scheme 5.13).

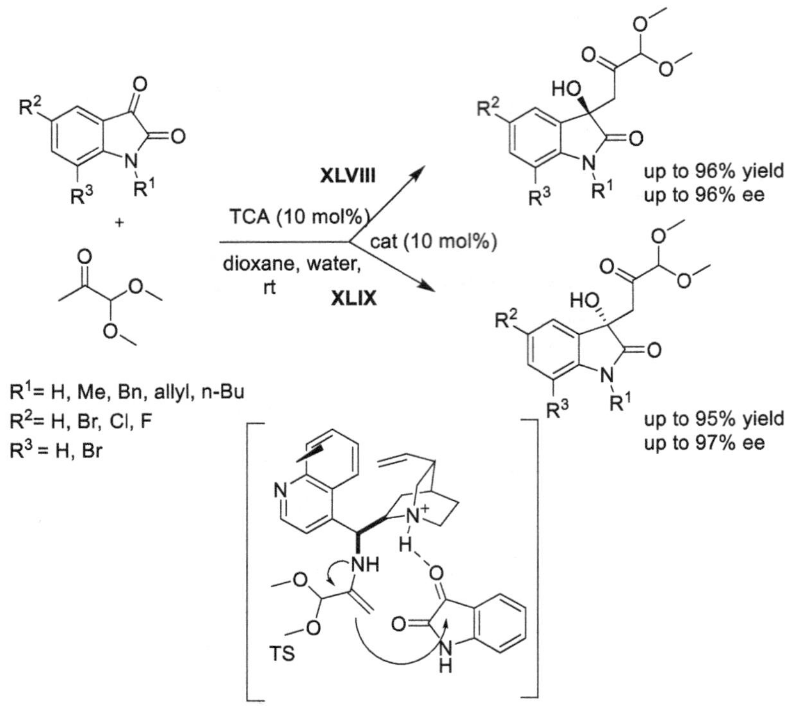
XLVIII
TCA (10 mol%)
cat (10 mol%)
dioxane, water,
rt
XLIX
up to 96% yield
up to 96% ee
up to 95% yield
up to 97% ee
R1= H, Me, Bn, allyl, n-Bu
R2= H, Br, Cl, F
R3 = H, Br
TS

Scheme 5.12. Organocatalytic reaction of pyruvic aldehydes dimethyl acetal with isatin derivatives.

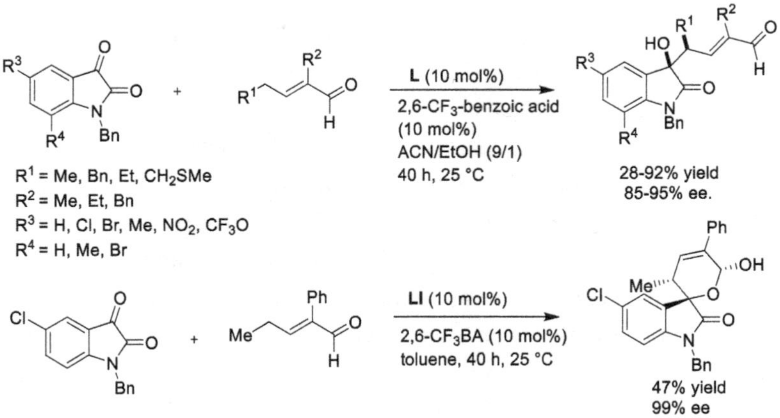
L (10 mol%)
2,6-CF3-benzoic acid
(10 mol%)
ACN/EtOH (9/1)
40 h, 25 °C
28-92% yield
85-95% ee.
R1 = Me, Bn, Et, CH2SMe
R2 = Me, Et, Bn
R3 = H, Cl, Br, Me, NO2, CF3O
R4 = H, Me, Br
LI (10 mol%)
2,6-CF3BA (10 mol%)
toluene, 40 h, 25 °C
47% yield
99% ee

Scheme 5.13. Asymmetric vinylogous aldol reaction of α-branched enals with isatin derivatives catalyzed by diarylprolinol.

Jiang and co-workers developed bifunctional tertiary amine thiourea **LII** catalyzed enantioselective vinylogous aldol reaction of allyl ketones with isatins to obtain the product in 57–95% yield, 89–99% ee and >99:1 d.r. (Scheme 5.14) [58]. The different derivatives of allyl ketones as well as isatin derivatives were well tolerated under the optimized conditions. The product could be further transformed to obtain precursors for the synthesis of biologically important spirooxindoles in 96% ee.

The reaction of allenic esters with isatins catalyzed by chiral *N,N′*-dioxide **LIII** provides carbinol allenoates in 53–99% yield, 57:43–90:10 d.r. and 68:32–96:4 e.r. (Scheme 5.15) [59].

Wang *et al.* reported asymmetric cross-aldol reaction of isatin derivatives with α,β-unsaturated ketones catalyzed by *Cinchona* thiourea **LIV/LV** to afford product in 18–98% yield with 55–97% ee (Scheme 5.16) [60].

The synthesis of hydroxyoxindole derivative has been reported *via* organocatalytic aldol reaction of acetylphosphonates with activated

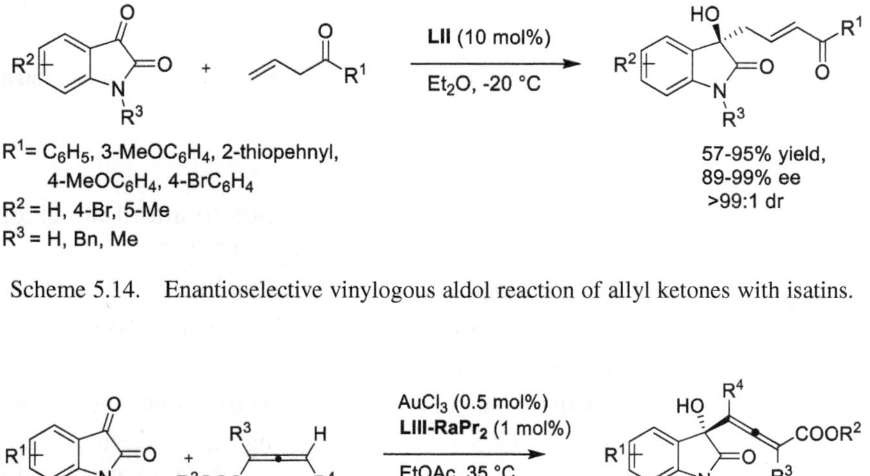

LII (10 mol%)
Et2O, -20 °C
R1= C6H5, 3-MeOC6H4, 2-thiopehnyl, 4-MeOC6H4, 4-BrC6H4
R2 = H, 4-Br, 5-Me
R3 = H, Bn, Me
57-95% yield,
89-99% ee
>99:1 dr

Scheme 5.14. Enantioselective vinylogous aldol reaction of allyl ketones with isatins.

AuCl3 (0.5 mol%)
LIII-RaPr2 (1 mol%)
EtOAc, 35 °C
3 Å MS
R1 = 4-Cl, 4-Br, 5-F, 5-Br, 5-I, 5-Me, 5-MeO, 5-CF3O, 6-Br, 7-F, 7-Br, 7-CF3
R2 = Me, Et, Bn, Ph, iPr
R3 = H, Bn, Et, iBu, i-pentyl, n-decyl, 2-MeC6H4CH2
R4 H, Me, Et, Bn, Ph, -decyl
53-99% yield
57:43-99:1 dr
68:32-96:4 er

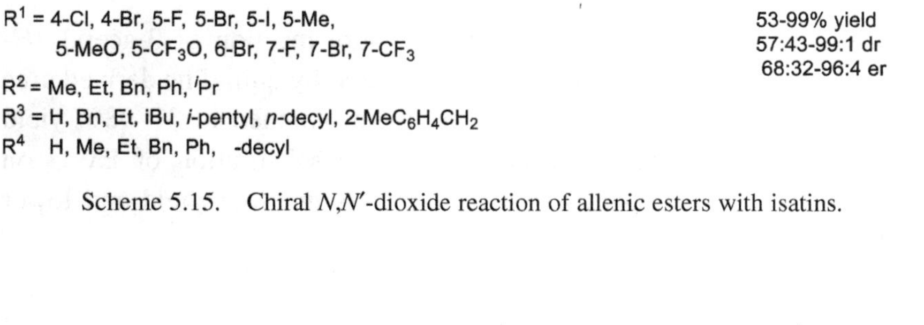

Scheme 5.15. Chiral *N,N′*-dioxide reaction of allenic esters with isatins.

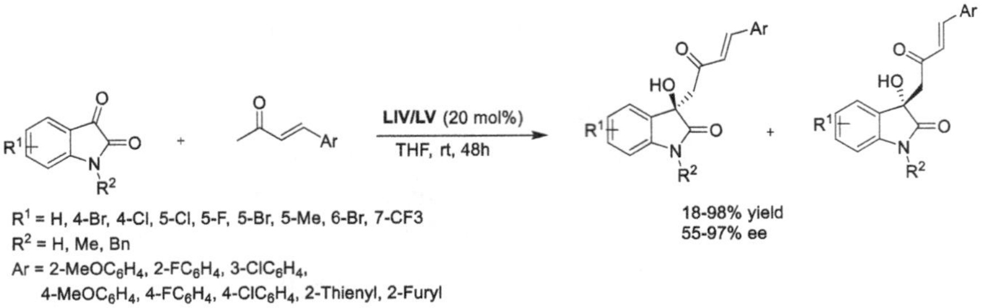
LIV/LV (20 mol%)
THF, rt, 48h
R1 = H, 4-Br, 4-Cl, 5-Cl, 5-F, 5-Br, 5-Me, 6-Br, 7-CF3
R2 = H, Me, Bn
Ar = 2-MeOC6H4, 2-FC6H4, 3-ClC6H4, 4-MeOC6H4, 4-FC6H4, 4-ClC6H4, 2-Thienyl, 2-Furyl
18-98% yield
55-97% ee

Scheme 5.16. Asymmetric cross-aldol reaction of isatin derivatives with α,β-unsaturated ketones.

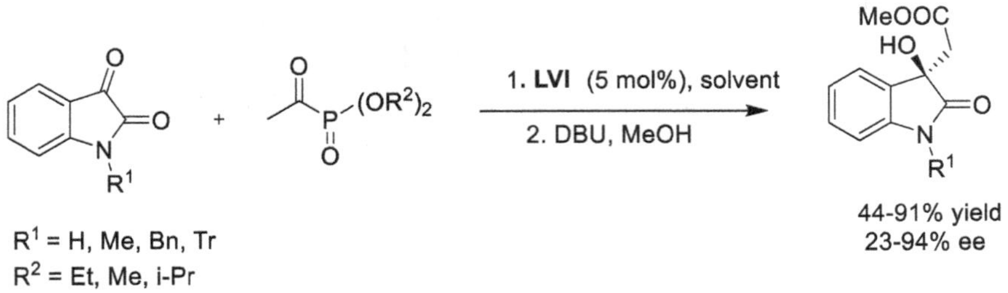
1. LVI (5 mol%), solvent
2. DBU, MeOH
MeOOC
HO
R1 = H, Me, Bn, Tr
R2 = Et, Me, i-Pr
44-91% yield
23-94% ee

Scheme 5.17. Organocatalytic aldol reaction of acetylphosphonates with activated carbonyl compounds.

carbonyl compounds. The **LVI** catalyzed the reaction to afford the aldol adduct in 44–91% yield with 23–94% ee (Scheme 5.17) [61].

The organocatalytic asymmetric C-1 functionalization of 1,3-dicarbonyl compound *via* reaction of isatin derivatives with 1,3-dicarbonyl compounds was disclosed. The quinidine thiourea **LIV** efficiently catalyzes the reaction to provide the 3-substituted-3-hydroxyoxindole derivatives in 58–93% yield with 83–93% ee. The resulting aldol adduct were further transformed into spirooxindoles in 52–92% ee and 88–97% yield (Scheme 5.18) [62].

Zhao *et al.* reported the asymmetric aldol reaction of 3-acetyl-2*H*-chromen-2-one with isatin derivatives catalyzed by quinidine derived urea as organocatalysts **LVII**. The products were obtained in 37–98% yield with 70–94% enantiomeric excess [63]. The substitution of EWG on 3-acetyl-2*H*-chromen-2-ones provide products in lower yield and lower

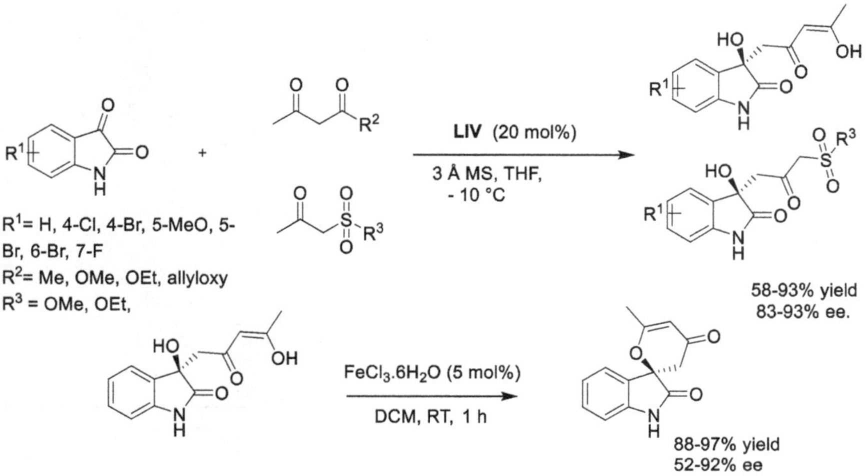

Scheme 5.18. Quinidine thiourea catalyzed reaction of isatin derivatives with 1,3-dicarbonyl compounds.

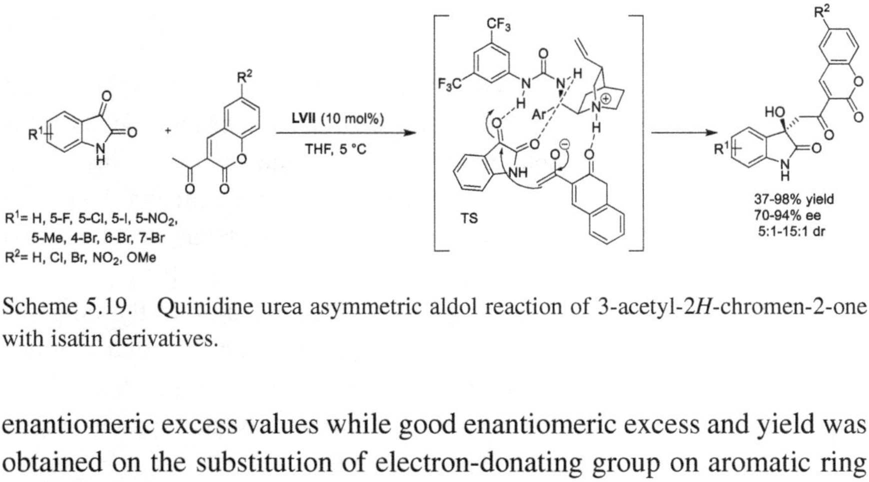

Scheme 5.19. Quinidine urea asymmetric aldol reaction of 3-acetyl-2*H*-chromen-2-one with isatin derivatives.

enantiomeric excess values while good enantiomeric excess and yield was obtained on the substitution of electron-donating group on aromatic ring (Scheme 5.19).

Miao *et al.* reported the quinidine **LVIII** catalyzed vinylogous Mukaiyama-aldol reaction of isatin derivatives with 2-(trimethylsilyloxy) furan in THF at −78°C to afford adduct in up to 94% yield and with good diastereoselectivities. The reactions proceeded smoothly within 15 min to give the desired products in good results regardless of the electronic properties of the substituents [64]. The *N*-protecting group of isatin derivatives

Table 5.8. *Cinchona* alkaloid catalyzed vinylogous Mukaiyama-aldol reaction of isatin derivatives with 2-(trimethylsilyloxy)furan and silyloxy dienes.

Sr. no.	Catalyst/reaction conditions	Yield (%)	ee (%)	d.r. (%)
1.	**Quinidine LVIII** (10 mol%), THF, −78°C, 15 min	51–94	—	Up to 96:4
2.	**Quinine thiourea LIV** (10 mol%), THF, −30°C	65–84	80:20–99:1	—

was found to have a certain influence on the reactivity. The synthesized adduct could be used for biological screening or serve as intermediates for further transformations (Table 5.8). Recently, organocatalytic vinylogous Mukaiyama-aldol reaction of silyloxy dienes and isatins using *Cinchona* thiourea **LIV** as bifunctional organocatalysts has been developed to provide the product in 65–84% yield with 80:20–99:1 e.r. [65].

Mikami and co-workers accomplished the application of the Pd complex catalyst **LXI** to the aldol reaction of isatin derivatives with trimethylsilyl ketene thioacetal. The aldol adducts were isolated in good yield and enantioselectivity and could be used as synthons for the synthesis of a variety of natural products (Scheme 5.20) [66].

A highly stereoselective Mukaiyama-aldol reaction of monofluorinated silyl enol ethers with isatins catalyzed by tertiary-amine catalyst **LIX** to afford 3-hydroxyoxindole derivatives in 37–98% yield, 70–94% ee and 7:1–15:1 d.r. (Table 5.9) [67] was also reported. In addition to this, *Cinchona* derived urea bearing *N*-alkyl group **LX** was also found to be efficient for this reaction (Table 5.9) [67]. The NH-isatins provide good

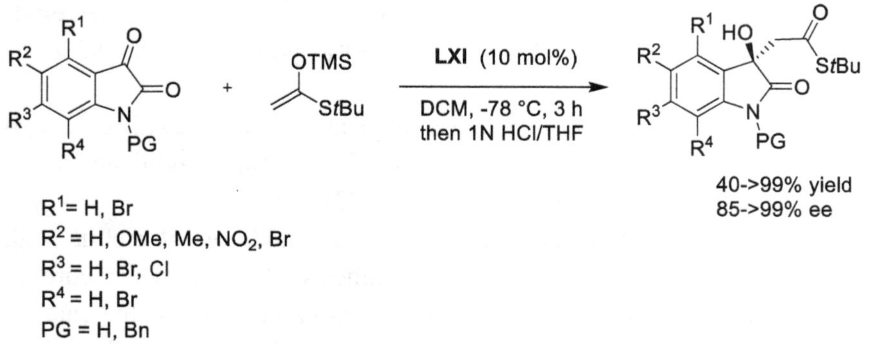

LXI (10 mol%)
DCM, -78 °C, 3 h
then 1N HCl/THF
40->99% yield
85->99% ee
R1= H, Br
R2 = H, OMe, Me, NO2, Br
R3 = H, Br, Cl
R4 = H, Br
PG = H, Bn

Scheme 5.20. Enantioselective aldol reaction of isatin derivatives with trimethylsilyl ketene thioacetal catalyzed by Pd complex catalyst.

Table 5.9. Organocatalytic Mukaiyama-aldol reaction of monofluorinated silyl enol ethers with isatins.

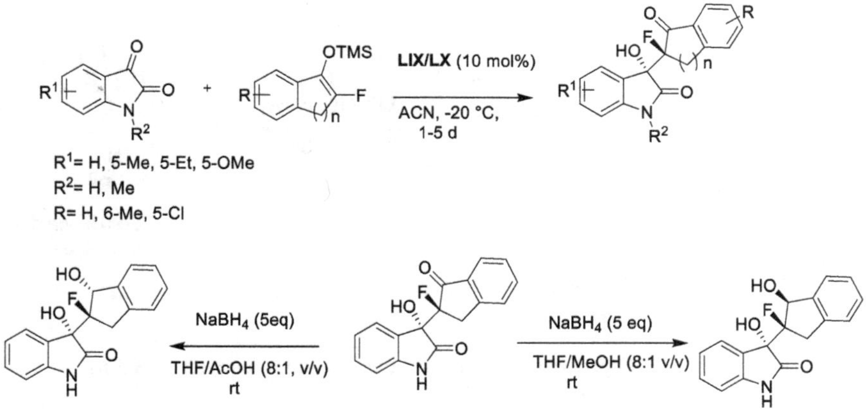

LIX/LX (10 mol%)
ACN, -20 °C,
1-5 d
R1= H, 5-Me, 5-Et, 5-OMe
R2= H, Me
R= H, 6-Me, 5-Cl
NaBH4 (5eq)
THF/AcOH (8:1, v/v)
rt
NaBH4 (5 eq)
THF/MeOH (8:1 v/v)
rt

Sr. no.	Catalyst/reaction conditions	Yield (%)	ee (%)	d.r. (%)
1.	**Cinchonidine urea LIX** (10 mol%), ACN, –20°C, 1–5 d	37–98	70–92	7:1–15:1
2.	**Quinine urea LX** (10 mol%), ACN, –20°C, 1–5 d	54–98	84–94	5:1–10:1

results as compared to *N*-methyl isatins. The formed product could be readily reduced to *syn*-diol and *anti*-diol by $NaBH_4$ using solvent system THF/MeOH and THF/AcOH [52].

The quinine urea **LIX** catalyzed synthesis of 3-difluoroalkenyl-substituted-3-hydroxyoxindoles in 78–90% yield with 88–95% ee (Scheme 5.21) [68].

The asymmetric aldol reaction of isatins with 2,2-difluoro-1,3-diketones catalyzed by quinine thiourea **LV** afforded aldol adduct in 42–95% yield with 58–98% ee (Scheme 5.22) [69].

Evano *et al.* developed asymmetric aldol addition of isatins with dihydroxyacetone derivatives for the synthesis of macrocyclic core of TMC-95A using H_2O as the cosolvent and amino alcohol as the catalyst **LXII** to afford the aldol products in good yields and high stereoselectivities (Scheme 5.23) [70].

Wu and co-workers explored the catalytic ability of *Cinchona* alkaloid catalyzed asymmetric aldol reaction of isatin derivatives with trifluoromethyl α-fluorinated β-keto gem-diols using AcOH as the additive to

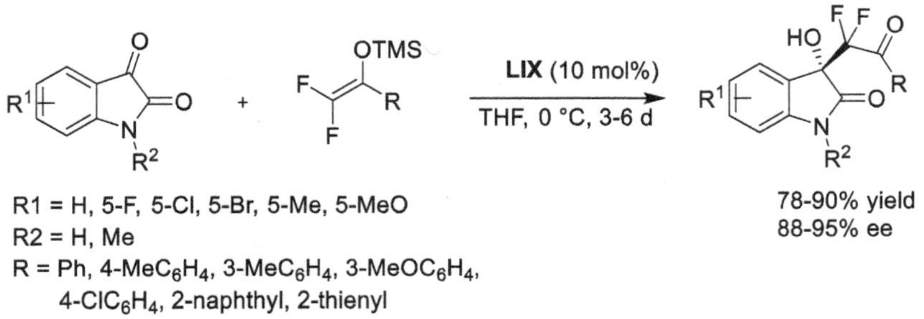

LIX (10 mol%)
THF, 0 °C, 3-6 d
OTMS
R1 = H, 5-F, 5-Cl, 5-Br, 5-Me, 5-MeO
R2 = H, Me
R = Ph, 4-MeC6H4, 3-MeC6H4, 3-MeOC6H4,
4-ClC6H4, 2-naphthyl, 2-thienyl
78-90% yield
88-95% ee

Scheme 5.21. Organocatalytic enantioselective aldol reaction of isatins.

LV (10 mol%)
MeOH, 25 °C
R1 = H, 4-Br, 5-Me, 5-OMe, 5-Cl, 5-F, 6-Cl, 6-Br, 7-F
R2 = Bn, Me
R3 = Ph, 4-MeC6H4, 3-MeC6H4,
4-ClC6H4, 4-BrC6H4, 4-FC6H4, 2-thienyl, Me
42-95% yield
58-98% ee

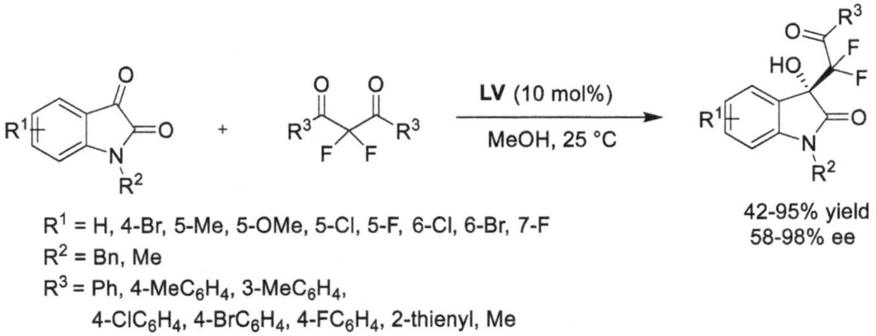

Scheme 5.22. Asymmetric aldol reaction of isatins with 2,2-difluoro-1,3-diketones.

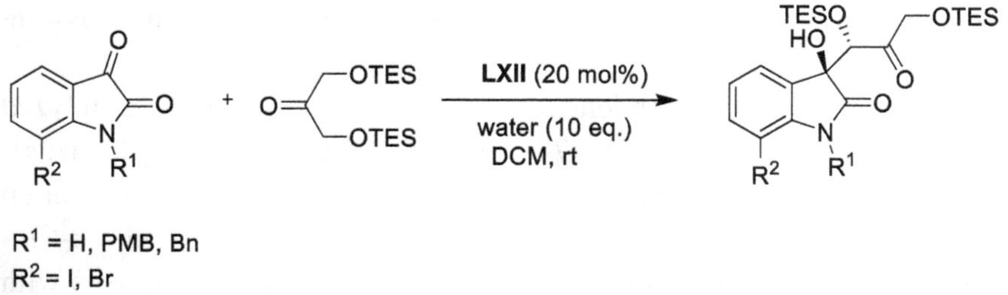

R^1 = H, PMB, Bn
R^2 = I, Br

Scheme 5.23.   Asymmetric aldol addition of isatins with dihydroxyacetone derivatives.

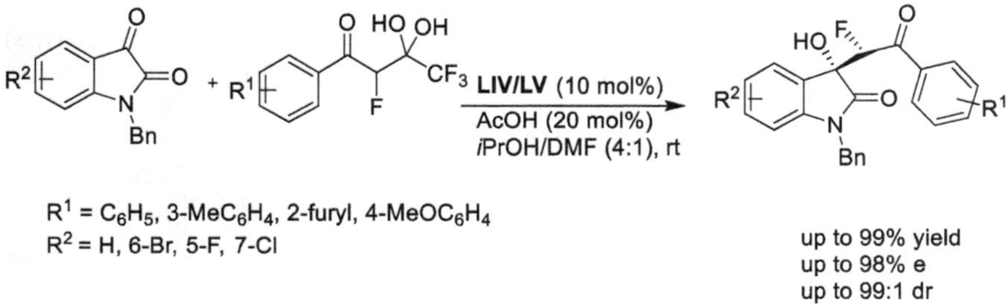

R^1 = C$_6$H$_5$, 3-MeC$_6$H$_4$, 2-furyl, 4-MeOC$_6$H$_4$
R^2 = H, 6-Br, 5-F, 7-Cl

up to 99% yield
up to 98% e
up to 99:1 dr

Scheme 5.24.   *Cinchona* alkaloid catalyzed asymmetric aldol reaction of isatin derivatives with trifluoromethyl α-fluorinated β-keto gem-diols.

afford 3-hydroxyoxindoles in up to 99% yield with enantiomeric excess up to 98% [71]. This protocol employed *gem*-diols as equivalents of fluorinated aryl/alkyl methyl ketone enolates for the carbon–carbon bond formation accompanied by the release of trifluoroacetate. The substrate scope includes the reaction of isatins with gem-diols with phenyl rings, heterocyclic groups and aliphatic groups (Scheme 5.24).

5.3. Asymmetric Henry Reaction of Isatins

The asymmetric Henry reaction is one of the important reactions to procure chiral β-nitro alcohols (for reviews on the catalytic asymmetric Henry reactions, see Ref. [72]). Wang *et al.* reported the first Henry reaction of isatin derivatives with nitroalkanes catalyzed by cupreine **LXIII**. The substituted-3-hydroxyoxindole derivatives were obtained in 92–99% yield with 74–95% enantiomeric excess [73]. The obtained product has been used for the synthesis of (+)-dioxibrassinin and (*S*)-(−)-spirobrassinin (Table 5.10).

The two other independent reports on the same reaction catalyzed by 9-*O*-benzylcupreine **LXIV** [74] and 9-*O*-3,5-trifluoromethylbenzoyl-cupreine **LXV** [75] has been disclosed to provide the Henry adduct in up to 98% yield and up to 97% ee. Rao and co-workers reported *bis-Cinchona* alkaloid **LXVI** catalyzed Henry reaction of isatin derivatives with nitroalkanes. The desired product was obtained in 79–94% yield with 64–97% ee (Table 5.10) [76].

Table 5.10. Asymmetric Henry reaction of isatin derivatives with nitroalkanes catalyzed by *Cinchona* alkaloid organocatalysts.

Sr. no.	Catalyst/reaction conditions	Yield (%)	*i* (%)
1.	**Cupreine LXIII** (10 mol%), PhCOOH (10 mol%), DMA, −15°C	92–99	74–95
2.	**9-*O*-benzylcupreine LXIV** (10 mol%), THF, −30°C, 4–5 d	95–97	76–92
3.	**9-*O*-3,5-trifluoromethyl benzoyl cupreine LXV** (10 mol%),THF, 5°C, 4d	79–98	53–97
4.	**(DHQD)₂PHAL LXVI** (4 mol%), THF, −10°C	79–94	64–97

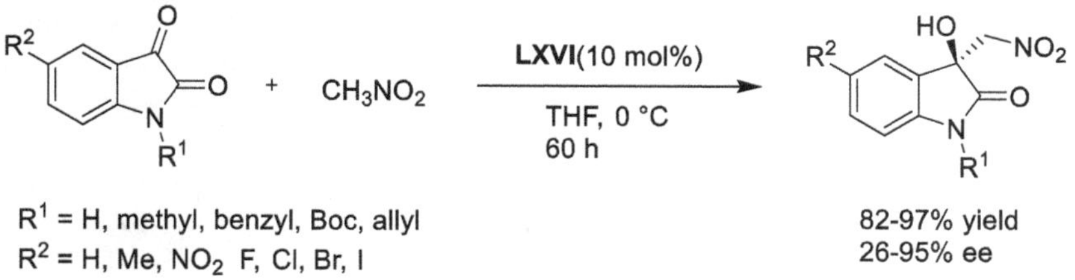

R^1 = H, methyl, benzyl, Boc, allyl
R^2 = H, Me, NO$_2$ F, Cl, Br, I

82-97% yield
26-95% ee

Scheme 5.25.   Hybrid-type squaramide fused amino alcohol catalyzed asymmetric Henry reaction of isatin derivatives.

Recently, hybrid-type squaramide fused amino alcohol catalysts **LXVI** have been developed and their catalytic ability has been explored for Henry reaction of isatin derivatives with nitroalkanes. The chiral Henry adduct was isolated in 82–97% yield with 26–95% ee (Scheme 5.25) [77].

5.4. Asymmetric Arylation Reaction of Isatins

Hayashi and co-workers reported Rh-catalyzed addition of arylboronic acid to isatin derivatives. The (*R*)-MeO-mop ligand turned out to be the best ligand to furnish the product in 79–98% yield and 72–93% ee. This Rh-(*R*)-MeO-mop complex **LXVIII** also catalyzes the addition of alkenylboronic acid to isatin derivatives to provide product in 91–93% yield with 88–93% ee [78]. The Rh-catalyzed asymmetric addition of phenylboronic acid to isatin derivatives using phosphoramidite ligand **LXIX** was also reported. The product was obtained in up to 99% yield and up to 55% ee [79]. The Rh–sulfur–alkene hybrid ligand complex **LXX** was used to catalyze the reaction of phenylboronic acid with isatin derivatives to afford product in up to 87% yield and with up to 76% ee (Table 5.11) [80].

In another report, the chiral sulfoxide phosphine ligand **LXXI** and Rhodium complex were used to catalyze the reaction of arylboronic acid with isatin derivatives. The chiral 3-aryl-3-hydroxy-2-oxindoles were obtained in 93–99% yield with 85–92% enantiomeric excess (Table 5.11) [81].

Qin and co-workers developed tetra-*ortho* substituted biphenyl phosphinoimine ligand **LXIV** using catalytic mixture of Pd(OAc)$_2$ to catalyze the addition of arylboronic acid to isatin derivatives. The 3-aryl-3-hydroxyoxindoles were obtained in 36–78% yield with 38–73% enantiomeric

Table 5.11.　Rh-catalyzed asymmetric addition of arylboronic acid to isatin derivatives.

R^1 = H, PMB, Bn
R^2 = H, 5-Cl, 5-F, 5-Me
R = Ar-, Ph-CH=CH-, n-pent-CH=CH-

Sr. no.	Catalyst/reaction conditions	Yield (%)	ee (%)
1.	**(R)-MeO-mop LXVIII** (10 mol%), $[RhCl(C_2H_4)_2]_2$ (5 mol%), KOH (15 mol%), THF/H_2O (20:1), 50°C,24 h	79–98	72–93 (S)
2.	**Phosphoramidite ligand LXIX** (9 mol%), $[(C_2H_4)_2Rh(acac)$ (3 mol%), $PB(OH)_2$ (2 eq.), dioxane, 40°C, 4h	99	55 (S)
3.	**Chiral *tert*-butanesulfinamide derived ligand LXX** (6 mol%),$PhB(OH)_2$ (2 eq.), $[Rh(C_2H_4)_2Cl]_2$ (2.5 mol%), KOH (7.5 mol%), toluene/H_2O (2:1), rt, 2h	87	76 (R)
4.	**Chiral sulfoxide phosphine ligand LXXI** (4.8 mol%), $[Rh(C_2H_4)_2Cl]_2$ (2 mol%), DIPEA (5 mol%), MeOH, 40°C, 3h	93–99	85–92 (R)

excess [82]. Shi *et al.* reported the enantioselective arylation of isatin derivatives with arylboronic acid employing chiral C_2-symmetric cationic NHC-Pd^2+ diaqua complex as a catalyst **LXV** to afford product in 79–92% yield with 60–80% ee (Table 5.12) [83].

The chiral ligand based on a chiral bridged biphenyl backbone **LXXIV** complexed with $[IrCodCl]_2$ catalyst catalyzed the asymmetric addition of arylboronic acid to isatin derivatives. The product was obtained in 48–98% yield with 77–95% ee (Scheme 5.26) [84].

A copper catalyzed addition of arylboronates to isatin derivatives by use of chiral *N*-heterocyclic carbene ligands **LXXV** has also been developed. The products were obtained in 67–94% yield with 67–92% ee (Scheme 5.27) [85].

Yamamoto *et al.* reported Ru-catalyzed addition of arylboronic acids to isatins using chiral *O*-linked C_2-symmetric bidentate phosphoramidite

Table 5.12. Pd-catalyzed asymmetric addition of arylboronic acid to isatin derivatives.

Pd catalysts
R-B(OH)2
HO
R
R2
R1
O
N
R1 = Bn, 1-anthracenylmethyl
R2 = H, 5-Cl, 5-Me, 5-OMe, 7-Me, 7-OMe
R = Ph, 4-MeOPh, 3-MeOPh, 4-FPh

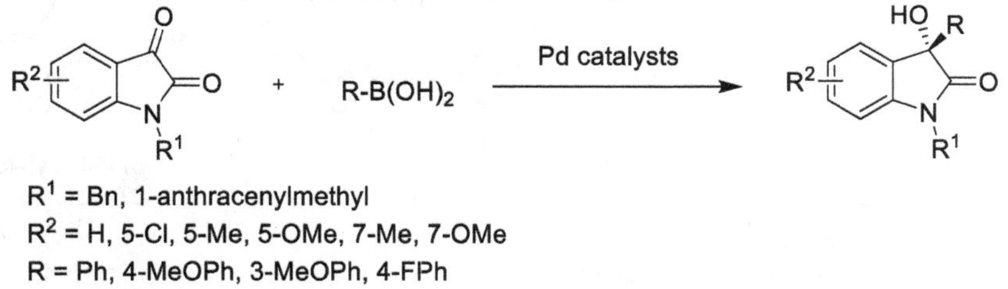

Sr. no.	Catalyst/reaction conditions	Yield (%)	ee (%)
1.	*tetra*-Ortho-substituted biphenyl phosphinoimine ligand LXXII (10 mol%), $Pd(OAc)_2$(5 mol%), $ArB(OH)_2$, $BF_3.Et_2O$ (4 eq.), THF, rt, 48h	36–78	38–73
2.	Chiral cationic C2-symmetric-NHC Pd^2+ diaqua complexes LXXIII (5 mol%), $LiOH.H_2O$ (1 eq.), Naphthalen-2-ol (0.15 eq.), THF, rt, 48 h	79–92	60–80

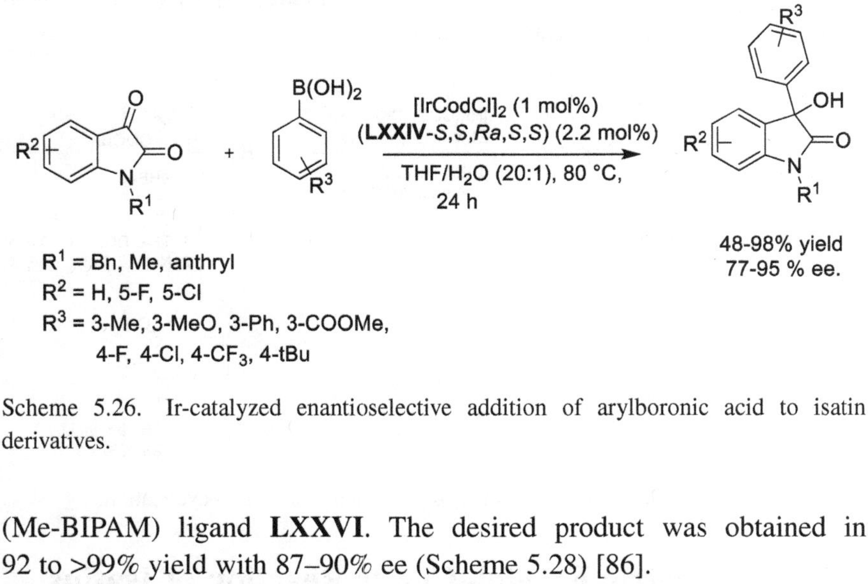

B(OH)2
[IrCodCl]2 (1 mol%)
(LXXIV-S,S,Ra,S,S) (2.2 mol%)
THF/H2O (20:1), 80 °C,
24 h
R3
OH
R2
R1
O
N
R1 = Bn, Me, anthryl
R2 = H, 5-F, 5-Cl
R3 = 3-Me, 3-MeO, 3-Ph, 3-COOMe,
4-F, 4-Cl, 4-CF3, 4-tBu
48-98% yield
77-95 % ee.

Scheme 5.26. Ir-catalyzed enantioselective addition of arylboronic acid to isatin derivatives.

(Me-BIPAM) ligand **LXXVI**. The desired product was obtained in 92 to >99% yield with 87–90% ee (Scheme 5.28) [86].

A rhodium-catalyzed asymmetric arylation–cyclization sequence was developed to provide 3-hydroxyoxindoles in 95–99% yield with 93–98% ee (Scheme 5.29) [87].

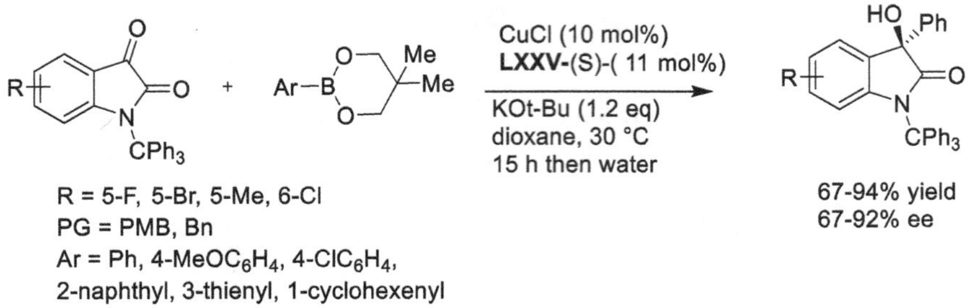

CuCl (10 mol%)
LXXV-(S)-(11 mol%)
KOt-Bu (1.2 eq)
dioxane, 30 °C
15 h then water
R = 5-F, 5-Br, 5-Me, 6-Cl
PG = PMB, Bn
Ar = Ph, 4-MeOC6H4, 4-ClC6H4,
2-naphthyl, 3-thienyl, 1-cyclohexenyl
67-94% yield
67-92% ee

Scheme 5.27. *N*-heterocyclic carbene ligands catalyzed asymmetric addition of arylboronic acid to isatin derivatives.

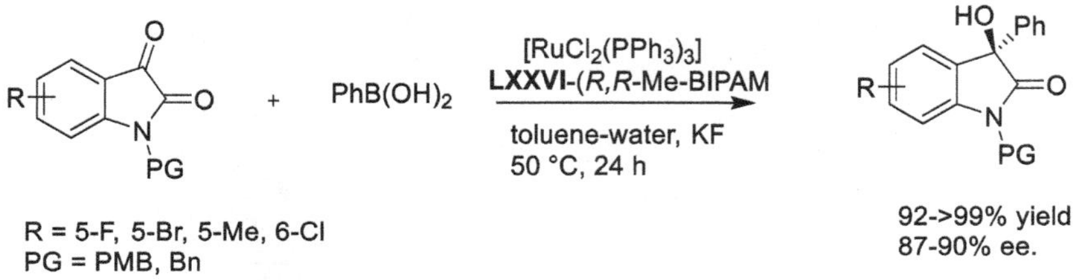

PhB(OH)2
[RuCl2(PPh3)3]
LXXVI-(R,R-Me-BIPAM
toluene-water, KF
50 °C, 24 h
R = 5-F, 5-Br, 5-Me, 6-Cl
PG = PMB, Bn
92->99% yield
87-90% ee.

Scheme 5.28. Ru-catalyzed addition of arylboronic acids to isatins using chiral *O*-linked C_2-symmetric bidentate phosphoramidite ligand.

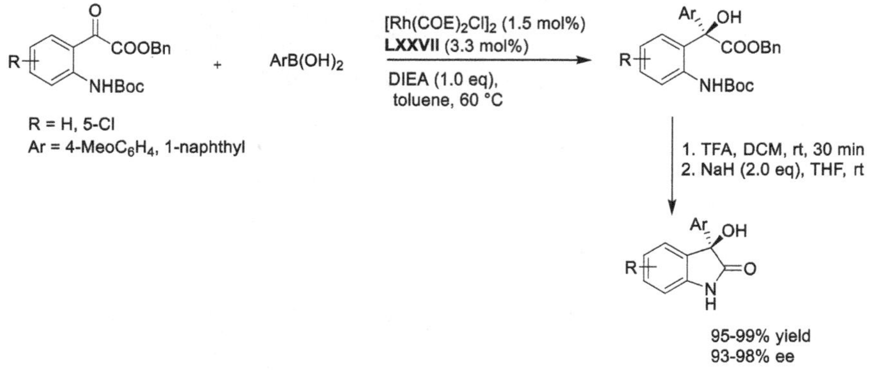

ArB(OH)2
[Rh(COE)2Cl]2 (1.5 mol%)
LXXVII (3.3 mol%)
DIEA (1.0 eq),
toluene, 60 °C
R = H, 5-Cl
Ar = 4-MeoC6H4, 1-naphthyl
1. TFA, DCM, rt, 30 min
2. NaH (2.0 eq), THF, rt
95-99% yield
93-98% ee

Scheme 5.29. Rhodium-catalyzed asymmetric arylation–cyclization.

5.5. Asymmetric Friedel–Crafts Reaction of Isatins

The asymmetric Friedel–Crafts reaction has emerged as a powerful carbon–carbon bond formation reaction (for selected reviews on

Table 5.13. 9-*O*-benzylcupreine catalyzed Friedel–Crafts reaction of isatin derivatives.

R^1 = H, Me
R^2 = H, 5-Cl, 5-Br, 5-MeO
R = H, 5-Br, 5-MeO, 5-NO$_2$, 2-Me

Sr. no.	Catalyst/reaction conditions	Yield (%)	ee (%)
1.	**Cupreine LXIII** (20 mol%), PhCOOH (20 mol%), 1,4-dioxane, rt, 4–5 d	68–97	76–91
2.	**9-*O*-benzylcupreine LXIV** (15 mol%), THF, MS, rt, 96 h	88–99	80–99

organocatalytic asymmetric Friedel–Crafts reaction, see Ref. [88]). Wang *et al.* developed cupreine **LXIII** catalyzed Friedel–Crafts reaction of isatin derivatives with different indoles to furnish 3-indolyl-3-hydroxyindoles in 68–97% yield with 76–91% ee [89]. At the same time, Chimni and co-workers reported 9-*O*-benzylcupreine **LXIV** catalyzed Friedel–Crafts reaction of isatin derivatives with different indoles to afford the product in 88–99% yield with 80–99% ee. The bifunctional mode of activation of catalyst was demonstrated on the basis of several experiments (Table 5.13) [90].

The chiral Sc(III) and Ir(III) complexes were also used as catalysts for the reaction of indole with isatin derivatives. The product was obtained in 72–99% yield with 88–99% ee [91]. This methodology was further extended for the nucleophilic addition reaction of isatins with pyrrole derivatives catalyzed by indium(III)-inda-pybox complex **LXXVIII**. The products were obtained in 45–98% yield with 94 to >99% ee (Scheme 5.30).

The Friedel–Crafts reaction of 1-naphthol with isatins catalyzed by Cinchonidine thiourea **CXIV** was also reported. A variety of naphthol derivatives reacted well with isatins to afford the chiral product in 70–84% yield with 37–77% ee (Scheme 5.31) [92].

The Friedel–Crafts reaction of activated phenols and sesamol with isatins has also been developed employing cinchonidine thiourea as

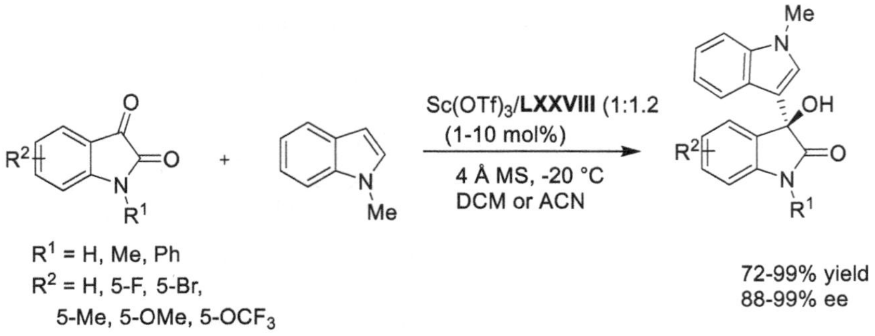

Scheme 5.30. Metal-catalyzed Friedel–Crafts reaction of indole with isatins.

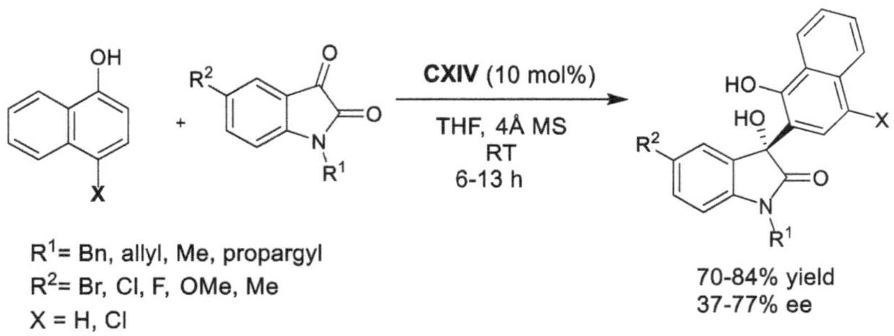

Scheme 5.31. CDT catalyzed Friedel–Crafts reaction of 1-naphthol with isatins.

catalyst **CXIV**. The chiral 3-aryl-3-hydroxy-2-oxindoles were obtained in 80–95% yield with 83–94% ee [93]. The obtained Friedel–Crafts adduct was further subjected to Cu(I) catalyzed azide-alkyne cycloaddition to afford biologically active 3-aryl-3-hydroxy-2-oxindoles having 1,2,3-triazole moiety in good enantiomeric excess (Scheme 5.32).

Wang and co-workers developed chiral tridentate Schiff-base/Cu complex **CXV** for the asymmetric Friedel–Crafts reaction of pyrrole with isatin derivatives. For *N*-substituted isatin derivatives, the Friedel–Crafts adduct were obtained in up to >99% ee while only moderate enantiomeric excess (56–69% ee) were obtained for *N*-H isatin derivatives (Scheme 5.33) [94].

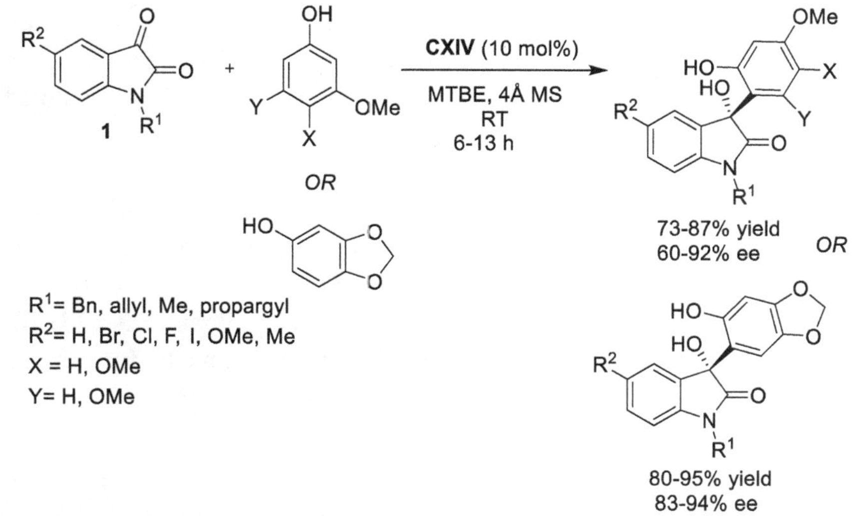

R2
O
O
N
1
R1
+
OH
Y
OMe
X
OR
HO
O
O
CXIV (10 mol%)
MTBE, 4Å MS
RT
6-13 h
OMe
HO
X
HO
R2
Y
O
N
R1
73-87% yield
60-92% ee
OR
R1= Bn, allyl, Me, propargyl
R2= H, Br, Cl, F, I, OMe, Me
X = H, OMe
Y= H, OMe
80-95% yield
83-94% ee

Scheme 5.32. Organocatalytic Friedel–Crafts reaction of activated phenols and sesamol with isatins.

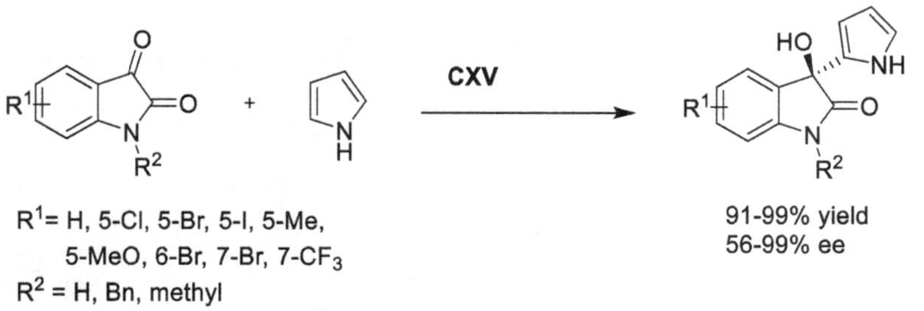

O
R1
O
N
R2
+
N
H
CXV
HO
NH
R1
O
N
R2
R1= H, 5-Cl, 5-Br, 5-I, 5-Me,
5-MeO, 6-Br, 7-Br, 7-CF3
R2 = H, Bn, methyl
91-99% yield
56-99% ee

Scheme 5.33. Chiral tridentate Schiff-base/Cu complex catalyzed asymmetric Friedel–Crafts reaction of pyrrole with isatin derivatives.

5.6. Asymmetric Hydrophosphonylation Reaction of Isatins

Phosphonic esters have received significant attention because of their important role as metabolic intermediates, regulatory switches for proteins and backbones for genetic information (for reviews on the enantioselective addition of P–H bond to C–C/C–X double bond, see Ref. [95]).

Wang *et al.* reported the first enantioselective hydrophosphonylation reaction of isatin derivatives with diphenylphosphite catalyzed by *Cinchona* alkaloids **CXIII** to provide bioactive 3-hydroxy-3-phospho-substituted oxindoles in 77–99% yield with 25–73% ee (Scheme 5.34) [96].

The Ti-salen complexes generated *in situ* using a series of chiral macrocyclic salen ligands **LXXXI** were also employed as catalysts for enantioselective hydrophosphonylation of isatins. The corresponding products were isolated in 80–90% yield with 74 to >99% ee [97]. Same reaction was reported using novel fluorous *bis*(oxazoline) as a ligand **LXXXII** for catalyzing asymmetric hydrophosphonylation of isatins to furnish the product in 69–91% yield with 66–92% ee (Table 5.14) [98].

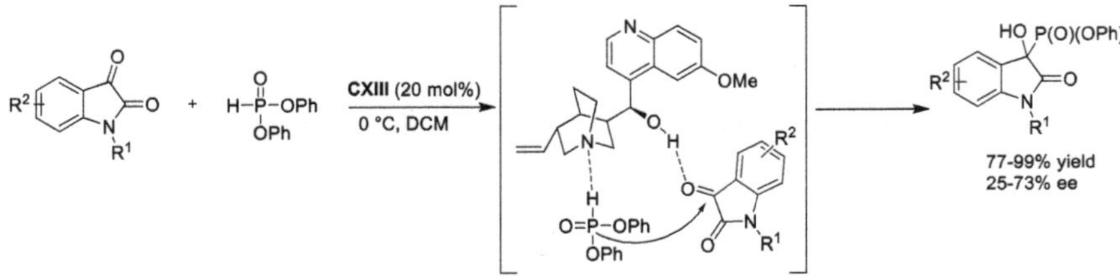

Scheme 5.34. *Cinchona* alkaloids catalyzed enantioselective hydrophosphonylation reaction of isatin derivatives.

Table 5.14. Chiral macrocyclic salen ligands catalyzed enantioselective hydrophosphonylation reaction of isatins.

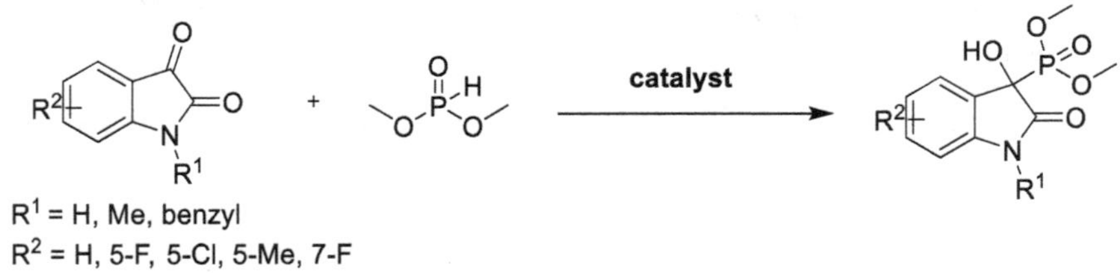

R^1 = H, Me, benzyl
R^2 = H, 5-F, 5-Cl, 5-Me, 7-F

Sr. no.	Catalyst/reaction conditions	Yield (%)	ee (%)
1.	**Macrocyclic Ti(IV)-salen complex as ligand LXXXI** (2.5 mol%), Ti(O*i*Pr)$_4$ (5 mol%), toluene, rt, 20–30 h	80–90	74 to >99
2.	**Fluorous *bis*(oxazoline) LXXXII** (10 mol%), Cu(OAc)$_2$ (10 mol%), KF (10 mol%), MeOH (1 mL)	69–91	66–92

5.7. Asymmetric Morita–Baylis–Hillman Reaction

The MBH reaction is one of the most important and useful carbon–carbon bond forming reactions for the synthesis of β-hydroxycarbonyl compounds with an α-alkylidene group [99]. Zhou et al. disclosed the first catalytic enantioselective MBH reaction of isatin derivatives with acrolein catalyzed by β-isocupreidine **LXXXIII** to furnish enantiomerically enriched 3-substituted-3-hydroxyoxindoles in 65–97% yield with 90–98% ee (Scheme 5.35) [100]. The synthesized product served as a precursor for the synthesis of hydroxyfuroindoline and chiral lactone.

The application of β-isocupreidine as a catalyst **LXXXIII** for asymmetric MBH reaction of acrylate with isatin derivatives to provide a quaternary hydroxylated stereocenter on the oxindole with 70–99% yield and 87–94% ee was also reported (Scheme 5.36) [101]. Lu and co-workers also reported the similar transformation that provides adduct in 61–96% yield with 86–96% ee catalyzed by 10 mol% of β-isocupreidine [102].

Wu and co-workers have also studied the MBH reaction of acrylates with isatin derivatives employing bifunctional phosphinothiourea organocatalyst based on a chiral cyclohexane scaffold **LXXVII**. The reaction

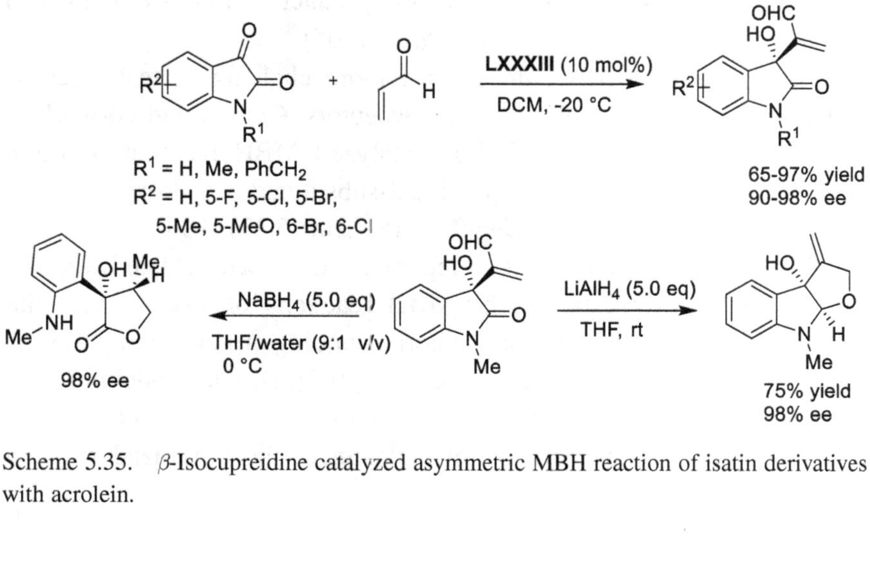

Scheme 5.35. β-Isocupreidine catalyzed asymmetric MBH reaction of isatin derivatives with acrolein.

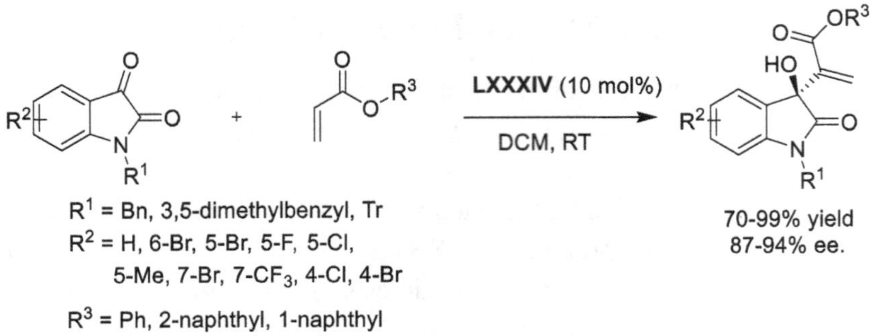

R^1 = Bn, 3,5-dimethylbenzyl, Tr
R^2 = H, 6-Br, 5-Br, 5-F, 5-Cl,
      5-Me, 7-Br, 7-CF$_3$, 4-Cl, 4-Br

R^3 = Ph, 2-naphthyl, 1-naphthyl

70-99% yield
87-94% ee.

Scheme 5.36. β-Isocupreidine catalyzed enantioselective MBH reaction of different isatins with acrylates.

proceeded smoothly in the presence of 10 mol% of catalyst to furnish MBH adduct 82–99% yield with 8–69% ee [103]. The authors have also developed a new bifunctional phosphine organocatalyst bearing squaramide as H-bond donor in order to increase the enantiomeric excess of the reaction. The phosphine-squaramide **LXXXV** catalyzed the reaction of isatins with acrylates providing adduct in 77–99% yield and 71–95% ee. The squaramide catalyst was found to be superior than the thiourea catalyst for this transformation (Table 5.15) [104].

The same reaction when catalyzed by tunable bifunctional phosphine squaramide catalyst **LXXXVI**, provided product in 74–93% yield with 87–95% enantiomeric excess (Table 5.15) [105].

Maleimide as a MBH donor is a more challenging task because maleimides are traditionally Michael acceptors. Chimni and co-workers reported β-isocupreidine **LXXXIII** catalyzed MBH reaction of isatin derivatives with maleimides to provide 3-substituted-3-hydroxyoxindoles in up to 96% yield with up to >99% ee (Scheme 5.37) [106].

Recently, Chen and co-workers reported that β-Isocupreidine **LXXXIII** can also efficiently catalyze the MBH reactions of 7-azaisatins with maleimide derivatives affording the corresponding products in up to 98% yield with up to 98% enantiomeric excess [107]. Besides maleimide, the other activated alkenes such as methyl and ethyl acrylates and acrolein were also found to be equally tolerable under the optimized reaction

Table 5.15. β-Isocupreidine catalyzed enantioselective MBH reaction of different isatins with acrylates.

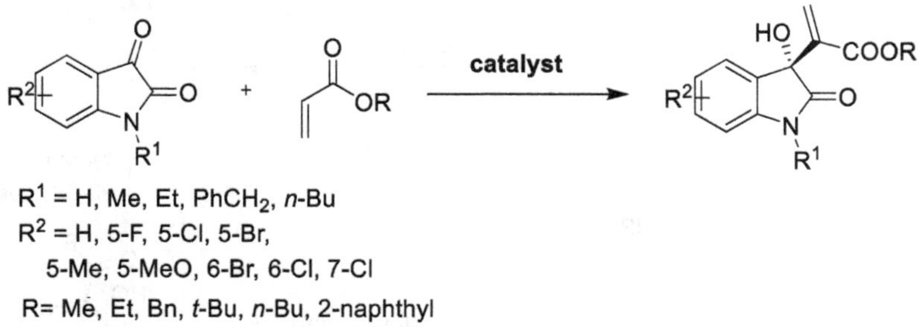

catalyst
COOR
R1 = H, Me, Et, PhCH2, n-Bu
R2 = H, 5-F, 5-Cl, 5-Br,
5-Me, 5-MeO, 6-Br, 6-Cl, 7-Cl
R= Me, Et, Bn, t-Bu, n-Bu, 2-naphthyl

Sr. no.	Catalyst/reaction conditions	Yield (%)	ee (%)
1.	**Phosphinothiourea LXXVII** (10 mol%), THF, 0°C	82–99	8–69
2.	**4-nitrosubstituted-Phosphine-squaramide LXXXV** (2 mol%), EtOAc, 25°C	77–99	71–95 (*S*)
3.	**Bifunctional phosphine-squaramide organocatalyst LXXXVI** (5 mol%), EtOAc, rt	74–93	87–95

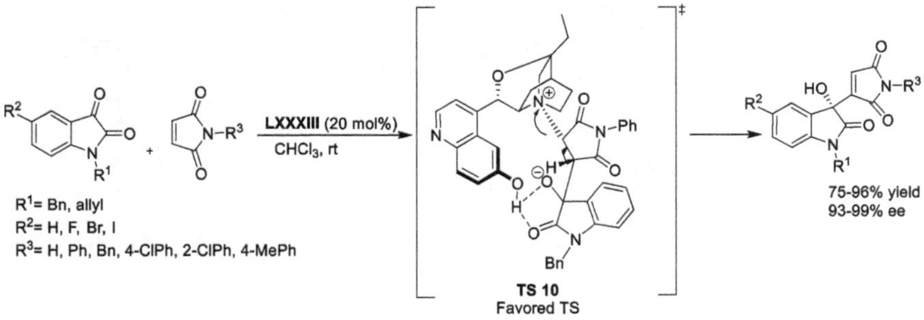

LXXXIII (20 mol%)
CHCl3, rt
R1= Bn, allyl
R2= H, F, Br, I
R3= H, Ph, Bn, 4-ClPh, 2-ClPh, 4-MePh
TS 10
Favored TS
75-96% yield
93-99% ee

Scheme 5.37. β-Isocupreidine catalyzed MBH reaction of isatin derivatives with maleimides.

conditions providing the corresponding products in excellent yields and with excellent enantiomeric excess. The authors found 7-azaisatins are better electrophiles than isatins that can be used to access bioactive molecules (Scheme 5.38).

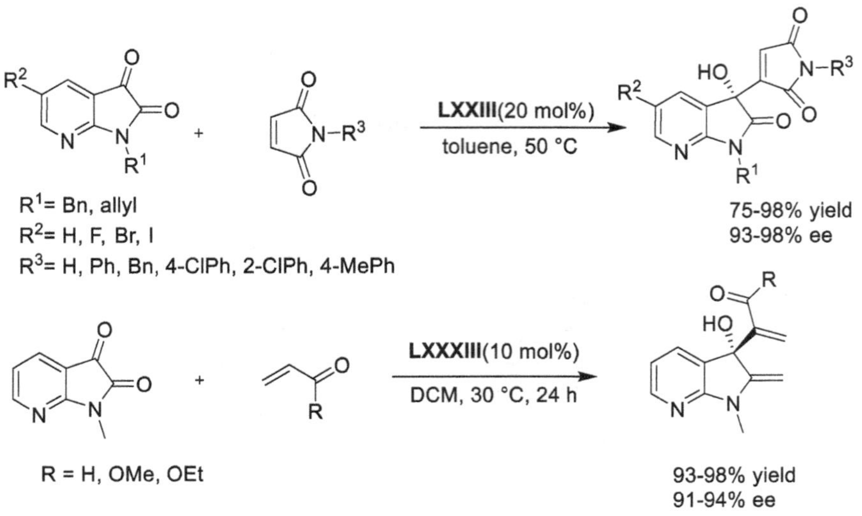

Scheme 5.38. *β*-isocupreidine catalyzed MBH reaction of 7-azaisatins with maleimide derivatives.

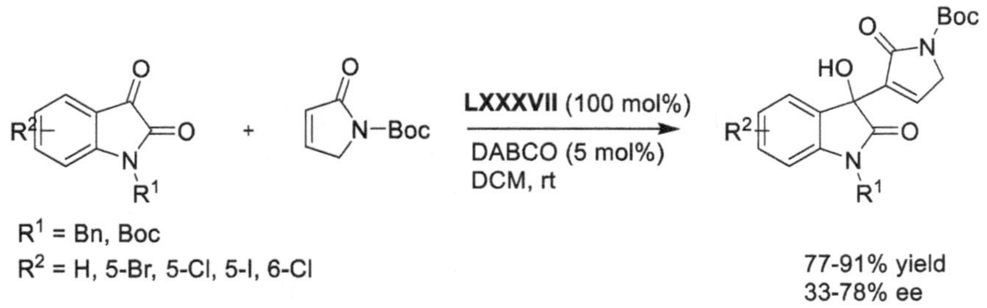

Scheme 5.39. *Bis*-thiourea catalyst catalyzed asymmetric MBH reaction of isatin derivatives with α,β-unsaturated γ-butyrolactam.

Pan *et al.* disclosed *bis*-thiourea catalyst **LXXXVII** catalyzed asymmetric MBH reaction of isatin derivatives with α,β-unsaturated γ-butyrolactam to afford the oxindole derivatives in up to 91% yield and 78% ee (Scheme 5.39) [108].

5.8. Asymmetric Michael Reaction of Isatin

The Michael addition reaction plays an important role among the numerous carbon–carbon bond forming reactions and has attracted much

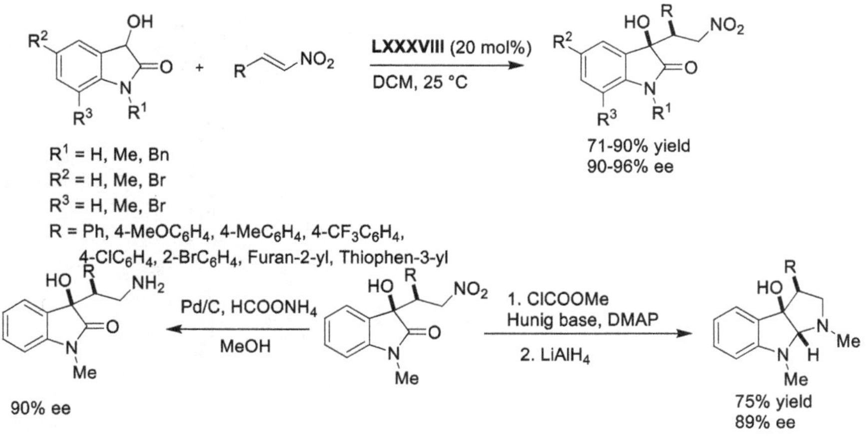

R^1 = H, Me, Bn
R^2 = H, Me, Br
R^3 = H, Me, Br
R = Ph, 4-MeOC$_6$H$_4$, 4-MeC$_6$H$_4$, 4-CF$_3$C$_6$H$_4$,
4-ClC$_6$H$_4$, 2-BrC$_6$H$_4$, Furan-2-yl, Thiophen-3-yl

71-90% yield
90-96% ee

90% ee

Pd/C, HCOONH$_4$
MeOH

1. ClCOOMe
Hunig base, DMAP
2. LiAlH$_4$

75% yield
89% ee

Scheme 5.40. Organocatalytic enantioselective Michael addition of dioxindoles to β-substituted nitroalkene derivatives.

attention (for recent reviews of asymmetric Michael addition reactions, see Ref. [109]). Melchiorre and co-workers reported an organocatalytic enantioselective Michael addition of dioxindoles to β-substituted nitroalkene derivatives that provides a new approach to the synthesis of optically active 3-substituted-3-hydroxy-2-oxindole derivatives. The bifunctional primary amine thiourea **LXXXVIII** catalyzed the reaction to furnish 3-substituted-3-hydroxy-2-oxindoles in 71–90% yield and up to 3.6:1 d.r. and up to 96% ee, while minimizing the formation of isatide [110]. The synthesized adduct can be further transformed into a product which bears the hexahydropyrrolo[2,3-b]indole unit found in many natural molecules (Scheme 5.40).

5.9. Decarboxylative Addition to Isatins

Shibata *et al.* reported organocatalytic enantioselective decarboxylative addition of malonic acid half thioesters to isatin derivatives employing squaramide as a catalyst **LXXXIX** providing products in 91–99% yield with 46–92% ee [111]. The synthetic utility of this methodology has been demonstrated in the first enantioselective synthesis of (2)-flustraminol B (Scheme 5.41).

Pan and co-workers disclosed Yb(OTf)$_3$/**XC**-catalyzed enantioselective decarboxylative addition reaction of isatin derivatives with different

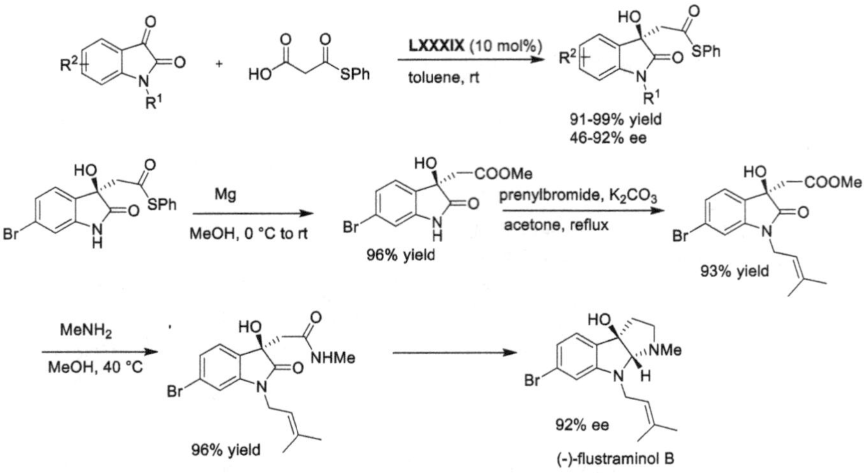

Scheme 5.41.   Organocatalytic enantioselective decarboxylative addition of malonic acid half thioesters to isatin derivatives.

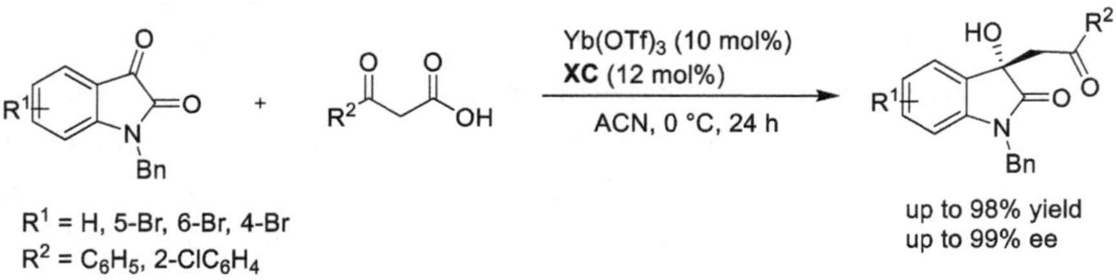

Scheme 5.42.   Yb(OTf)$_3$-catalyzed enantioselective decarboxylative addition reaction of isatin derivatives with β-ketoacids.

β-ketoacids. The chiral 3-hydroxyoxindoles were obtained in up to 98% yield and up to 99% ee. The use of 4 Å molecular sieves were found to enhance the stereoselectivity. The aliphatic β-ketoacids has also been used in the reaction that furnish the product in good results under the optimized conditions (Scheme 5.42) [112].

Lu research group reported enantioselective decarboxylative addition reaction of *N*-Boc isatins with β-ketoacids employing 10 mol% of *Cinchona* derived thiourea as catalyst **CXIV** to afford 3-substituted-3-hydroxy-2-oxindole derivatives in 82–96% yield with 82–97% enantiomeric excess [113]. The reaction with unprotected isatin gave the product

Table 5.16.   *Cinchona* thiourea catalyzed enantioselective decarboxylative addition reaction of *N*-Boc isatins with β-ketoacids.

R^1 = H, 5-Me, 5-OMe, 5-Cl, 5-Br, 5-NO$_2$, 7-F
R^2 = C$_6$H$_5$, 2-ClC$_6$H$_4$, 4-FC$_6$H$_4$, 3-ClC$_6$H$_4$, 2-OMeC$_6$H$_4$, 2-naphthyl, 2-thiophenyl

Sr. no.	Catalyst/reaction conditions	Yield (%)	ee (%)
1.	**Cinchonidine thiourea CXIV** (10 mol%), CHCl$_3$, 0°C, 24h	82–96	82–97
2.	**Binaphthyl-modified squaramide XCI** (1 mol%), CHCl$_3$, 0°C	75–98	89–97

in 93% yield, but with 5% enantiomeric excess. Kim and co-workers also reported the same reaction catalyzed by chiral bifunctional squaramide as organocatalyst **XCI** that provides adduct in 75–98% yield with 89–97% enantiomeric excess (Table 5.16) [114].

The asymmetric decarboxylative cyanomethylation of isatins with cyanoacetic acid using the L-Proline-derived thiourea as the catalyst **XCII** was also reported that provides the products in up to 82% yield and up tp 99% ee [115]. The different isatins were able to react with cyanoacetic acid to afford adduct in good stereoselectivities (Table 5.17). The copper catalyzed asymmetric reaction of acetonitrile with isatin derivatives using the Cu(OTf)$_2$ as the catalyst in the presence of K$_2$CO$_3$ and a fluorinated *bis*(oxazoline) **XCIII** as the ligand to afford the products in up to 66% yield and with up to 92% ee [116]. The authors recovered the fluorinated *bis*(oxazoline) **XCIII** and reused it for a few times without any activity loss.

5.10. Hydroxylation Reaction

Itoh and co-workers reported organocatalytic hydroxylation reaction of oxindoles using molecular oxygen as an oxidant catalyzed by

Table 5.17. L-Proline-derived thiourea and copper catalyzed asymmetric cyanomethylation of isatins.

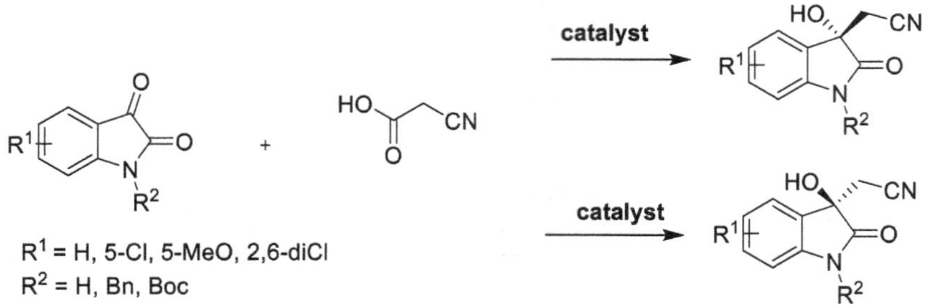

catalyst
catalyst
HO
CN
R1
R2
O
R1 = H, 5-Cl, 5-MeO, 2,6-diCl
R2 = H, Bn, Boc

Sr. no.	Catalyst/reaction conditions	Yield (%)	ee (%)
1.	**L-Proline derived thiourea XCII** (5 mol%), MTBE, 25°C	72–82	70–99
2.	**Fluorous *bis*(oxazoline) as ligand XCIII** (10 mol%), K_2CO_3, $Cu(OTf)_2$ (10 mol%), CAN (1 mL), 100°C	40–66	64–92

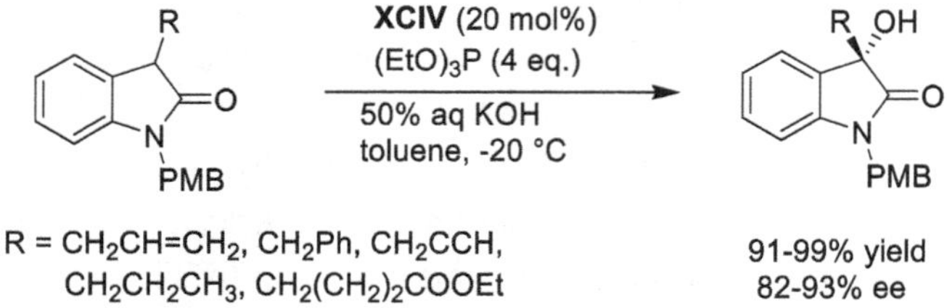

XCIV (20 mol%)
(EtO)3P (4 eq.)
50% aq KOH
toluene, -20 °C
PMB
R = CH2CH=CH2, CH2Ph, CH2CCH,
CH2CH2CH3, CH2(CH2)2COOEt
91-99% yield
82-93% ee

Scheme 5.43. PTC catalyzed organocatalytic hydroxylation reaction of oxindoles.

phase-transfer catalyst (PTC) **XCIV** to provide the adduct in 91–99% yield and 82–93% ee [117]. The reaction with 3-propyl-2-oxindole derivative resulted in the formation of adduct in lower enantiomeric excess of 67%. The obtained adduct was further transformed into optically active 3-allyl-3-hydroxy-2-oxindole, a synthetic precursor of alkaloid CPC-1 (Scheme 5.43).

The first catalytic enantioselective hydroxylation reaction of 3-aryl-2-oxindoles using the DBFOX-Zn(II) (cat-Zn(II)) **CXVI** catalyst with oxaziridine as oxidant was also reported that furnished the chiral 3-aryl-3-hydroxy-2-oxindoles in 76–95% yield and 90–97% ee. The use of

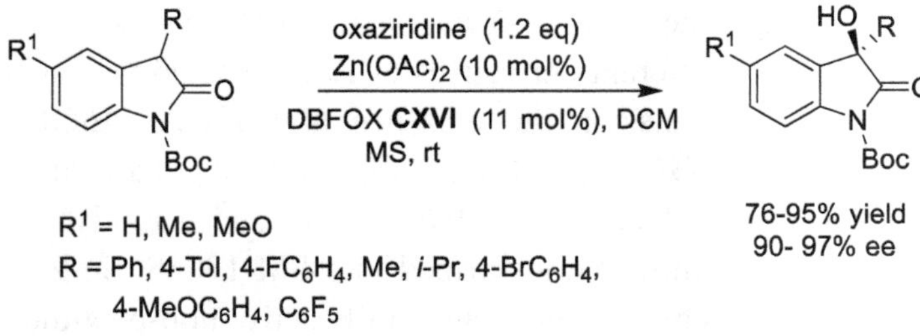

R
R1
O
N
Boc
oxaziridine (1.2 eq)
Zn(OAc)2 (10 mol%)
DBFOX CXVI (11 mol%), DCM
MS, rt
HO
R
R1
O
N
Boc
R1 = H, Me, MeO
R = Ph, 4-Tol, 4-FC6H4, Me, i-Pr, 4-BrC6H4,
4-MeOC6H4, C6F5
76-95% yield
90- 97% ee

Scheme 5.44. Enantioselective hydroxylation reaction of 3-aryl-2-oxindoles.

Table 5.18. Enantioselective hydroxylation of 3-methyl-2-oxindole using TMPO as the oxygen source.

R
R1
O
N
Boc
catalyst
HO
R
R1
O
N
Boc
R1 = H, Me, OMe
R = H, Ph, 4-CH3C6H4, 4-FC6H4,
4-BrC6H4CH2, 4-ClC6H4CH2, 4-OMeC6H4CH2

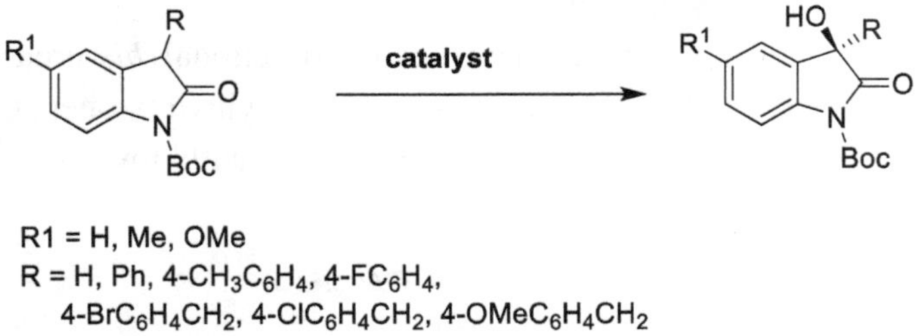

Sr. no.	Catalyst/reaction conditions	Yield (%)	ee (%)
1.	**Sc^{III}/*N,N'*-dioxide complex XCVI** (5 mol%), CH_3CCl_3, 30°C	66–92	88–95

3-alkyl-2-oxindoles provides chiral 3-alkyl-3-hydroxy-2-oxindoles in 68–90% yield and 83–86% ee (Scheme 5.44) [118].

Feng *et al.* have developed the enantioselective hydroxylation of 3-methyl-2-oxindole using TMPO as the oxygen source. The 3-methyl-3-hydroxy-2-oxindole was obtained in 92% yield and up to 95% ee by using $Sc(OTf)_3$/**XCVI** complex as a catalyst (Table 5.18) [119].

5.11. Addition of Organotrimethoxysilanes to Isatins

Shibasaki *et al.* developed intermolecular CuF-catalyzed enantioselective alkenylation and arylation of isatins using silicon-based nucleophiles that are generally more facile than boron-based nucleophiles, under CuF catalysis [120]. The different derivatives of isatin reacted well with alkenyl-/aryl-trimethoxysilanes in the presence of CuF_3PAr_22EtOH, chiral phosphine ligand **CXVII**, and additive ZnF_2 in toluene to afford tetrasubstituted-3-hydroxyoxindoles in 90–99% yield and 81–97% ee. The obtained adduct was further transformed into SM-130686 in 26% yield and 44% ee which is a highly potent and orally active non-peptidic growth hormone secretagogue (Scheme 5.45).

Shibasaki *et al.* disclosed enantioselective synthesis of 3-alkyl-3-hydroxy-2-oxindole *via* addition of diethyl zinc {$(CH_3)_2Zn$} to *N*-methylisatin catalyzed by **CXVIII** to furnish 3-methyl-3-hydroxy-1-methyl-2-oxindole in 82% yield with 76% ee (Scheme 5.46) [121].

Lassaletta and co-workers developed bifunctional *bis*-urea catalyst **XCVI** catalyzed reactions of isatins with *tert*-butylhydrazones to furnish 3-hydroxy-2-oxindoles in up to 99% yield but with low enantiomeric

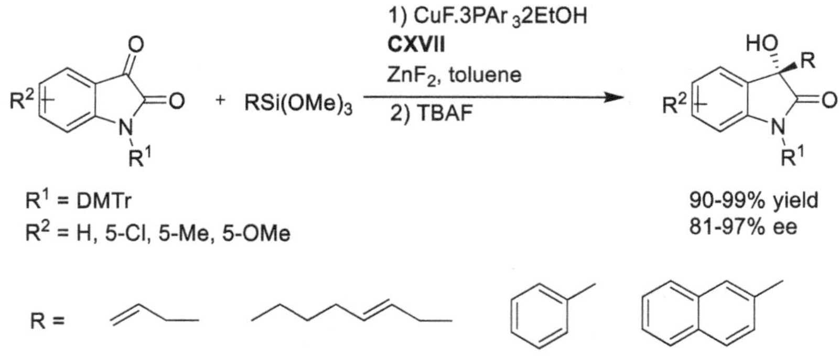

Scheme 5.45.   Intermolecular CuF-catalyzed enantioselective alkenylation and arylation of isatins.

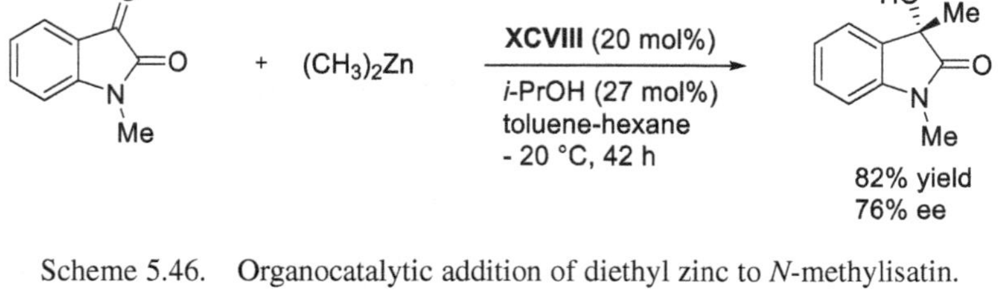

Scheme 5.46.   Organocatalytic addition of diethyl zinc to *N*-methylisatin.

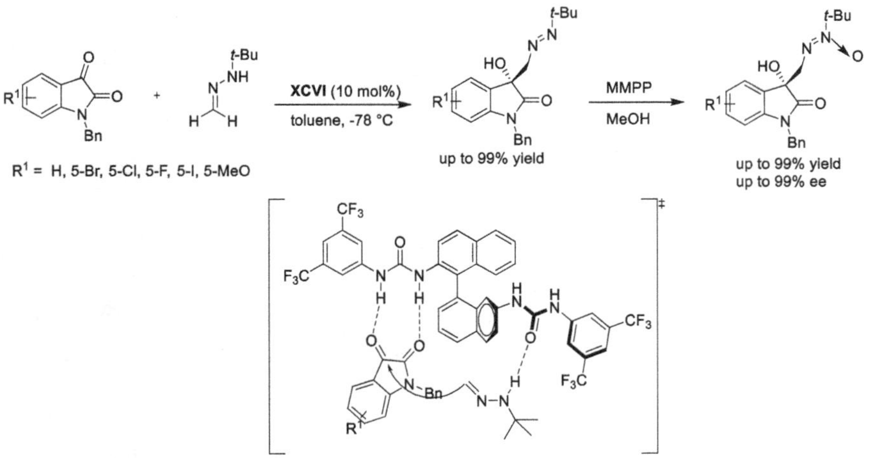

Scheme 5.47. Asymmetric reaction of tert-butyl hydrazones with isatins followed by oxidation.

excess, due to racemization of the adduct under the reaction conditions [122]. The treatment of the obtained adducts with magnesium monoperoxyphthalate (MMPP) provided the corresponding azoxy compounds in up to 99% yield and with up to 99% ee. A transition state model was proposed to gain insight into the reaction mechanism (Scheme 5.47).

5.12. Hetero-ene Reaction

Feng *et al.* reported hetero-ene reaction of isatin derivatives with alkyl enol ether employing a rare earth metal complex as catalyst to furnish chiral 3-substituted-3-hydroxyoxindole derivatives. The *in situ* generated complex by $Mg(OTf)_3$ along with ligand catalyzed the reaction to afford the product in 52–98% yield and 94–99% ee. The allylated product could be further hydrolyzed into product in 96% yield with 97% ee [123]. The reduction of adduct with $LiAlH_4$ furnished adduct in 92% yield with >99% ee, which is a common motif found in pyrrolidinoindoline-type natural products (Scheme 5.48).

The chiral dicationic Pd-complex catalyzed enantioselective hetero-ene of isatin derivatives was reported to afford optically active tertiary alcohols. The isopropenyloxy(tri-isopropyl)silane reacted efficiently with isatins

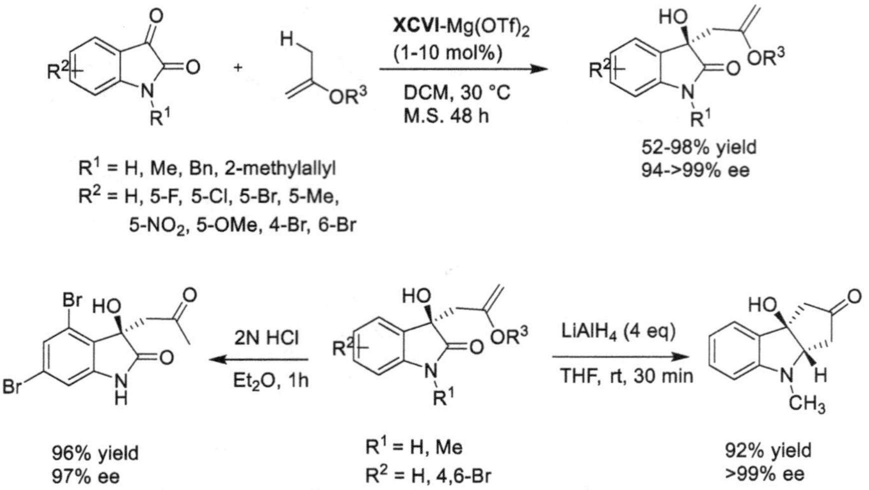

XCVI-Mg(OTf)2
(1-10 mol%)
DCM, 30 °C
M.S. 48 h
52-98% yield
94->99% ee
R1 = H, Me, Bn, 2-methylallyl
R2 = H, 5-F, 5-Cl, 5-Br, 5-Me,
5-NO2, 5-OMe, 4-Br, 6-Br
2N HCl
Et2O, 1h
LiAlH4 (4 eq)
THF, rt, 30 min
96% yield
97% ee
R1 = H, Me
R2 = H, 4,6-Br
92% yield
>99% ee

Scheme 5.48. Hetero-ene reaction of isatin derivatives with alkyl enol ether employing a rare earth metal complex as catalyst.

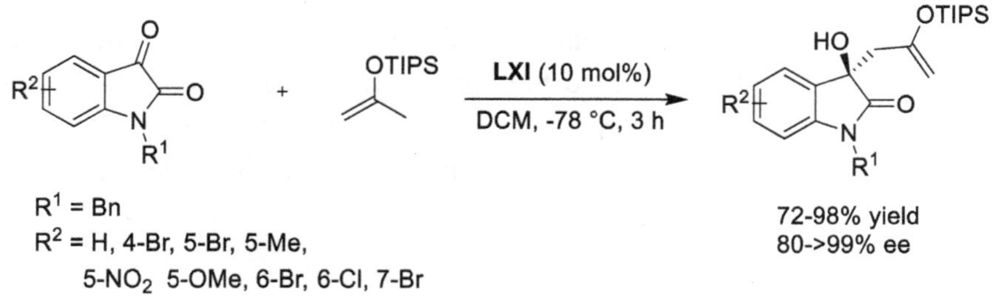

OTIPS
LXI (10 mol%)
DCM, -78 °C, 3 h
R1 = Bn
R2 = H, 4-Br, 5-Br, 5-Me,
5-NO2 5-OMe, 6-Br, 6-Cl, 7-Br
72-98% yield
80->99% ee

Scheme 5.49. Chiral dicationic Pd-complex catalyzed enantioselective hetero-ene of isatin derivatives.

in the presence of 10 mol% of Pd complex to provide the product in 72–98% yield and 80–99% ee. Further, this methodology was extended for the synthesis of adduct in 87% yield and 99% ee (Scheme 5.49).

5.13. Allylation, Crotylation and Prenylation of Isatins

Takayama and co-workers reported the allylation of isatin derivatives with tetraallylstannane employing catalytic mixture of (*R*)-(+)-BINOL

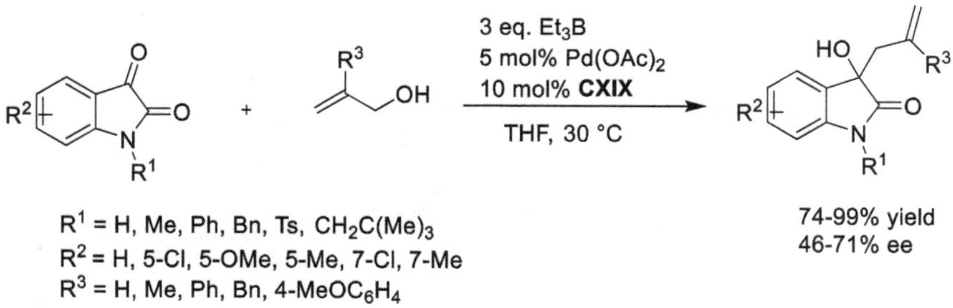

Scheme 5.50.   Asymmetric allylation of isatin derivatives with allylic alcohols.

and $Ti(OiPr)_4$ to provide allylated compounds in 89% yield with 42% ee [124].

Zhou *et al.* have developed the palladium catalyzed asymmetric allylation of isatin derivatives with allylic alcohols as an allyl donor in presence of chiral spiro phosphoramidite ligand **CXIX** to furnish chiral tertiary homoallylic alcohols in up to 99% yield and with 46–71% ee (Scheme 5.50) [125].

The highly enantioselective allylations, crotylations, and prenylations of isatin derivatives catalyzed by an Ir complex in the absence of stoichiometric amounts of allylmetal reagents was reported [126]. The allylations, crotylations, and prenylations of *N*-benzylisatin derivatives have been performed using cyclometalated C and O benzoate complex generated *in situ* from [{Ir(cod)Cl}$_2$]. The allylation of isatin derivatives using allyl acetate provided tertiary homoallyllic alcohols in 65–92% yield and 91–96% ee. The crotylation of different derivatives of isatin using α-methyl allyl acetate as the crotyl donor furnished the chiral product in 64–87% yield 80–92% ee. The reverse prenylation of isatin derivatives has been performed using 1,1-dimethylallene. The product with two contiguous quaternary carbon centers was obtained in 70–90% yield and 90–96% ee (Scheme 5.51).

Franz and co-workers developed metal catalyzed enantioselective Hosomi–Sakurai allylation of isatin derivatives with substituted allylic silanes [127]. The $Sc(OTf)_3$-indapybox catalyst **XC** with a TMSCl activator catalyzed the reaction to provide 3-substituted- 3-hydroxy-2-oxindoles in 72–99% yield and 80–99% ee (Scheme 5.52).

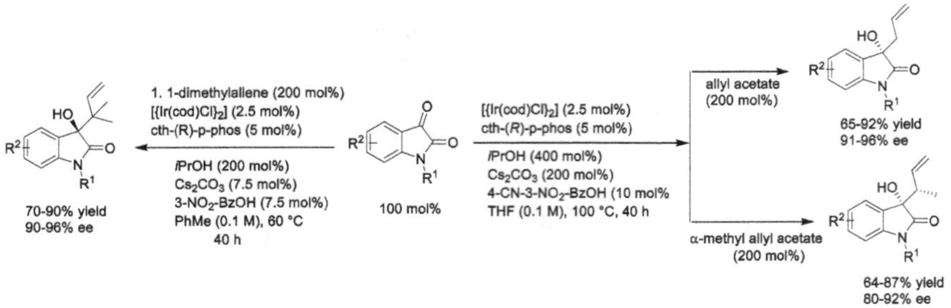

Scheme 5.51. Enantioselective allylations, crotylations, and prenylations of isatin derivatives.

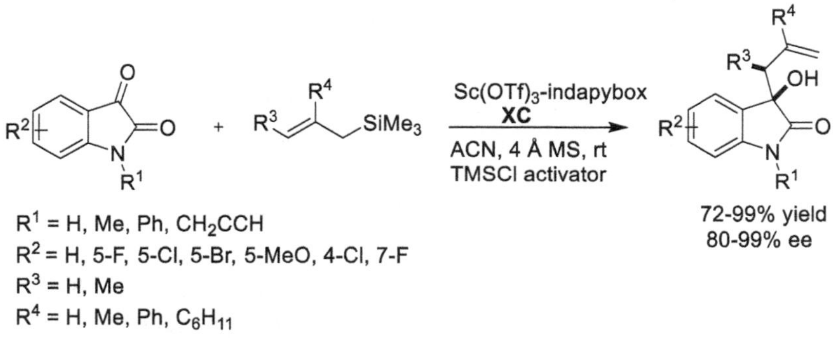

Scheme 5.52. Enantioselective Hosomi–Sakurai allylation of isatin derivatives.

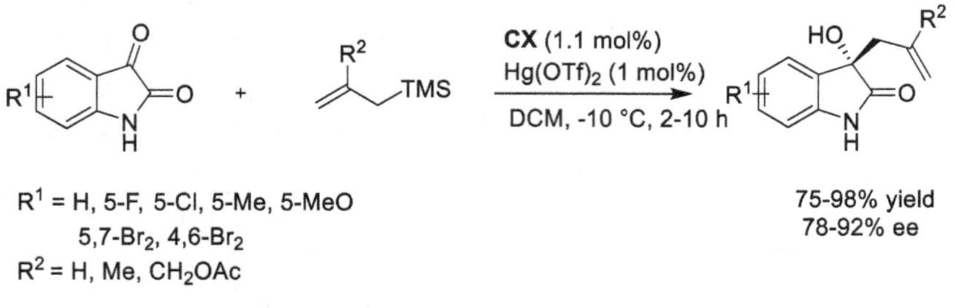

Scheme 5.53. Enantioselective allylation reaction of isatins.

Synthesis of 3-allyl-3-hydroxyoxindoles from the allylation reaction of isatins was reported using chiral complex derived from (S)-difluorophos **CX** and $Hg(OTf)_2$. The product was isolated in 75–98% yield with 78–92% ee (Scheme 5.53) [128].

Table 5.19. Enantioselective allylation reaction of isatins.

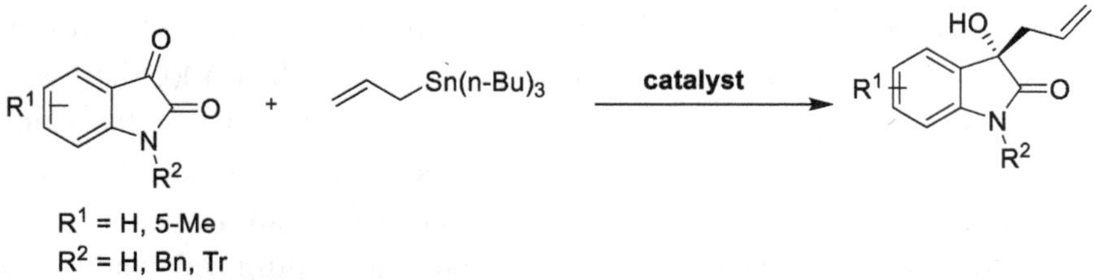

Sn(n-Bu)3
catalyst
HO
R1
R2
R1 = H, 5-Me
R2 = H, Bn, Tr

Sr. no.	Catalyst/reaction conditions	Yield (%)	ee (%)
1.	**Pincer Pd(II) complex CII** (5 mol%), DCM (1 mL), −60°C, 12h	89–95	32–85 (*S*)
2.	**(*R*)-2-amino-3-methyl-1,1-diphenylbutanol derived squaramide CIII** (2.5 mol%), DCM, −20°C	20–95	62–98 (*R*)

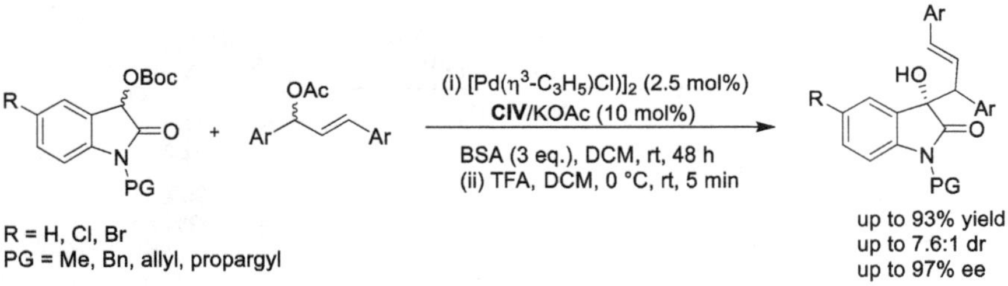

OBoc
R
PG
OAc
Ar
Ar
(i) [Pd(η3-C3H5)Cl)]2 (2.5 mol%)
CIV/KOAc (10 mol%)
BSA (3 eq.), DCM, rt, 48 h
(ii) TFA, DCM, 0 °C, rt, 5 min
HO
R = H, Cl, Br
PG = Me, Bn, allyl, propargyl
up to 93% yield
up to 7.6:1 dr
up to 97% ee

Scheme 5.54. Pd/*bis*(oxazoline) complex catalyzed asymmetric allylation of 3-O-Boc-oxindole.

Song and co-workers employed the chiral CNN (three atoms attached to the palladium) pincer Pd complex **CII** to catalyze the allylation of isatins with allyltributyltin. The product was obtained in 89–95% yield with 32–85% ee (Table 5.19) [129]. Same reaction was performed using squaramide **CIII** as a catalyst to furnish the allylation products in up to 95% and up to 98% ee under mild conditions [130].

Kesavan and co-workers disclosed Pd/*bis*(oxazoline) (**CIV**) complex catalyzed asymmetric allylation of 3-*O*-Boc-oxindole to furnish 3-allyl-3-hydroxyoxindoles in up to 93% yield with up to 97% ee and up to 7.6:1 d.r. (Scheme 5.54) [131].

An indium/PyBox complex-catalyzed enantioselective allylation of isatins using stannylated reagents was reported. The allylated product was formed in up to >99% and with up to 99% ee. The [In(**CV**)(OTf)$_3$] was used as a precatalyst in this reaction. The author found that the stannylated reagents that do not contain the secondary amide moiety gave low enantiomeric excess (≈22% ee). In addition to this, silyl reagents also failed to afford the desirable products under the optimized conditions (Scheme 5.55) [132].

The DuanPhos catalyst **CVI** in presence of $CoBr_2$ catalyzed the vinylation of isatins by vinyl boronic acids to afford the adduct in 50–90% yield with 90–97% ee (Scheme 5.56) [133].

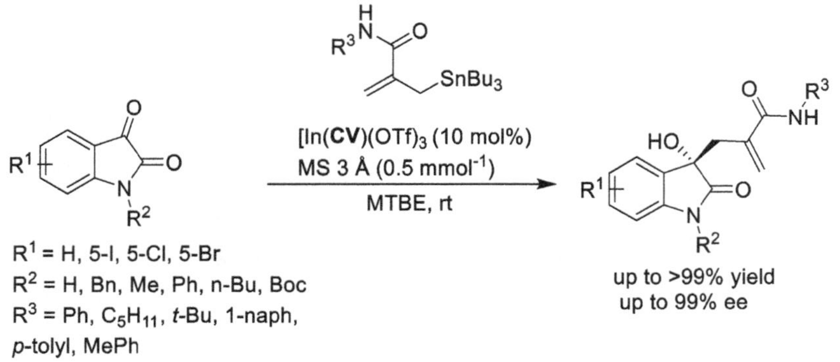

SnBu3
[In(CV)(OTf)3 (10 mol%)
MS 3 Å (0.5 mmol-1)
MTBE, rt
R1 = H, 5-I, 5-Cl, 5-Br
R2 = H, Bn, Me, Ph, n-Bu, Boc
R3 = Ph, C5H11, t-Bu, 1-naph,
p-tolyl, MePh
up to >99% yield
up to 99% ee

Scheme 5.55. In-catalyzed asymmetric allylation of isatins.

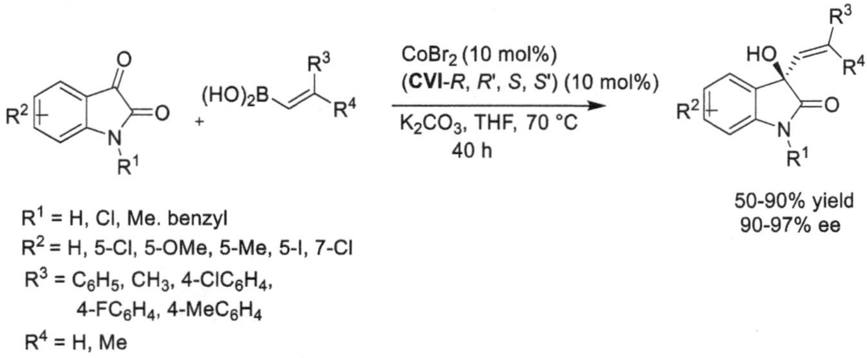

(HO)2B
CoBr2 (10 mol%)
(CVI-R, R', S, S') (10 mol%)
K2CO3, THF, 70 °C
40 h
R1 = H, Cl, Me. benzyl
R2 = H, 5-Cl, 5-OMe, 5-Me, 5-I, 7-Cl
R3 = C6H5, CH3, 4-ClC6H4,
4-FC6H4, 4-MeC6H4
R4 = H, Me
50-90% yield
90-97% ee

Scheme 5.56. Enantioselective vinylation of isatins by vinyl boronic acids.

The asymmetric synthesis of homoallylic alcohols *via* addition of allenylboron reagents to isatins was also reported. The 0.5% of catalyst loading and 0.5% NaO*t*-Bu were used for the reaction to afford the product in up to 98% yield and with up to 98.5:1.5 e.r. [134]. The author found that the phenolic hydroxy group of the catalyst played an important role in controlling the enantioselectivity by forming hydrogen bonds with reactants (Scheme 5.57) [135].

The alkynylation of isatin derivatives with alkynes was also reported. The bifunctional guanidine **CVIII**/CuI catalyzed the reaction to furnish products in 70–99% yield and 26–96% ee. Same reaction was reported using CuI and chiral phosphine ligand as catalyst to afford the product in 78–97% yield with 80–93% ee (Scheme 5.58) [136].

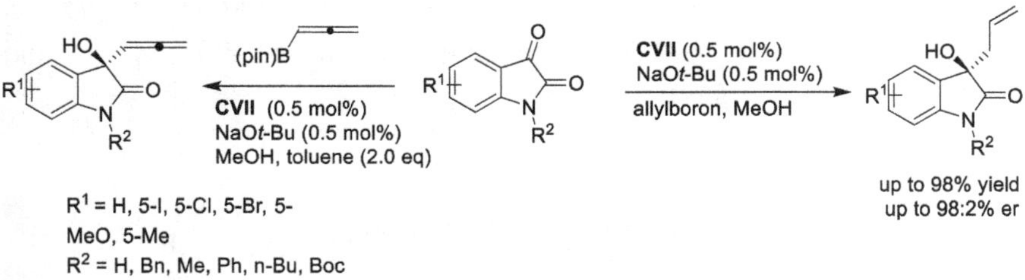

(pin)B
CVII (0.5 mol%)
NaOt-Bu (0.5 mol%)
MeOH, toluene (2.0 eq)
CVII (0.5 mol%)
NaOt-Bu (0.5 mol%)
allylboron, MeOH
up to 98% yield
up to 98:2% er
R1 = H, 5-I, 5-Cl, 5-Br, 5-MeO, 5-Me
R2 = H, Bn, Me, Ph, n-Bu, Boc

Scheme 5.57. Asymmetric addition of allenylboron reagents to isatins.

CVIII (2.5 mol%), CuI (3 mol%)
2,4,6-collidine (20 mol%)
Toluene, 30 °C, 3-4 days
70-99% yield
26-96% ee.
R1 = H, Bn
R2 = H, 5-Cl, 5-OMe, 5-Me, 5-I, 7-Cl, 7-I, 7-CF3, 3-ClC6H4, 4-MeC6H4, CH2C6H4, cyclopropyl
R3 = 2-FC6H4, 3-FC6H4, 4-FC6H4 3-MeC6H4, CH2OCOPh, cyclopropyl, cyclohexyl, propenyl

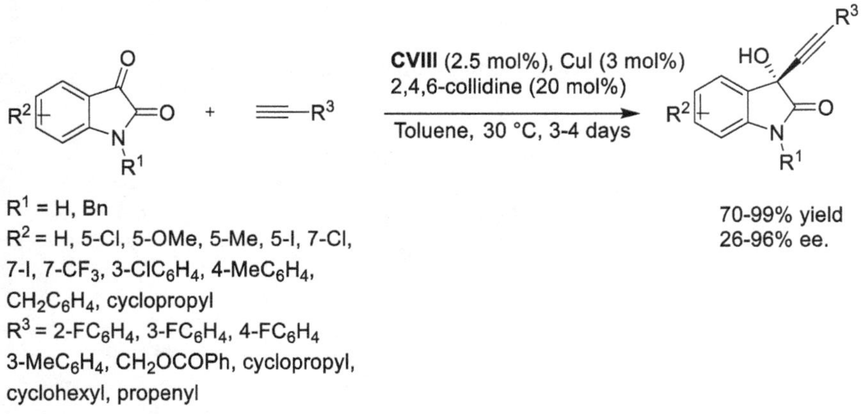

Scheme 5.58. Guanidine/CuI catalyzed alkynylation of isatin.

5.14. Anilide Cyclizations

The highly enantioselective intermolecular arylation of isatins was developed to obtain 3-hydroxy-3-aryl-2-oxindoles. This methodology suffered from limitation in the case of isatin derivatives with CF_3 group at the C-4 position that led to a dramatic decrease in the reactivity and enantioselectivity of the reaction. To overcome this limitation, the author developed an alternative strategy where a catalytic enantioselective intramolecular arylation of α-keto amides was carried out using Ph-BPE as a ligand to obtain product in 70–90% yield and 77–87% enantiomeric excess (Scheme 5.59).

The same group reported the palladium–difluorphos complex catalyzed intramolecular arylation of α-keto amides to furnish chiral 3-substituted-3-hydroxy-2-oxindoles [137]. The reaction was catalyzed by a chiral catalyst derived from (*R*)-DifluorPhos **CX** and $[Pd(CH_3CN)_4](BF_4)_2$ providing tetrasubstituted-3-hydroxyoxindoles in 55–93% yield and 82–99% ee (Scheme 5.60).

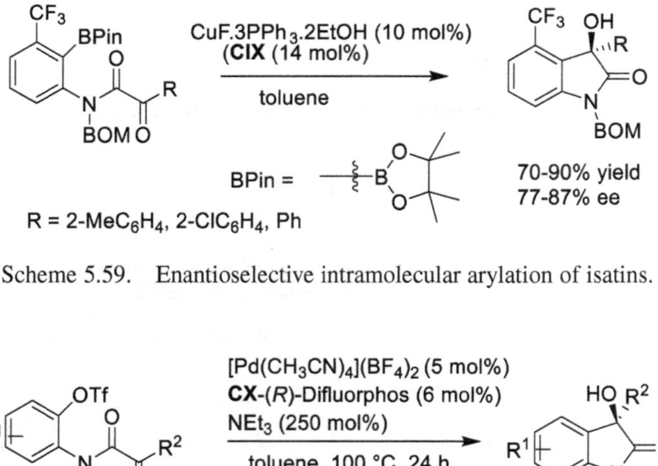

Scheme 5.59. Enantioselective intramolecular arylation of isatins.

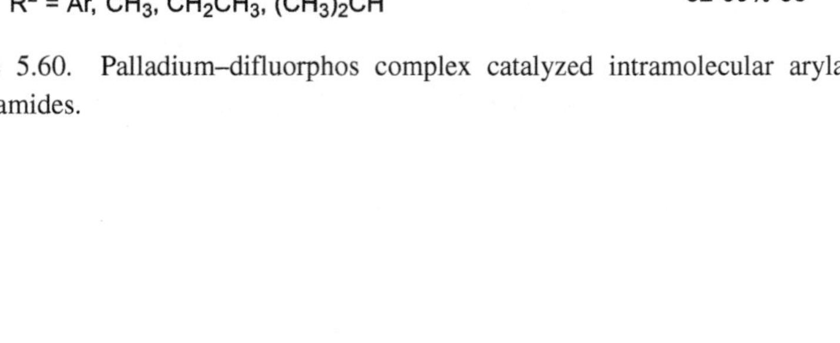

Scheme 5.60. Palladium–difluorphos complex catalyzed intramolecular arylation of α-keto amides.

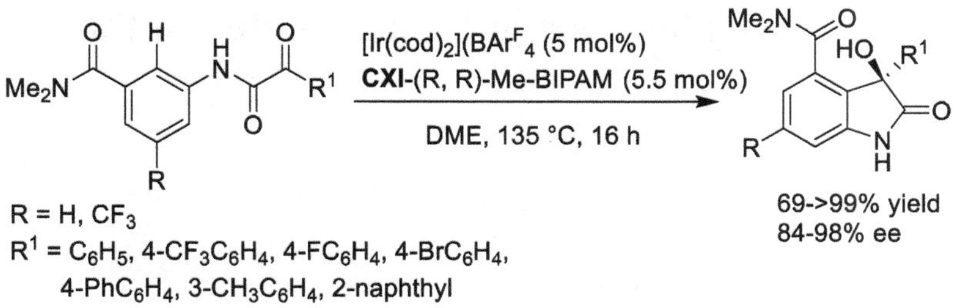

[Ir(cod)2](BArF4 (5 mol%)
CXI-(R, R)-Me-BIPAM (5.5 mol%)
DME, 135 °C, 16 h
R = H, CF3
R1 = C6H5, 4-CF3C6H4, 4-FC6H4, 4-BrC6H4, 4-PhC6H4, 3-CH3C6H4, 2-naphthyl
69->99% yield
84-98% ee

Scheme 5.61. Me-BIPAM catalyzed asymmetric intramolecular direct hydroarylation of α-ketoamides.

CXII (10 mol%)
THF, -78 °C
R1 = H, 5-F, 5-Cl, 5-Br, 5-MeO, 5-Me, 6-F
Ar = 4-MeC6H4, 4-FC6H4, 4-IC6H4, 4-SMeC6H4
R = Et, Me
59-98% yield
92:8-99:1 er
>20:1 dr

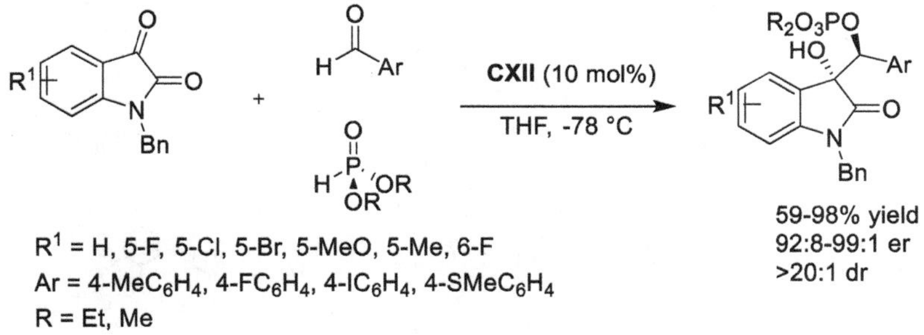

Scheme 5.62. Chiral iminophosphoranes catalyzed stereoselective reductive coupling reaction between isatins and aldehydes.

Yamamoto developed iridium complex $[Ir(cod)_2](BArF_4)$ and the chiral *O*-linked bidentate phosphoramidite (R,R)-Me-BIPAM **CXI** catalyzed asymmetric intramolecular direct hydroarylation of α-ketoamides to afford optically active 3-substituted-3-hydroxy-2-oxindoles in 69 to >99% yield with 84–98% ee (Scheme 5.61) [138].

A metal-free stereoselective reductive coupling reaction between isatins and aldehydes has been reported. Chiral iminophosphoranes **CXII** were used as organocatalyst to provide the product in 59–98% yield, 92:8–99:1 e.r. and >20:1 dr (Scheme 5.62) [139].

Appendix A. Catalysts and Ligands Cited in this Chapter

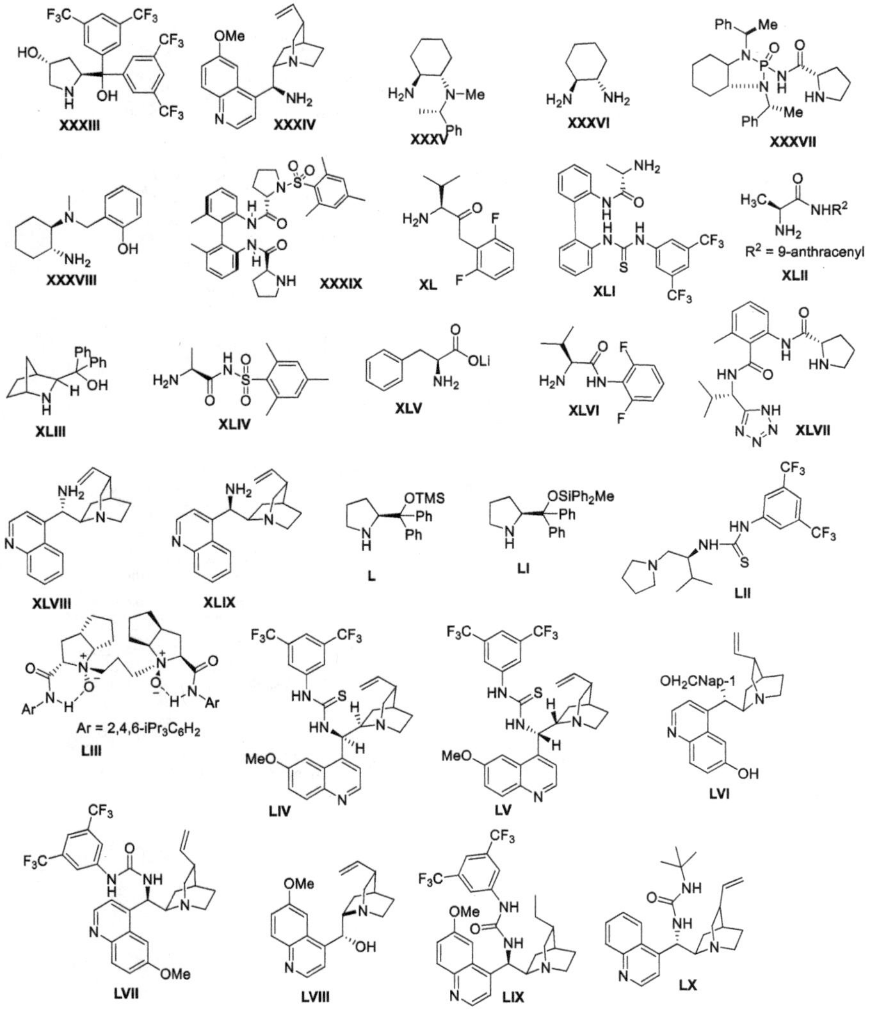

XXXIII

XXXIV

XXXV

XXXVI

XXXVII

XXXVIII

XXXIX

XL

XLI

XLII

R² = 9-anthracenyl

XLIII

XLIV

XLV

XLVI

XLVII

XLVIII

XLIX

L

LI

LII

LIII

Ar = 2,4,6-iPr₃C₆H₂

LIV

LV

LVI

LVII

LVIII

LIX

LX

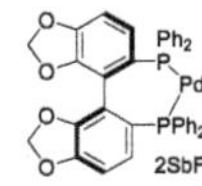

LXI   2SbF^{6-}

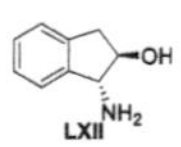

LXII

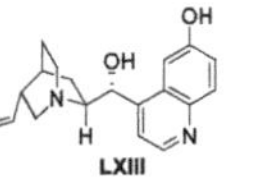

LXIII

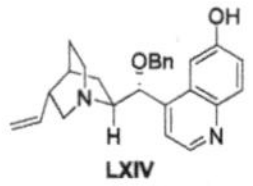

LXIV

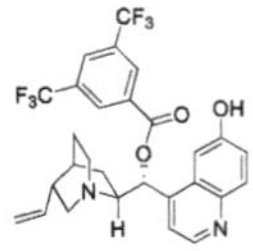

LXV

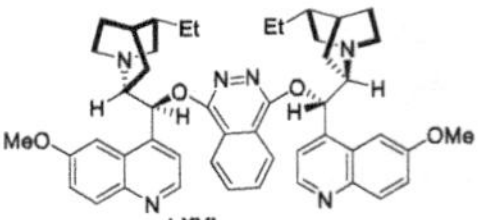

LXVI

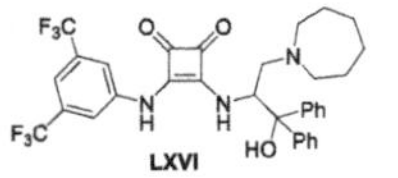

LXVII

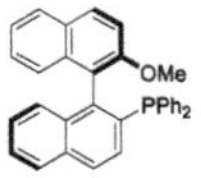

LXVIII

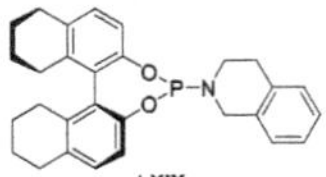

LXIX

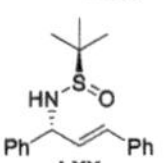

LXX

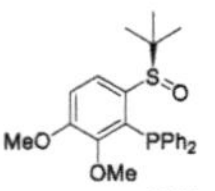

LXXI

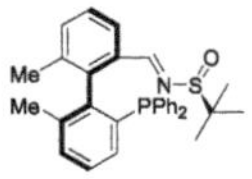

LXXII

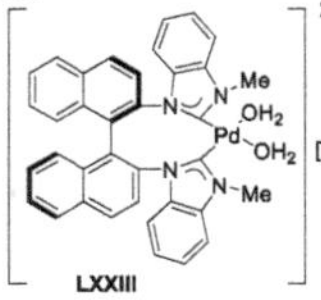

LXXIII   [TfOH]$^{2-}$

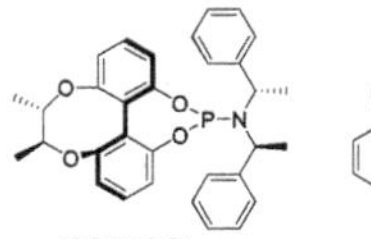

(*S*,*S*,*Ra*,*S*,*S*) LXXIV

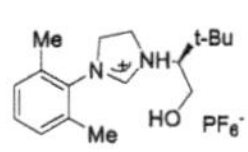

LXXV

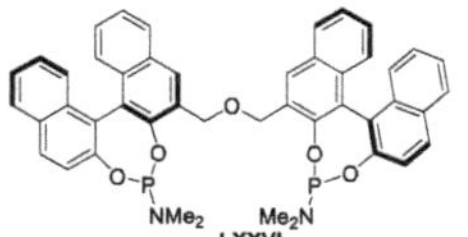

LXXVI

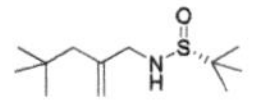

LXXVII

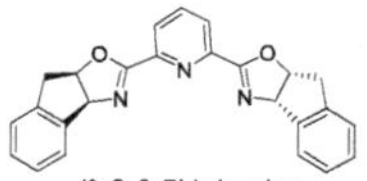

(3a*S*, 8a*R*)-inda-pybox LXXVIII

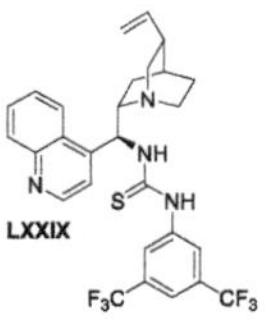

LXXIX

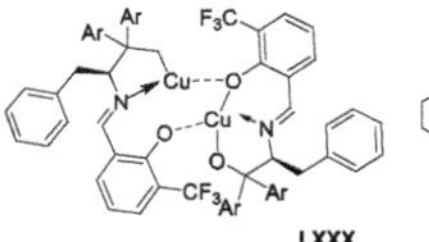

LXXX   Ar = *p*-Tolyl

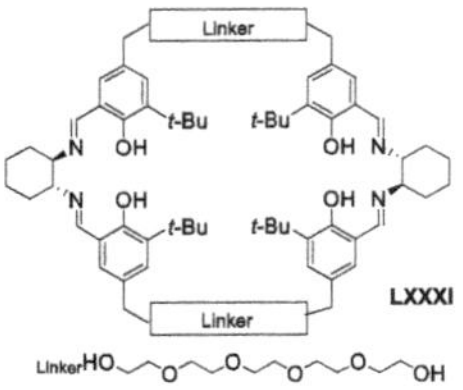

LXXXI

Linker HO⌒O⌒O⌒O⌒OH

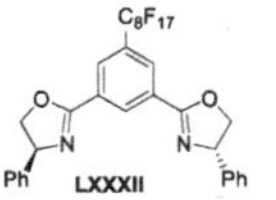

LXXXII

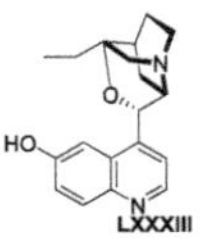

LXXXIII

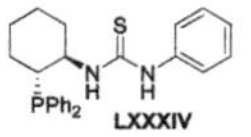

LXXXIV

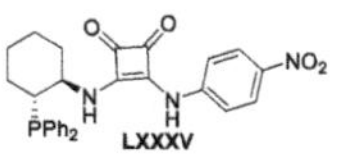

LXXXV

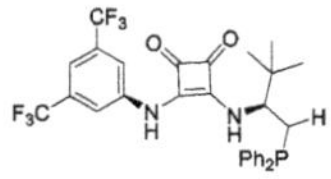

LXXXVI

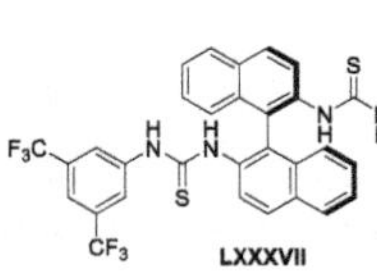

LXXXVII

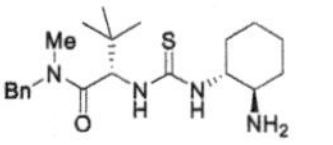

LXXXVIII

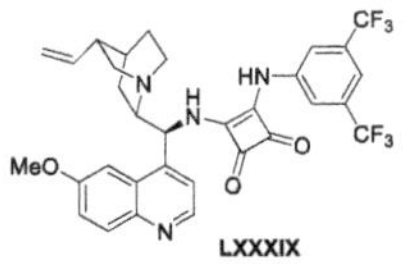

LXXXIX

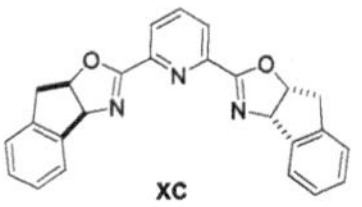

XC

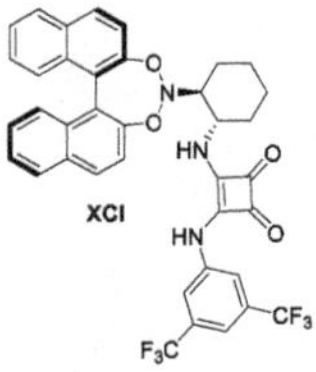

XCI

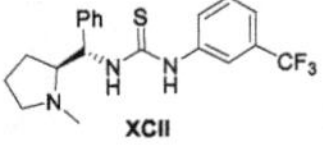

XCII

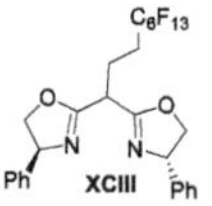

XCIII

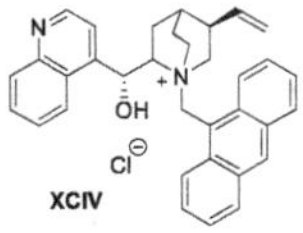

XCIV

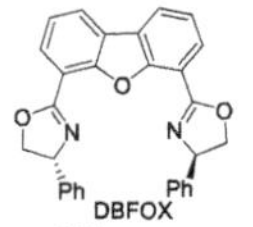

DBFOX
XCV

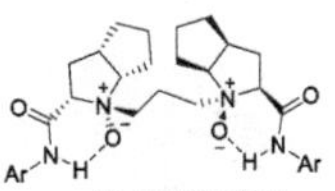

Ar = 2,4,6-iPr$_3$C$_6$H$_2$
XCVI

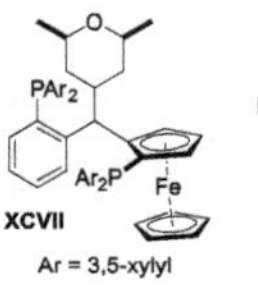

Ar = 3,5-xylyl
XCVII

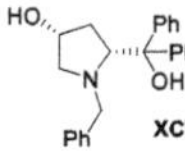

XCVIII

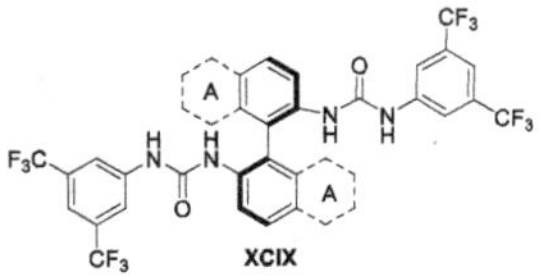

XCIX
A = phenyl ring, cyclohexyl ring

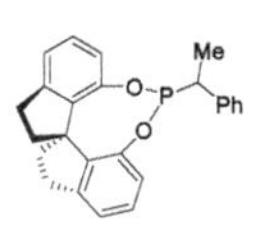

C

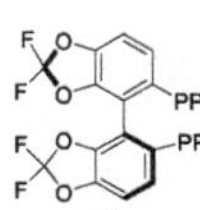

CI

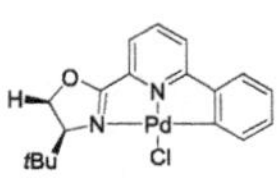

CII

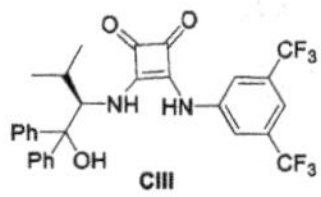

CIII

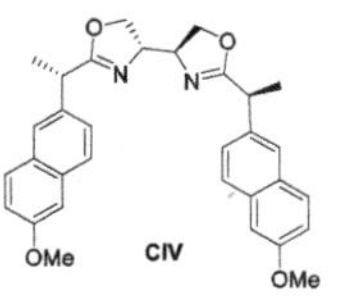

CIV

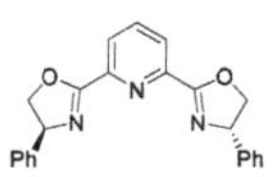

CV

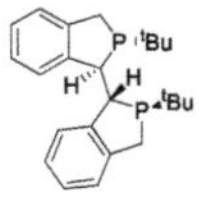

CVI

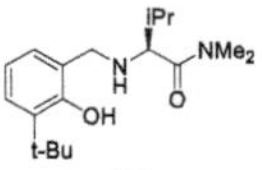

CVII

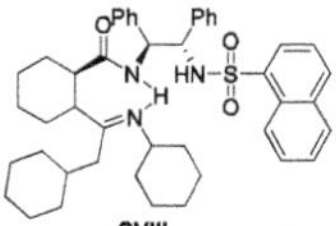

CVIII

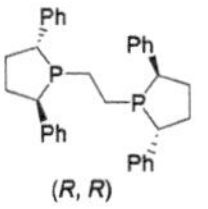

(R, R)
CIX

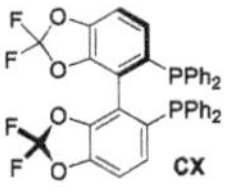

(R)-DifluorPhos
CX

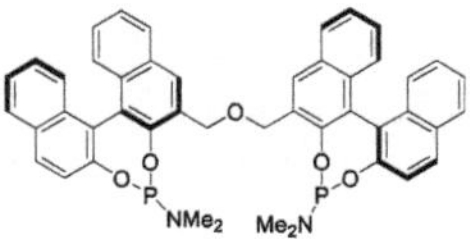

(R, R)-Me-BIPAM
CXI

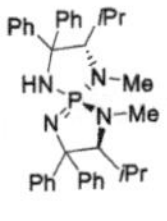

CXII

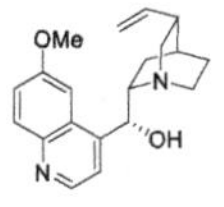

CXIII

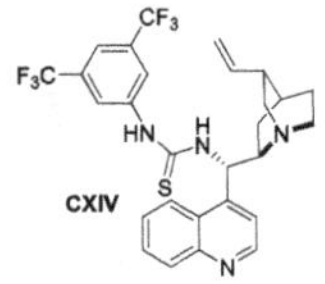

CXIV

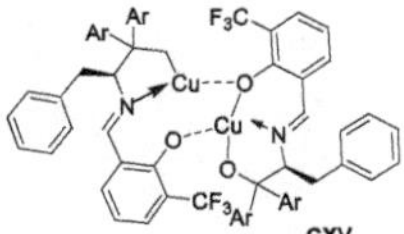

Ar = p-Tolyl   CXV

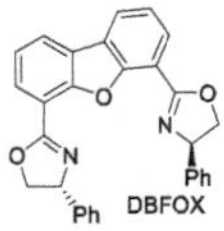

DBFOX
CXVI

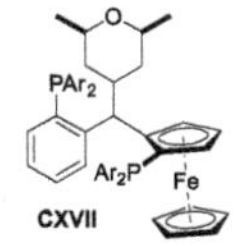

Ar$_2$P
CXVII
Ar = 3,5-xylyl

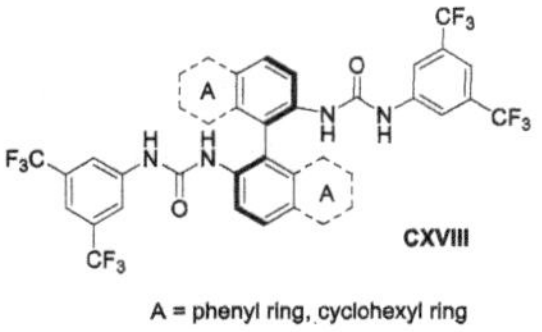

CXVIII
A = phenyl ring, cyclohexyl ring

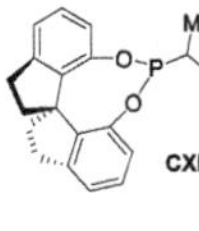

CXIX

Appendix B. List of Abbreviation

AcOH	Acetic acid
BINAP	2,2′-*Bis*(diphenylphosphino)-1,1′-binaphthyl
BIPAM	Bidentate phosphoramidite
Bn	Benzyl
Boc	*tert*-Butoxycarbonyl
Bz	Benzoyl (PhCO)
CD	Cinchonidine
CDT	Cinchonidine thiourea
CN	Cinchonine
CPN	Cupreine
dba	Dibenzylideneacetone
DACH	1,2-diaminocyclohexane
DBAD	Di-*tert*-butyl azodicarboxylate
DFT	Density functional theory
DIPEA	*N,N*-Diisopropylethaneamine
d.r.	Diastereomeric ratio
ee	Enantiomeric excess
ICD	Isocupreidine
MeOH	Methanol
MS	Molecular sieves
NFSI	*N*-Fluorobenzenesulfonimide
NHC	*N*-Heterocyclic carbene
Npt	Naphthyl
PG	Protecting group
PMP	4-Methoxyphenyl
PTC	Phase-transfer catalyst
QN	Quinine
TFA	Trifluoroacetic acid
TfO	Triflate (trifluoromethanesulfonate)
THF	Tetrahydrofuran
TMSCl	Trimethylsilyl chloride

References

1. (a) S. Peddibhotla, *Curr. Bioact. Compd.* **2009**, *5*, 20; (b) C. V. Galliford, K. A. Scheidt, *Angew. Chem. Int. Ed.* **2007**, *46*, 8748; (c) C. Marti, E. M. Carreira, *Eur. J. Org. Chem.* **2003**, 2209.

2. Y. Kamano, H.-P. Zhang, Y. Ichihara, H. Kizu, K. Komiyama, G. R. Pettit, *Tetrahedron Lett.* **1995**, *36*, 2783; (b) S. Miah, C. J. Moody, I. C. Richards, A. M. Z. Slawin, *J. Chem. Soc. Perkin Trans.* **1997**, *1*, 2405.

3. T. Kagata, S. Saito, H. Shigemori, A. Ohsaki, H. Ishiyama, T. Kubota, J. Kobayashi, *J. Nat. Prod.* **2006**, *69*, 1517.

4. Y. Q., Tang, I. Sattler, R. Thiericke, S. Grabley, X.-Z. Feng, *Eur. J. Org. Chem.* **2001**, 261.

5. M. Kitajima, I. Mori, K. Arai, N. Kogure, H. Takayama, *Tetrahedron Lett.* **2006**, *47*, 3199.

6. K. C. Nicolaou, P. B. Rao, J. L. Hao, M. V. Reddy, G. Rassias, X. H. Huang, D. Y. K. Chen, S. A. Snyder, *Angew. Chem. Int. Ed.* **2003**, *42*, 1753.

7. V. U. Khuzhaev, I. Zhalolov, K. K. Turgunov, B. Tashkhodzhaev, M. G. Levkovich, S. F. Aripova, A. S. Shashkov, *Chem. Nat. Compd.* **2004**, *40*, 269.

8. J. Kohno, Y. Koguchi, M. Nishio, K. Nakao, M. Juroda, R. Shimizu, T. Ohnuki, S. Komatsubara, *J. Org. Chem.* **2000**, *65*, 990.

9. (a) R. Mahrwald, Reactions, *Modern Aldol,* Wiley-VCH, Weinheim, **2004**. Vols. 1 and 2; (b) B. List, *Tetrahedron*, **2002**, *58*, 5573; (c) C. Palomo, M. Oiarbide, J. M. Garcia, *Chem. Soc. Rev.* **2004**, *33*, 65.

10. B. List, R. A. Lerner, C. F. Barbas, *J. Am. Chem. Soc.* **2000**, *122*, 2395.

11. (a) Z. Tang, F. Jiang, L. T. Yu, X. Cui, L. Z. Gong, A. Q. Mi, Y. Z. Jiang, Y. D. Wu, *J. Am. Chem. Soc.* **2003**, *125*, 5262; (b) Z. Tang, F. Jiang, X. Cui, L. Z. Gong, A. Q. Mi, Y. Z. Jiang, Y. D. Wu, *Proc. Natl. Acad. Sci. U.S.A.* **2004**, *101*, 5755.

12. F. Braude, H. G. Lindwall, *J. Am. Chem. Soc.* **1933**, *55*, 325.

13. G. Luppi, M. Monari, R. J. Corrêa, F. A. Violante, A. C. Pinto, B. Kaptein, Q. B. Broxterman, S. J. Garden, C. Tomasini, *Tetrahedron*, **2006**, *62*, 12017.

14. A. V. Malkov, M. A. Kabeshov, M. Bella, O. Kysilka, D. A. Malyshev, K. Pluháčková, P. Kočovský, *Org. Lett.* **2007**, *9*, 5473.

15. S. Nakamura, N. Hara, H. Nakashima, K. Kubo, N. Shibata, T. Toru, *Chem.–Eur. J.* **2008**, *14*, 8079.

16. S. Wei, B. Schmid, F. Z. Macaev, S. N. Curlat, A. V. Malkov, S. B. Tsogoeva, *Asymmetric Catal.* **2015**, *2*, 1.

17. G. Luppi, P. G. Cozzi, M. Monari, B. Kaptein, Q. B. Broxterman, C. Tomasini, *J. Org. Chem.* **2005**, *70*, 7418.

18. J. G. Hernández, V. G-López, E. Juaristi, *Tetrahedron*, **2012**, *68*, 92.

19. B. Xu, L. Li, S. Gou, *Tetrahedron: Asymmetry*, **2013**, *24*, 1556.

20. J. T. Guo, B. Q. Zhang, Y. Luo, Z. Guan, Y. H. He, *Asian J Org. Chem.* **2017**, *6*, 605.

21. G. D. Yadav, S. Singh, *Tetrahedron: Asymmetry*, **2016**, *27*, 463.

22. A. S. Saiyed, A.V. Bedekar, *Tetrahedron Asymmetry*, **2013**, *24*, 1035.

23. A. Ricci, L. Bernardi, C. Gioia, S. Vierucci, M. Robitzer, F. Quignard, *Chem. Commun.* **2010**, *46*, 6288.

24. F. Zhang, C. Li, C. Qi, *Tetrahedron: Asymmetry*, **2013**, *24*, 380.

25. M. Kinsella, P. G. Duggan, C. M. Lennon, *Tetrahedron: Asymmetry*, **2011**, *22*, 1423.

26. J.-R. Chen, X.-P. Liu, X.-Y. Zhu, L. Li, Y.-F. Qiao, J.-M. Zhang, W.-J. Xiao, *Tetrahedron*, **2007**, *63*, 10437.

27. Q. Guo, M. Bhanushali, C. G. Zhao, *Angew. Chem. Int. Ed.* **2010**, *49*, 9460.

28. V. Erkizan, Y. Kong, M. Merchant, S. Schlottmann, J. S. Barber- Rotenberg, L. Yuan, O. D. Abaan, T.-H. Chou, S. Dakshanamurthy, M. L. Brown, A. Uren, J. A. Toretsky, *Nat. Med.* **2009**, *15*, 750.

29. S. Allu, N. Molleti, R. Panem, V. K. Singh, *Tetrahedron Lett.* **2011**, *52*, 4080.

30. Y. Lu, Y. Ma, S. Yang, M. Ma, H. Chu, C. Song, *Tetrahedron: Asymmetry*, **2013**, *24*, 1082.

31. C. Shen, F. Shen, H. Xia, P. Zhang, X. Chen, *Tetrahedron: Asymmetry*, **2011**, *22*, 708.

32. H. Lu, J. Bai, J. Xu, T. Yang, X. Lin, J. Li, F. Ren, *Tetrahedron*, **2015**, *71*, 2610.

33. T. P. Kumar, N. Manjula, K. Katragunta, *Tetrahedron: Asymmetry*, **2015**, *26*, 1281.

34. F. Xue, S. Zhang, L. Liu, W. Duan, W. Wang, *Chem. Asian J.* **2009**, *4*, 1664.

35. T. Itoh, H. Ishikawa, Y. Hayashi, *Org. Lett.* **2009**, *11*, 3854.

36. W. B. Chen, X. L. Du, L. F. Cun, X. M. Zhang, W. C. Yuan, *Tetrahedron*, **2010**, *66*, 1441.

37. Q. Guo, J. C.-G. Zhao, *Tetrahedron Lett.* **2012**, *53*, 1768.

38. G. Pandey, J. Khamrai, *Asian J. Org. Chem.* **2016**, *5*, 621.

39. S. Hu, L. Zhang, J. Li, S. Luo, J. P. Cheng, *Eur. J. Org. Chem.* **2011**, 3347.

40. Q. Guo, J. C. G. Zhao, *Tetrahedron Lett.* **2012**, *53*, 1768.

41. M. Raj, N. Veerasamy, V. K. Singh, *Tetrahedron Lett.* **2010**, *51*, 2157.

42. (a) M. Raj, V. Maya, S. K. Ginotra, V. K. Singh, *Org. Lett.* **2006**, *8*, 4097; (b) V. Maya, M. Raj, V. K. Singh, *Org. Lett.* **2007**, *9*, 2593.

43. Y. Liu, P. Gao, J. Wang, Q. Sun, Z. Ge, R. Li, *Synlett*, **2012**, *23*, 1031.

44. C. C. Hernández, P. E. H. Gonzáleza, E. Juaristi, *Synth.* **2018**, DOI: 10.1055/s-0036-1591916.

45. F. Zhang, C. Li, C. Qi, *Tetrahedron: Asymmetry*, **2013**, *24*, 380.

46. Y. Tanimura, K. Yasunaga, K. Ishimaru, *Eur. J. Org. Chem.* **2013**, 6535.

47. H. Zhao, W. Meng, Z. Yang, T. Tian, Z. Sheng, H. Li, X. Song, Y. Zhang, S. Yang, B. Li, *Chin. J. Chem.* **2014**, *32*, 417.

48. A. Kumar, S. S. Chimni, *Tetrahedron*, **2013**, *69*, 5197.

49. J. Kimura, U. V. S. Reddy, Y. Kohari, C. Seki, Y. Mawatari, K. Uwai, Y. Okuyama, E. Kwon, M. Tokiwa, M. Takeshita, T. Iwasa, H. Nakano, *Eur. J. Org. Chem.* **2016**, 3748.

50. A. Ogasawara, U. V. S. Reddy, C. Seki , Y. Okuyama, K. Uwai, M. Tokiwa, M. Takeshita, H. Nakano, *Tetrahedron: Asymmetry*, **2016**, *27*, 1062.

51. J. Wang, Q. Liu, Q. Hao, Y. Sun, Y. Luo, H. Yang, *Chirality*, **2015**, *27*, 314.

52. G. Chen, Y. Ju, T. Yang, Z. Li, W. Ang, Z. Sang, J. Liu, Y. Luo, *Tetrahedron: Asymmetry*, **2015**, *26*, 943.

53. H. Dong, J. Liu, L. Ma, L. Ouyang, *Catal.* **2016**, *6*, 186.

54. Y. Tanimura, K. Yasunaga, K. Ishimaru, *Tetrahedron*, **2014**, *70*, 2816.

55. A. J. Pearson, S. Panda, S. D. Bunge, *J. Org. Chem.* **2013**, *78*, 9921.

56. A. Kumar, S. S. Chimni, *Eur. J. Org. Chem.* **2013**, 4780.

57. C. Cassani, P. Melchiorre, *Org. Lett.* **2012**, *14*, 5590.

58. B. Zhu, W. Zhang, R. Lee, Z. Han, W. Yang, D. Tan, K. W. Huang, Z. Jiang, *Angew. Chem. Int. Ed.* **2013**, *52*, 6666.

59. G. Wang, X. Liu, Y. Chen, J. Yang, J. Li, L. Lin, X. Feng, *ACS Catal.* **2016**, *6*, 2482.

60. G. G. Liu, H. Zhao, Y. B. Lan, B. Wu, X. F. Huang, J. Chen, J. C. Tao, X. W. Wang, *Tetrahedron*, **2012**, *68*, 3843.

61. J. Guang, Q. Guo, J. C. G. Zhao, *Org. Lett.* **2012**, *14*, 3174.

62. H. Liu, H. Wu, Z. Luo, J. Shen, G. Kang, B. Liu, Z. Wan, J. Jiang, *Chem. Eur. J.* **2012**, *18*, 11899.

63. S. Abbaraju, J. C. G. Zhao, *Adv. Synth. Catal.* **2014**, *356*, 237.

64. S. Kong, W. Fan, H. Lyu, J. Zhan, X. Miao, Z. Miao, *Synth. Commun.* **2014**, *44*, 936.

65. V. L. Martín, J. H. Martín, J. A. Fernández-Salas, J. Alemán, *Chem. Commun.* **2018**, *54*, 2781.

66. K. Aikawa, S. Mimura, Y. Numata, K. Mikami, *Eur. J. Org. Chem.* **2011**, 62.

67. Y. L. Liu, F. M. Liao, Y. F. Niu, X. L. Zhao, J. Zhou, *Org. Chem. Front.* **2014**, *1*, 742.

68. Y. L. Liu, J. Zhou, *Chem. Commun.* **2012**, *48*, 1919.

69. J. Qian, W. Yi, X. Huang, J. P. Jasinski, W. Zhang, *Adv Synth. Catal.* **2016**, *358*, 2811.

70. A. Coste, A. Bayle, J. Marrot, G. Evano, *Org. Lett.* **2014**, *16*, 1306.

71. I. Saidalimu, X. Fang, X.-P. He, J. Liang, X. Yang, F. Wu, *Angew. Chem. Int. Ed.* **2013**, *52*, 5566.

72. (a) C. Palomo, M. Oiarbide, A. Laso, *Eur. J. Org. Chem.* **2007**, 2561; (b) J. Boruwa, N. Gogoi, P. P. Saikia, N. C. Barua, *Tetrahedron: Asymmetry*, **2006**, *17*, 3315; (c) C. Palomo, M. Oiarbide, A. Mielgo, *Angew. Chem. Int. Ed.* **2004**, *43*, 5442.

73. L. Liu, S. Zhang, F. Xue, G. Lou, H. Zhang, S. Ma, W. Duan, W. Wang, *Chem. Eur. J.* **2011**, *17*, 7791.

74. M. Q. Li, J. X. Zhang, X. F. Huang, B. Wu, Z. M. Liu, J. Chen, X. D. Li, X.W. Wang, *Eur. J. Org. Chem.* **2011**, 5237.

75. Y. Zhang, Z. J. Li, H. S. Xu, Y. Zhang, W. Wang, *RSC Adv.* **2011**, *1*, 389.

76. P. S. Prathima, K. Srinivas, K. Balaswamy, R. Arundhathi, G. N. Reddy, B. Sridhar, M. M. Rao, P. R. Likhar, *Tetrahedron: Asymmetry*, **2011**, *22*, 2099.

77. M. Chennapuram, U. V. S. Reddy, C. Seki, Y. Okuyama, E. Kwon, K. Uwai, M. Tokiwa, M. Takeshita, H. Nakano, *Eur. J. Org. Chem.* **2017**, 1638.

78. R. Shintani, M. Inoue, T. Hayashi, *Angew. Chem. Int. Ed.* **2006**, *45*, 3353.

79. P. Y. Toullec, R. B. C. Jagt, J. G. de Vries, B. L. Feringa, A. J. Minnaard, *Org. Lett.* **2006**, *8*, 2715.

80. X. Feng, Y. Nie, J. Yang, H. Du, *Org. Lett.* **2012**, *14*, 624.

81. J. Gui, G. Chen, P. Cao, J. Liao, *Tetrahedron: Asymmetry*, **2012**, *23*, 554.

82. H. Lai, Z. Huang, Q. Wu, Y. Qin, *J. Org. Chem.* **2009**, *74*, 283.

83. Z. Liu, P. Gu, M. Shi, P. McDowell, G. Li, *Org. Lett.* **2011**, *13*, 2314.

84. Y. Zhuang, Y. He, Z. Zhou, W. Xia, C. Cheng, M. Wang, B. Chen, Z. Zhou, J. Pang, L. Qiu, *J. Org. Chem.* **2015**, *80*, 6968.

85. R. Shintani, K. Takatsu and T. Hayashi, *Chem. Commun.* **2010**, *46*, 6822.

86. Y. Yamamoto, M. Yohda, T. Shirai, H. Ito, N. Miyaura, *Chem. Asian J.* **2012**, *7*, 2446.

87. Y. Li, D. X. Zhu, M. H. Xu, *Chem. Commun.* **2013**, *49*, 11659.

88. (a) S. L. You, Q. Cai, M. Zeng, *Chem. Soc. Rev.* **2009**, *38*, 2190; (b) V. Terrasson, R. M. de Figueiredo, J. M. Campagne, *Eur. J. Org. Chem.* **2010**, 2635; (c) M. Zen, S. L. You, *Synlett*, **2010**, 1289; (d) P. Chauhan,

S. S. Chimni, *RSC Adv.* **2012**, *2*, 6117; (e) M. Bandini, A. U. Ronchi, *Catalytic Asymmetric Friedel–Crafts Alkylations,* Wiley-VCH, Weinheim, **2009**.

89. J. Deng, S. Zhang, P. Ding, H. Jiang, W. Wang, J. Li, *Adv. Synth. Catal.* **2010**, *352*, 833.

90. P. Chauhan, S. S. Chimni, *Chem. Eur. J.* **2010**, *16*, 7709.

91. N. V. Hanhan, A. H. Sahin, T. W. Chang, J. C. Fettinger, A. K. Franz, *Angew. Chem. Int. Ed.* **2010**, *49*, 744.

92. J. Kaur, A. Kumar, S. S. Chimni, *Tetrahedron Lett.* **2014**, *55*, 2138.

93. (a) A. Kumar, J. Kaur, P. Chauhan, S. S. Chimni, *Chem. Asian J.* **2014**, *9*, 1305; (b) J. Kaur, A. Kumar, S. S. Chimni, *RSC Adv.* **2014**, *4*, 62367.

94. C. Li, F. Guo, K. Xu, S. Zhang, Y. Hu, Z. Zha, Z. Wang, *Org. Lett.* **2014**, *16*, 3192.

95. (a) Y. Zhu, J. P. Malerich, V. H. Rawal, *Angew. Chem. Int. Ed.* **2010**, *49*, 153; (b) H. Groger, B. Hammer, *Chem.–Eur. J.* **2000**, 943; (c) L. Albrecht, A. Albrecht, H. Krawczyk, K. A. Jorgensen, *Chem.–Eur. J.* **2010**, *16*, 28; (d) D. Zhao, R. Wang, *Chem. Soc. Rev.* **2012**, *41*, 2095.

96. L. Peng, L. L. Wang, J. F. Bai, L. N. Jia, Q. C. Yang, Q. C. Huang, X.Y. Xu, L. X. Wang, *Tetrahedron Lett.* **2011**, *52*, 1157.

97. M. Nazish, A. Jakhar, N. H. Khan, S. Verma, R. I. Kureshy, S. H. R. Abdi, H. C. Bajaj, *App. Cat. A: General,* **2016**, *515*, 116.

98. T. Deng, H. Wang, C. Cai, *Org. Biomol. Chem.* **2014**, *12*, 5843.

99. For recent reviews on the Morita–Baylis–Hillman (MBH) reaction, see: (a) D. Basavaiah, B. S. Reddy, S. S. Badsara, *Chem. Rev.* **2010**, *110*, 5447; (b) C. G. Lima- Junior, M. L. A. A. Vasconcellos, *Bioorg. Med. Chem.* **2012**, *20*, 3954; (c) Y. Wei, M. Shi, *Chem. Rev.* **2013**, *113*, 6659; (d) G. Masson, C. Housseman and J. Zhu, *Angew. Chem. Int. Ed.* **2007**, *46*, 4614; (e) V. Declerck, J. Martinez and F. Lamaty, *Chem. Rev.* **2009**, *109*, 1; (f) Y. Wei, M. Shi, *Acc. Chem. Res.* **2010**, *43*, 1005; (g) Y. Wei, M. Shi, *Chin. Sci. Bull.* **2010**, *55*, 1699.

100. Y. L. Liu, B. L. Wang, J. J. Cao, L. Chen, Y. X. Zhang, C. Wang, J. Zhou, *J. Am. Chem. Soc.* **2010**, *132*, 15176.

101. X. Y. Guan, Y. Wei, M. Shi, *Chem. Eur. J.* **2010**, *16*, 13617.

102. F. Zhong, G.Y. Chen, Y. Lu, *Org. Lett.* **2011**, *13*, 82.

103. C. C. Wang, X.Y. Wu, *Tetrahedron,* **2011**, *67*, 2974.

104. J.Y. Qian, C. C. Wang, F. Sha, X.Y. Wu, *RSC Adv.* **2012**, *2*, 6042.

105. Z. Dong, C. Yan, Y. Gao, C. Dong, G. Qiu, H. B. Zhou, *Adv. Synth. Catal.* **2015**, *357*, 2132.

106. P. Chauhan, S. S. Chimni, *Asian J. Org. Chem.* **2013**, *2*, 586.

107. Q. He, G. Zhan, W. Du, Y.C. Chen *Beilstein, J. Org. Chem.* **2016**, *12*, 309.

108. Z. Duan, Z. Zhang, P. Qian, J. Han, Y. Pan, *RSC Adv.* **2013**, *3*, 10127.

109. (a) J. Christoffers, A. Baro, *Angew. Chem. Int. Ed.* **2003**, *42*, 1688; (b) O. M. Berner, L. Tedeschi, D. Enders, *Eur. J. Org. Chem.* **2002**, 1877; (c) N. Krause, A. Hoffmann-Röder, *Synthesis*, **2001**, 171; (d) D. Almasi, D. A. Alonso, C. Najera, *Tetrahedron: Asymmetry*, **2007**, *18*, 299; (e) S. B. Tsogoeva, *Eur. J. Org. Chem.* **2007**, 1701; (f) S. Sulzer-Mosse, A. Alexakis, *Chem. Commun.* **2007**, 3123; (g) J. L. Vicario, D. Badía, L. Carrillo, *Synthesis*, **2007**, 2065; (h) S. Jautze, R. Peters, *Synthesis*, **2010**, 365.

110. M. Retini, G. Bergonzini, P. Melchiorre, *Chem. Commun.* **2012**, *48*, 3336.

111. N. Hara, S. Nakamura, Y. Funahashi, N. Shibata, *Adv. Synth. Catal.* **2011**, *353*, 2976.

112. Z. Duan, J. Han, P. Qian, Z. Zhang, Y. Wang, Y. Pan, *Org. Biomol. Chem.* **2013**, *11*, 6456.

113. F. Zhong, W. Yao, X. Dou,Y. Lu, *Org. Lett.* **2012**, *14*, 4018.

114. C. W. Suh, C. W. Chang, K. W. Choi, Y. J. Lim, D. Y. Kim, *Tetrahedron Lett.* **2013**, *54*, 3651.

115. V. P. R. Gajulapalli, P. Vinayagam, V. Kesavan, *Org. Biomol. Chem.* **2014**, *12*, 4186.

116. T. Deng, H. Wang, C. Cai, *Eur. J. Org. Chem.* **2014**, 7259.

117. D. Sano, K. Nagata, T. Itoh, *Org. Lett.* **2008**, *10*, 1593.

118. T. Ishimaru, N. Shibata, J. Nagai, S. Nakamura, T. Toru, S. Kanemasa, *J. Am. Chem. Soc.* **2006**, *128*, 16488.

119. K. Shen, X. Liu, G. Wang, L. Lin, X. Feng, *Angew. Chem. Int. Ed.* **2011**, *50*, 4684.

120. D. Tomita, K. Yamatsugu, M. Kanai, M. Shibasaki, *J. Am. Chem. Soc.* **2009**, *131*, 6946.

121. K. Funabashi, M. Jachmann, M. Kanai, M. Shibasaki, *Angew. Chem. Int. Ed.* **2003**, *42*, 5489.

122. D. Monge, A. M. Crespo-Peña, E. Martin-Zamora, E. Álvarez, R. Fernández, J. M. Lassaletta, *Chem. Eur. J.* **2013**, *19*, 8421.

123. K. Zheng, C. Yin, X. Liu, L. Lin, X. Feng, *Angew. Chem. Int. Ed.* **2011**, *50*, 2573.

124. M. Kitajima, I. Mori, K. Arai, N. Kogure, H. Takayama, *Tetrahedron Lett.* **2006**, *47*, 3199.

125. X. C. Qiao, S. F. Zhu, Q. L. Zhou, *Tetrahedron: Asymmetry*, **2009**, *20*, 1254.

126. J. Itoh, S. B. Han, M. J. Krische, *Angew. Chem. Int. Ed.* 2009, *48*, 6313.

127. N. V. Hanhan,Y. C. Tang, N. T. Tran, A. K. Franz, *Org. Lett.* **2012**, *14*, 2218.

128. Z. Y. Cao, J. S. Jiang, J. Zhou, *Org. Biomol. Chem.* **2016**, *14*, 5500.

129. T. Wang, X. Q. Hao, J. J. Huang, K. Wang, J.-F. Gong, M.-P. Song, *Organometallics*, **2014**, *33*, 194.

130. D. Ghosh, N. Gupta, S. H. R. Abdi, S. Nandi, N. H. Khan, R. I. Kureshy, H. C. Bajaj, *Eur. J. Org. Chem.* **2015**, 2801.

131. S. Jayakumar, N. Kumarswamyreddy, M. Prakash, V. Kesavan, *Org. Lett.* **2015**, *17*, 1066.

132. M. Takahashi, Y. Murata, F. Yagishita, M. Sakamoto, T. Sengoku, H. Yoda, *Chem. – Eur. J.* **2014**, *20*, 11091.

133. Y. Huang, R. Z. Huang, Yu Zhao, *J. Am. Chem. Soc.* **2016**, *138*, 6571.

134. H. Wu, K. Haeffner, A. H. Hoveyda, *J. Am. Chem. Soc.* **2014**, *136*, 3780.

135. D. L. Silverio, S. Torker, T. Pilyugina, E. M.Vieira, M. L. Snapper, F. Haeffner, A. H. Hoveyda, *Nature*, **2013**, *494*, 216.

136. Q. Chen, Y. Tang, T. Huang, X. Liu, L. Lin, X. Feng, *Angew. Chem. Int. Ed.* **2016**, *55*, 5286.

147. L. Yin, M. Kanai, M. Shibasaki, *Angew. Chem. Int. Ed.* **2011**, *50*, 7620.

138. T. Shirai, H. Ito, Y. Yamamoto, *Angew. Chem. Int. Ed.* **2014**, *53*, 2658.

139. M. A. Horwitz, N. Tanaka, T. Yokosaka, D. Uraguchi, J. S. Johnson, T. Ooi, *Chem. Sci.* **2015**, *6*, 6086.

Chapter 6

3-Hetero-3-Substituted Oxindoles

6.1. Introduction

In the previous chapters, the synthesis of oxindoles containing C–O or C–N bond have been described, but other carbon–heteroatom bonds are important in organic synthesis, especially if they can be easily manipulated again. Moreover, 3-hetero-3-substituted oxindoles, such as 3-halo, 3-sulphur, 3-phosphorus derivatives, can show themselves biological activity.

For instance, 3-alkyl- or 3-aryl-3-fluorooxindoles are drugs, thus their synthesis stimulated the development of both enantioselective fluorination methods on 3-substituted oxindoles as well as C–C bond formation in 3-fluorooxindoles. The corresponding 3-chloro and 3-bromooxindoles are not useful in drug discovery chemistry, thus their asymmetric synthesis is less studied. On the other hand, they are good substrates for S_N2 reactions, but in these cases, the enantioenrichment is generally produced during the nucleophilic substitution reaction, as it was widely reported in this book.

Many pharmaceutical and biological compounds bear a carbon–sulfur bond and its formation constitutes a very important general reaction [1]. In particular, oxindoles bearing a sulfur atom at the carbon stereocentre have been found to have anti-cancer, antifungal, or antitubercular activities.

The enantioselective hydrophosphonylation of imines is a well-explored reaction, because α-aminophosphonates are both important synthetic intermediates and exhibit themselves a broad spectrum of biological activities often connected with the absolute stereochemistry. Thus, the addition of diphenyl phosphonate to ketimines derived from isatins is widely studied.

6.2. 3-Fluoro-3-Substituted Oxindoles

Although compounds, in which the halogen acts as an electrophile, are not very common owing to the high electronegativity of halogens, the first reactions affording 3-fluoro-3-substituted oxindoles employed stoichiometric amounts of a combination between *bis*-cinchona alkaloids [mainly $(DHQD)_2PYR$ but also with $(DHQ)_2AQN$ or $(DHQ)_2PHAL$] and an electrophilic fluoro donor such as Selectfluor. Yields were good, enantiomeric excesses sometimes poor, but, above all, the configuration was not determined (Scheme 6.1) [2]. Unfortunately, when Shibata revisited this reaction in organocatalytic conditions, he found that Selectfluor reacted more easily with the substrate than the cinchona alkaloid, thus the background reaction was faster and a racemate was recovered [3]. Some better results were obtained by a chiral C2-symmetric NHC–palladium complex (Scheme 6.2) [4]. However, ee values are still moderate and, when the C-3 was substituted by electron-rich or -deficient aromatic rings as well as a benzyl group, the ee's are low or very low.

The prevalence of (*S*)-products allowed authors to envisage the transition state depicted in Scheme 6.2, in which the *N*-Boc moiety is involved in hydrogen bond interactions. In fact, protecting groups unable to give hydrogen bonds did not give enantio-enrichment. Then, the enolized

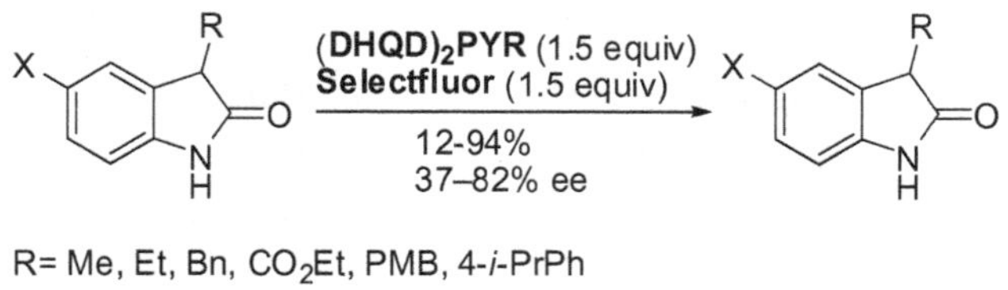

(DHQD)₂PYR (1.5 equiv)
Selectfluor (1.5 equiv)
12-94%
37–82% ee

R= Me, Et, Bn, CO₂Et, PMB, 4-*i*-PrPh

Scheme 6.1.   First fluorination reaction of prostereogenic oxindoles.

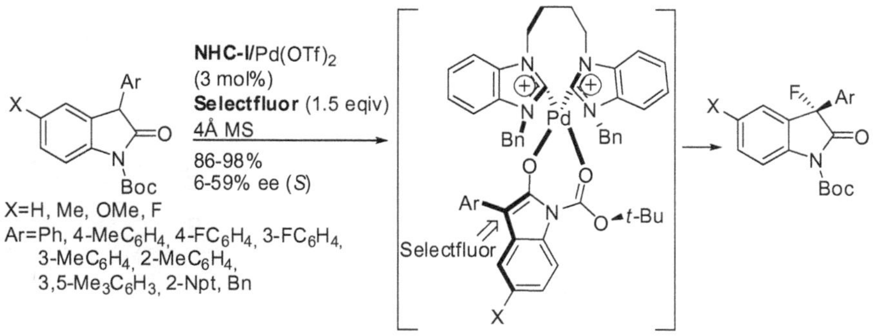

Scheme 6.2.   Palladium-catalyzed asymmetric fluorination of oxindoles.

3-substituted oxindoles could react with Selectfluor from the less hindered *Re*-face. It should be noted that another classical fluorinating agent such as NFSI was ineffective in affording products.

The importance of the drug BMS-204352 (MaxiPost, Appendix C), a potent and effective opener of the maxi-K channel, stimulated the research of asymmetric syntheses of 3-fluoro-3-substituted oxindoles. Shibata's group described one of the first examples and MaxiPost was prepared in 94% yield with 84% ee under their reaction conditions (stoichiometric (DHQ)$_2$AQN/Selectfluor combination) [5]. Moreover, they obtained also the enantiomer in 96% yield and with 68% ee using quinidine/Selectfluor stoichiometric combination.

Actually the stoichiometric cinchona alkaloid/Selectfluor is nothing but an *in situ*-prepared *N*-fluorinated cinchona alkaloid derivative. Cahard demonstrated that many other *N*-fluorinated cinchona alkaloids were able to give MaxiPost in almost quantitative yield with moderate ee (57–88%) [3]. However, the very high cost of these fluorinating agents and the moderate ee's made this reaction not useful for scaling up.

Then, Shibata's group used NFSI in the *bis*-cinchona alkaloid promoted reaction and was able to set up both catalytic and stoichiometric reactions providing products with the same enantioselectivity (Table 6.1) [3]. However, the paper reported only five examples. Addition of a base such as CsOH·H$_2$O greatly increased the enantioselectivity. It should be noted that this base is not supposed by authors to favor enolization of oxindole, unlike all the other research groups, working in the field.

Table 6.1.   Fluorination reactions of prostereogenic oxindoles under asymmetric metal catalysis (envisaged transition state are depicted in Figure 6.1).

R, X, PG	Catalyst/transition state	Yield (%)	ee (%)	Ref.
R=Ph, 4-MeC$_6$H$_4$, 4-FC$_6$H$_4$ X=H, 5-Me, 5-MeO PG=Boc	**(DHQD)$_2$AQN** (5 mol%) CsOH (6 equiv) by **TS-I**	77–92	79–87 (*S*)	[3]
	(DHQ)$_2$AQN (5 mol%) CsOH (6 equiv) by **TS-I**	86–99	84–86 (*R*)	[6]
R=Ph, Me, Et, Bn, *i*-Bu, CH$_2$COMe, 4-MeC$_6$H$_4$, 4-FC$_6$H$_4$, 2-MeOC$_6$H$_4$, 2-MeO-5-ClC$_6$H$_3$ X=H, 6-CF$_3$ PG=Boc	**PHOS-I** (5 mol%) Pd(OTf)$_2$ (2.5 mol%) by **TS-II**	72–97	71–96 (*S*)[a]	[7]
R=Me, Ph, 2-MeO-5-ClC$_6$H$_3$ X=H, 6-CF$_3$ PG=Boc	**BOX-I** (11 mol%), Ni(OAc)$_2$·4H$_2$O (10 mol%)	71–73	93–96 (*S*)[a]	[8]
R=Ph, Me, 4-FC$_6$H$_4$, PMP, 4-MeC$_6$H$_4$, 2-MeO-5-ClC$_6$H$_3$ X=H, 4-F, 5-F, 5-MeO, 5-Me, 6-CF$_3$, 7-F PG=Boc	**BOX-II** (12 mol%) Ni(ClO$_4$)$_2$·6H$_2$O (10 mol%), 4Å MS	89–95	95 to >99% (*S*)	[9]

R=Bn, Me, Et, Pr, Bu, *i*-Bu, (2-Et)Bu, Allyl, 4-MeC$_6$H$_4$CH$_2$, PMB, 4-ClC$_6$H$_4$CH$_2$, 4-FC$_6$H$_4$CH$_2$, 4-BrC$_6$H$_4$CH$_2$, 4-NO$_2$C$_6$H$_4$CH$_2$, 3-MeC$_6$H$_4$CH$_2$, 3-ClC$_6$H$_4$CH$_2$, 3-NO$_2$C$_6$H$_4$CH$_2$, 2-MeC$_6$H$_4$CH$_2$, 2-ClC$_6$H$_4$CH$_2$, 2,4-Cl$_2$C$_6$H$_3$CH$_2$, 3,4-(OCH$_2$O)C$_6$H$_3$CH$_2$, 2-thienylCH$_2$, 2-pyridylCH$_2$, 1-NptCH$_2$, 2-NptCH$_2$ X=H, 5-Cl PG=H	**NNO-Ia** (5 mol%) Sc(OTf)$_3$ (5 mol%) by **TS-III**	80–98	89–99 (*R*)	[10]
R=Ph, 4-ClC$_6$H$_4$ X=H, 5-Cl PG=H	**NNO-II** (5 mol%) Sc(OTf)$_3$ (5 mol%) by **TS-II**	85–98	89–93 (*R*)	
R=Me, Et, Pr, *i*-Bu, Bn X=H, 5-Me, 5-MeO, 5-F PG=Boc	**SAL-FEa** (5 mol%) AgClO$_4$ (5 mol%), 4Å MS by **TS-IV**	84–94	85–96 (*R*)	[11]
R=Ph, 4-MeC$_6$H$_4$, 4-FC$_6$H$_4$ X=H, 5-Me, 5-MeO, 5-F PG=Boc	**SAL-FEb** (5 mol%) AgClO$_4$ (5 mol%), 4Å MS by **TS-IV**	86–93	82–95 (*R*)	

Notes: [a] The stereochemistry was determined only for the MaxiPost precursor, it can be supposed that the other products were (*R*)-configured, because the different Cahn–Ingold–Prelog priorities in MaxiPost with respect to the other products made (*S*) this product and (*R*) the other ones.

Authors envisaged the formation of an *N*-fluoroammonium salt of the cinchona alkaloid in an open conformation (one of the two more stable conformation of (DHQ)$_2$AQN). Then this conformation accommodates in a transition state, where fluoronium ion is transferred to give the fluorinated products in the *R* configuration.

NFSI revealed then a much better fluorinating reagent in association with asymmetric metal complexes. By comparison of the different procedures, it should be noted that the *N*-Boc protection is present in most cases, because the metal cation is surmised to be also bound to the carbonyl group of the Boc protecting group.

The enolate is generally formed by means of molecular sieves, but either the basic nature of the catalyst or the intervention of an external base is reported. Sodeoka [7] and Feng [10] prepared MaxiPost in 38% yield with >99% ee after two steps and recrystallization and in 81% yield with 96% ee, respectively, while other authors prepared only the *N*-Boc protected MaxiPost. Sodeoka [7] was also able to obtain enriched 3-fluorooxindole in 53% yield and 93% ee (stereochemistry not reported), notwithstanding racemization of the product was before considered unavoidable owing to the easy enolization of the 3-fluorooxindole. In the *N,N'*-dioxide-Sc(OTf)$_3$ (**NNO-Ia** and **NNO-II**) [10] and iron(III)-salan (**SAL-FEa** and **b**) [11] catalyzed reaction, both 3-aryl- and 3-alkyloxindoles afforded products with good results only with a little change in the catalyst. Interestingly, Feng's procedure was the sole reported to be scalable until gram scale and working on unprotected indoles; authors surmised a transition state in which the Sc(III) ion was coordinated with four oxygens, only one of these belonging to the indole (Figure 6.1) [10]. Iron catalyst reaction was accelerated by the addition of silver salts to favor the *in situ* formation of a cationic iron complex [11].

Yang's group tested various substituted NFSIs [12], in order to tune their reactivity and selectivity in this reaction, in particular 4-substituted NFSIs (F, *t*-Bu, OMe, CF$_3$, and OCF$_3$) with (DHQD)$_2$PHAL (5 mol%) as the catalyst, in the presence of K$_2$CO$_3$ as the base (25–99% yields with 21–96% ee, (*S*)-isomer, 24 examples). Under these reaction conditions, 5-fluoro-3-aryindoles and 4-CF$_3$- or 4-CF$_3$O-NFSI were unreactive, while 4-*t*-Bu-NFSI lowered the reaction rate, but increased the enantioselectivity. It should be noted that authors supposed a transition state similar to

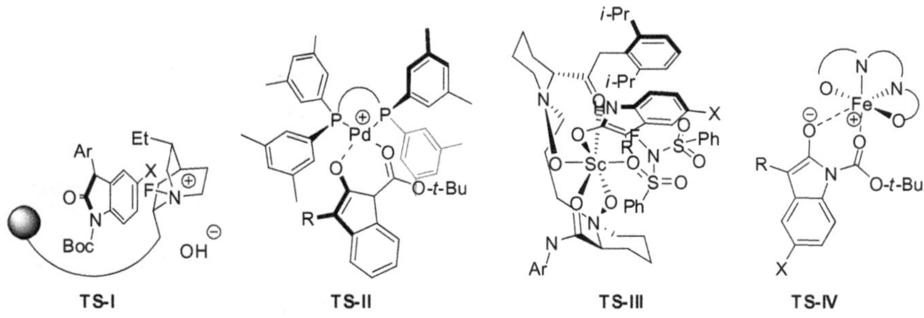

Figure 6.1.   Envisaged transition state of reaction in Table 6.1.

that envisaged by Shibata [3] that did not explain the role of the NFSI substituents on the enantioselectivity. In fact, the benzenesulfonimide ion is not involved in such a transition state. Authors wrote only a cryptic sentence: "Naturally, the different substituents on the phenyl ring of $(ArSO_2)_2N^-$ anion influence the ee value."

As much as Yang's group tested various substituted NFSIs with organocatalysts [12], He's research groups did in the enantioselective fluorination of 2-oxindoles catalyzed by (*R*)-**BINAP**-palladium complexes (0.5 mol%) [13]. The best results were obtained with 4,4′-difluoro-NFSI, with which 3-fluoro-2-oxindoles were recovered in 90–98% yield and 88–99% ee ((*S*)-isomer, 15 examples). In comparison with the Yang's screening:

(i) the influence of the different aryl groups of the NFSI on enantiose-lectivity was better explained, because the *bis*-benzenesulfonimide moiety is always complexed with the various hypothesized palladium intermediates;

(ii) 5-fluorooxindole gave the expected product in high yield and enantioselectivity, while 4-CF_3- or 4-CF_3O-NFSI were not tested. Moreover, Yang also found the highest yields with 4,4′-difluoro-NFSI, but with very low enantioselectivity.

The C–C bond formation onto prochiral 3-fluorooxindole by their conjugate addition to vinyl sulfones and nitroalkenes, γ-addition to allenes,

Scheme 6.3.   Enantioselective conjugate addition of 3-fluorooxindoles to vinyl sulfone.

aldol condensation, or arylation and allylic alkylation was developed only after 2013, much more recently with respect to fluorination methods.

The conjugate addition of 3-fluorooxindoles to 1,1-*bis*(benzenesulfonyl) ethylene was described by two different research groups and, interestingly, each one obtained a different enantiomer, thus now both enantiomers are available (Scheme 6.3).

Dou and Lu reported the synthesis of the (*S*)-isomer confirmed by X-ray analysis [14]. Kim's group prepared the (*R*)-isomer [15], as they affirmed comparing their optical rotations with those reported by Dou and Lu. Only Kim envisaged a transition state for the reaction, in which the tertiary amine deprotonated the 3-fluorooxindole and the urea moiety activated the vinyl sulfone. A similar transition state is very likely operative also in the other reaction, supposing that the faces shielded by the cinchona or the binaphthyl moieties might be opposite. Under Dou and Lu conditions, mono-sulfone was unreactive and 1,1-*bis*(benzenesulfonyl)-2-phenylethylene afforded the expected product in 97% yield, 9:1 dr and 98% ee, with a slightly different catalyst. Finally, also nitrostyrene led to the desired adduct in 95% yield, 3.5 (45% ee): 1 (28% ee) dr, under Dou and Lu conditions.

Later, Lu also carried out the reaction of 3-fluorooxindoles with 2,3-butadienoates under phosphine catalysis at 0.05 mmol scale (Scheme 6.4)

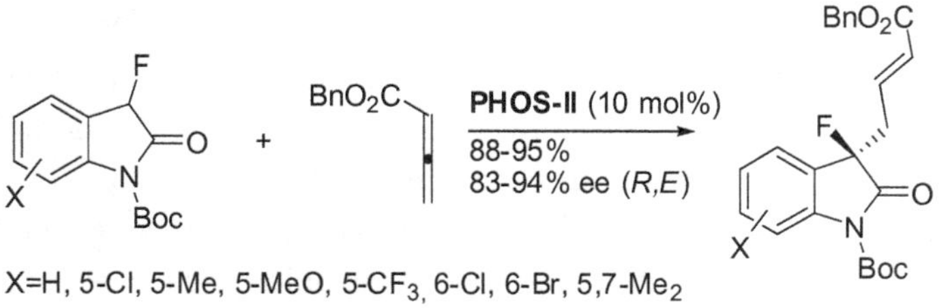

PHOS-II (10 mol%)
88–95%
83–94% ee (*R,E*)

X=H, 5-Cl, 5-Me, 5-MeO, 5-CF$_3$, 6-Cl, 6-Br, 5,7-Me$_2$

Scheme 6.4.   Asymmetric γ-addition of 3-fluoro-oxindoles to 2,3-butadienoates.

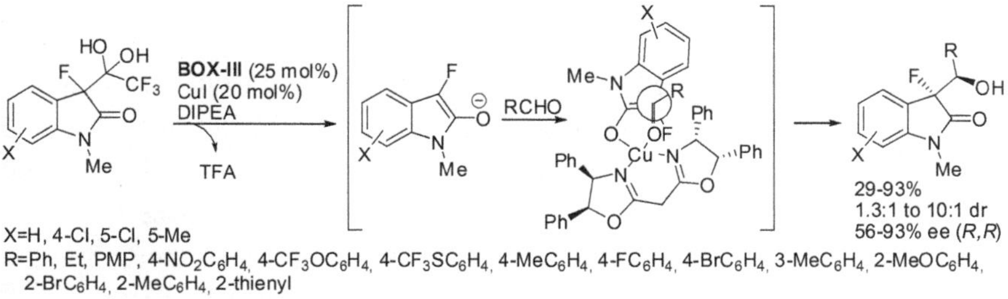

BOX-III (25 mol%)
CuI (20 mol%)
DIPEA
TFA

29–93%
1.3:1 to 10:1 dr
56–93% ee (*R,R*)

X=H, 4-Cl, 5-Cl, 5-Me
R=Ph, Et, PMP, 4-NO$_2$C$_6$H$_4$, 4-CF$_3$OC$_6$H$_4$, 4-CF$_3$SC$_6$H$_4$, 4-MeC$_6$H$_4$, 4-FC$_6$H$_4$, 4-BrC$_6$H$_4$, 3-MeC$_6$H$_4$, 2-MeOC$_6$H$_4$, 2-BrC$_6$H$_4$, 2-MeC$_6$H$_4$, 2-thienyl

Scheme 6.5.   Asymmetric aldol condensation of 3-fluorooxindoles.

[16], but the scale-up to 1 mmol did not substantially modify yield and enantioselectivity. Authors provided also procedures for the easy modification of the 4-(benzyloxy)-4-oxobut-2-enyl side chain, leading to 3-allyl or 3-alkyl substituted 3-fluorooxindoles in high yield. Once more, Lu did not provide a tentative mechanism; however, *N*–H indole as well as *N*-methylated catalyst dramatically decreased enantioselectivity, thus surmising interactions between *N*-Boc group of indole and the amide moiety of the catalyst.

The aldol condensation of 3-fluorooxindole and aldehydes was performed using Cu(I)/*bis*-oxazoline ligand as the catalyst, and a haloform-type reaction (Scheme 6.5) [17]. Unprotected *N*–H oxindole afforded the expected product in 74% yield with 3.5:1 dr and 79% ee, whereas 5-trifluoromethoxy-3-fluoro-*N*-methylindolin-2-one gave the prevalence of the (*S*,R**)-isomer (1.3:1) in only 30% ee. A curious and interesting feature of this reaction was that usual gravity-driven silica–gel column chromatography of one of the products showed an enantiomeric enrichment of the first samples (80% ee) and a depletion of the last ones (76% ee), demonstrating an unusually very little different interaction of two enantiomers toward achiral silica.

Then a diasteroselective Mannich reaction was performed, starting from chiral sulfinimides. It should be noted that this is not an asymmetric catalytic reaction, but it is a chiral auxiliary promoted reaction. Authors surmised a close chair-like transition state (Scheme 6.6, Eq. (6.1)) [18].

Recently another Mannich reaction was performed (Scheme 6.6, Eq. (6.2)) [19]. α-Amidosulfones were the stable precursors of imines, a chiral oligoethylene glycol and KF acted as a cation-binding catalyst and base, respectively. The reaction was scalable and could be extended to 3-chloro- (96% yield, >20:1 dr, 98% ee), 3-bromooxindole (48% yield, >20:1 dr, 98% ee).

3-Fluoro-3-substituted oxindoles were also obtained from the asymmetric allylic alkylation of 3-fluoroxoindoles (Scheme 6.7) in a nice scalable reaction (up to 1.15 g of product in 92% yield with 24:1 dr and >99% ee) [20]. Also non-symmetrically substituted allylic acetates can be used, but the ligand had to be changed, in order to favor the attack of the C-3 at the less hindered position of the η^3-palladium complex from allyl acetate.

It is worth noting that the allyl moiety always assumed a (R,E)-configuration, thus only $(3R,\alpha R,E)$ and $(3S,\alpha R,E)$ diastereomers are

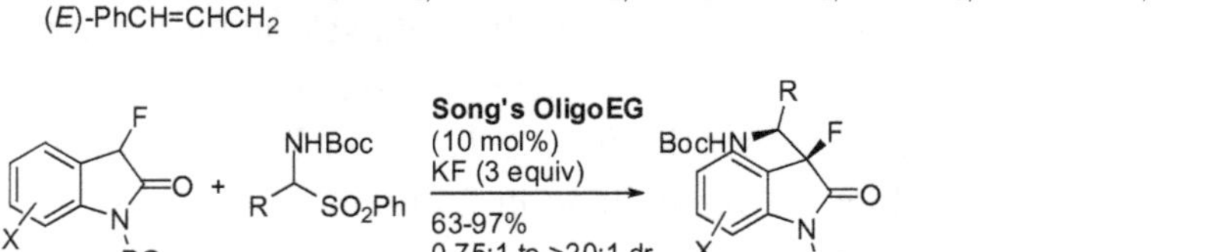

PG=Me, Bn, Propargyl, Cynnamyl
X=H, 5-MeO, 5-Cl
R=i-Bu, Bn, PMB, 4-MeC$_6$H$_4$CH$_2$, 4-CNC$_6$H$_4$CH$_2$, 4-NO$_2$C$_6$H$_4$CH$_2$, 1-NptCH$_2$, 4-pyridylCH$_2$, 2-furylCH$_2$,
 (E)-PhCH=CHCH$_2$

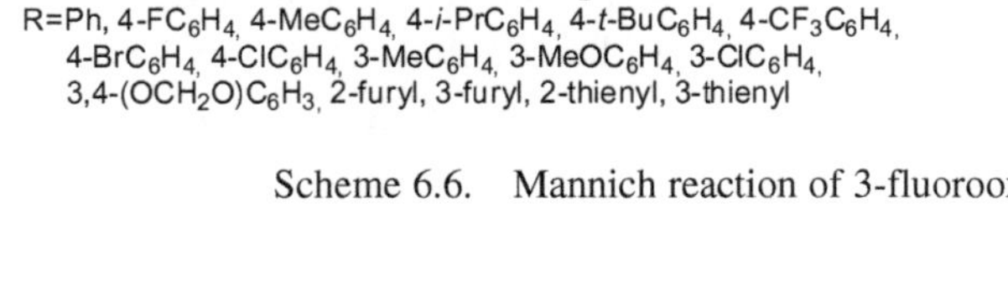

(6.1)

PG=Me, Et, Boc
X=H, 5-Me, 5-MeO, 5-F, 5-I, 6-Cl, 5,7-Me$_2$
R=Ph, 4-FC$_6$H$_4$, 4-MeC$_6$H$_4$, 4-i-PrC$_6$H$_4$, 4-t-BuC$_6$H$_4$, 4-CF$_3$C$_6$H$_4$,
 4-BrC$_6$H$_4$, 4-ClC$_6$H$_4$, 3-MeC$_6$H$_4$, 3-MeOC$_6$H$_4$, 3-ClC$_6$H$_4$,
 3,4-(OCH$_2$O)C$_6$H$_3$, 2-furyl, 3-furyl, 2-thienyl, 3-thienyl

(6.2)

Scheme 6.6. Mannich reaction of 3-fluorooxindoles.

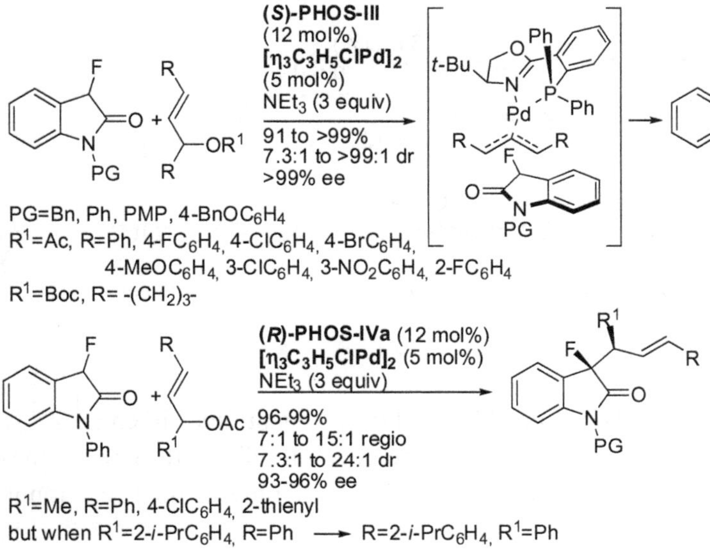

(S)-PHOS-III (12 mol%)
[η₃C₃H₅ClPd]₂ (5 mol%)
NEt₃ (3 equiv)

91 to >99%
7.3:1 to >99:1 dr
>99% ee

PG=Bn, Ph, PMP, 4-BnOC₆H₄
R¹=Ac, R=Ph, 4-FC₆H₄, 4-ClC₆H₄, 4-BrC₆H₄,
4-MeOC₆H₄, 3-ClC₆H₄, 3-NO₂C₆H₄, 2-FC₆H₄
R¹=Boc, R= -(CH₂)₃-

(R)-PHOS-IVa (12 mol%)
[η₃C₃H₅ClPd]₂ (5 mol%)
NEt₃ (3 equiv)

96-99%
7:1 to 15:1 regio
7.3:1 to 24:1 dr
93-96% ee

R¹=Me, R=Ph, 4-ClC₆H₄, 2-thienyl
but when R¹=2-*i*-PrC₆H₄, R=Ph ⟶ R=2-*i*-PrC₆H₄, R¹=Ph

Scheme 6.7.   Asymmetric allylic alkylation of 3-fluoroxoindoles.

(R)-PHOS-IVb (12 mol%)
Pd(dba)₂ (10 mol%)
Cs₂CO₃ or
K₃PO₄ (3 equiv)

50-97%
78-98% ee (*R*)

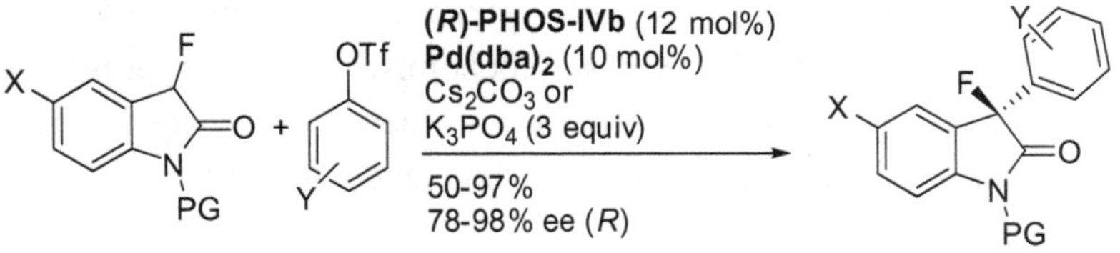

X=H, Me, MeO, F
PG=Me, Bn, Ph
Y=H, 4-CN, 4-CO₂Me, 4-NO₂, 4-Cl, 3-CN, 3,5-(MeO)₂

Scheme 6.8.   Asymmetric arylation of 3-fluoroxoindoles.

recovered, with prevalence of the former. Finally, either addition to the double bond or substitution of the fluorine atom can be performed without affecting the obtained enantioenrichment.

In the enantioselective arylation of 3-fluoroxoindoles (Scheme 6.8) [21], triflate leaving group was chosen, because it is much more labile than halides in an arylpalladium complex. Once more, scale-up of the reaction to 1.0 mmol was allowed (70% yield with 96% ee) even with less catalyst loading (5.0 mol% of Pd(dba)₂ and 6.0 mol% of catalyst with

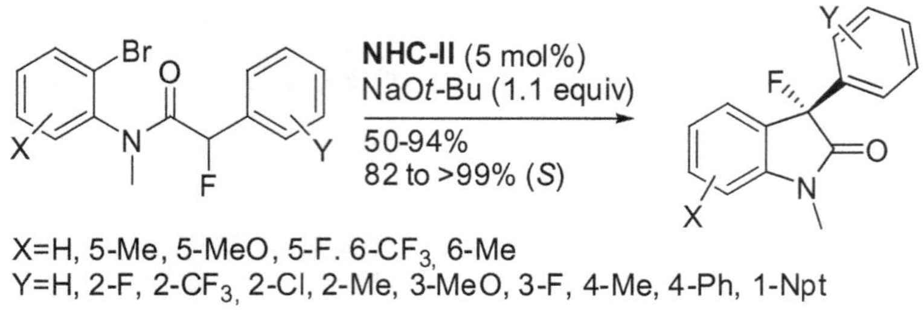

Br
O
N
F
X
Y
NHC-II (5 mol%)
NaOt-Bu (1.1 equiv)
50-94%
82 to >99% (S)
F
O
N
X=H, 5-Me, 5-MeO, 5-F. 6-CF3, 6-Me
Y=H, 2-F, 2-CF3, 2-Cl, 2-Me, 3-MeO, 3-F, 4-Me, 4-Ph, 1-Npt

Scheme 6.9. *De novo* asymmetric construction of 3-fluoroxoindoles.

K_3PO_4 as the base). The (*S*)-catalyst led to the enantiomer in comparable enantioselectivity. These compounds are generally difficult to be synthesized in an enantioselective manner and only one other similar reaction is reported [7].

Finally, a *de novo* asymmetric construction of the indole ring under palladium complex catalysis should be mentioned (Scheme 6.9) [22]. An excess of the base was found to decompose the stereogenic carbon in the substrate. The quantitative quadrant-diagram representations of the catalysts well explained the observed stereochemistry: in this representation, the most stable intermediate has the two aromatic rings placed in the less buried quadrants below the naphthyl ring, and thus led to the favored *S*-enantiomer.

6.3. Other Halogens

If 3-fluorooxindoles are drug candidates and MaxiPost synthesis largely prompted the research of asymmetric ways to them, the asymmetric synthesis of other halo-derivatives is much less studied. The most used chlorinating agent was NCS (Table 6.2). An influence of the C-3 substituents on the reaction course was observed by Shi and Zhao [23]: an electron-donating group on the phenyl ring lowered the reaction rate; a bulky aryl or 3-alkyl substituents lowered the enantioselectivities (*o*-tolyl group led to only 7% ee). Moreover, bulkier *N*-protecting groups also increased the stereochemical outcome. In addition, Antilla found lesser enantiomeric excesses with 3-alkyloxindoles [24].

Table 6.2.   Chlorination reactions of prostereogenic oxindoles with NCS.

R, X, PG	Catalyst	Yield (%)	ee (%)	Ref.
R=Ph, 4-FC$_6$H$_4$, 4-MeC$_6$H$_4$, 3-Me C$_6$H$_4$, 2-Me C$_6$H$_4$, 2-Npt, Me, Bn X=H, 5-F, 5-Me PG=Boc, Bn, Ph, Fmoc	**9-Bz-Quinine** (20 mol%)	81–99	7–93 (+)[a]	[23]
R=Ph, 4-FC$_6$H$_4$, 4-MeC$_6$H$_4$, 2-Npt, Me X=H, 5-MeO, 5-F, 5-Me,7-F PG=Boc, CO$_2$Me	**Ca[(*S*)-VAPOL-PA]$_2$** (2.5 mol%)	99	86 to >99 (*S*)	[24]
R=Ph, Me, 4-FC$_6$H$_4$, 4-MeC$_6$H$_4$, 3-MeC$_6$H$_4$, 2-Npt X=H, 5-Me, 5-MeO, 5-F PG=Boc[b]	**BIN** (10 mol%), 4Å MS Ni(BF$_4$)$_2$•6H$_2$O (10 mol%)	78–99	7–88 (*R*)	[25]
R=Bn, Me, Ph, 4-FC$_6$H$_4$, 4-MeC$_6$H$_4$, 2-Npt X=H, 5-F, 5-Me, 5-MeO, 7-F PG=Boc	**PHOS-Va** (5 mol%)	93–99	90–99 (*S*)	[26]

Notes: [a]Only a positive α_D is reported, but by comparison with the data of this table a (*S*)-configuration may be assumed. [b]Other protecting groups afforded racemic mixtures. When X=H, MeO, Me and R=4-MeC$_6$H$_4$ the lowest ee's were found.

Wang's group was able to scale-up the reaction to 1.24 g (99% yields with 99% ee) and to reuse the catalyst six times without loss of efficiency or enantioselectivity [26]. Moreover, the chiral *N*-Boc protected 3-chlorooxindole was submitted to S$_N$2 reaction with sodium *p*-toluenesulfinate in the presence of NEt$_3$ and the desired product was recovered in 87% yield with 96% ee.

Shi proposed that Ni(II) complex coordinated the indole enolate through the oxygen atoms on C-2 and of the Boc carbonyl group, in order to justify the absence of stereoselectivity with *N*-phenyl and *N*-benzyl protected indoles. In such an arrangement, NCS attacked from the less hindered *Re*-face [25].

Shibata attempted the synthesis of BMS-225113, the chloro analogue of MaxiPost (Appendix C) under the conditions reported in Table 6.1 [8], but NCS provided poor results. Chlorination was instead performed with CF_3SO_2Cl in 93% yield with 60–61% ee (two examples).

The other way to give 3-chloro-3-susbtituted oxindoles is the addition of 3-chlorooxindole enolate to electrophiles. For instance, two examples of conjugate addition to nitroalkenes with two different squaramides as the catalyst are reported (Scheme 6.10, Eqs. (6.3) and (6.4)) [27, 28]. A bifunctional catalysis was envisaged in which the tertiary basic nitrogen atom favored the enolization of the 3-chlorooxindole, and H-bonding donor sites activated the nitroalkene in the transition state. Moreover, in the transition state the *Re*-face of the oxindole enolate and the *Re*-face of the nitroalkene are faced because both catalysts gave a (*R*)-configured C-3 and a (*S*)-configured C-α. Kanger reported a failure in the reaction with aliphatic nitroalkenes [27], while Kim did not attempt the reaction [28].

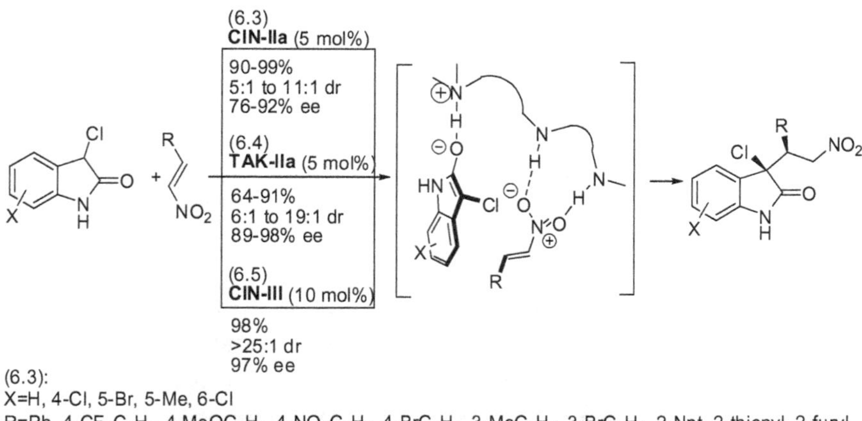

(6.3):
X=H, 4-Cl, 5-Br, 5-Me, 6-Cl
R=Ph, 4-CF$_3$C$_6$H$_4$, 4-MeOC$_6$H$_4$, 4-NO$_2$C$_6$H$_4$, 4-BrC$_6$H$_4$, 3-MeC$_6$H$_4$, 3-BrC$_6$H$_4$, 2-Npt, 2-thienyl, 2-furyl, 2-pyridyl
(6.4):
X=H, 4-Cl, 6-Cl
R=Ph, 4-MeOC$_6$H$_4$, 4-MeC$_6$H$_4$, 4-FC$_6$H$_4$, 3-FC$_6$H$_4$, 3-NO$_2$C$_6$H$_4$, 2-FC$_6$H$_4$, 3-NO$_2$C$_6$H$_4$, 2-thienyl, 2-furyl
(6.5):
X=H, R=Ph

Scheme 6.10.   Asymmetric addition of nitroalkenes to oxindoles.

Finally, another example of this reaction was reported by Lu and co-workers with a thiourea catalyst (Scheme 6.10, Eq. (6.5)) [29]. However, the aim of the paper was the one-pot formation of spirocyclopropaneoxindoles (see Chapter 2), a reaction not attempted with squaramides.

Feng and co-workers reported the enantioselective haloamination of 3-alkenyl-oxindoles with the simultaneous formation of C–N and C–halogen bonds (Scheme 6.11, Eq. (6.6)) [30]. These products could be the source of the potential drug spiro-aziridine oxindoles, but the reaction was not attempted. Evidence reported in the paper supported a transition state, in which the iron complex must be coordinated to the carbonyl group of oxindoles so that the observed (R,R)-configured product is formed.

The Mannich reaction of *N*-unprotected-3-bromooxindoles with *N*-Ts-imines provided the diastereomer of these compounds (Scheme 6.11, Eq. (6.7)) [31]. Notably, a source of protons in the *N*-Ts-imine deactivated the catalyst, because these acidic sites reacted with the tertiary nitrogen in the catalyst counteracting its basicity. The reaction could be successfully scaled up to 3 mmol, and intramolecular S_N2 reaction promoted by

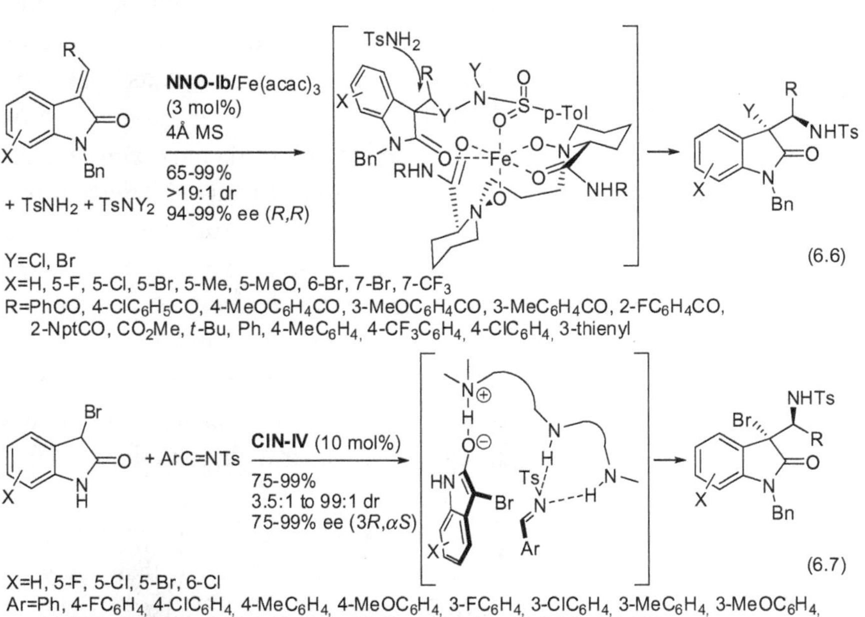

Y=Cl, Br

X=H, 5-F, 5-Cl, 5-Br, 5-Me, 5-MeO, 6-Br, 7-Br, 7-CF$_3$

R=PhCO, 4-ClC$_6$H$_5$CO, 4-MeOC$_6$H$_4$CO, 3-MeOC$_6$H$_4$CO, 3-MeC$_6$H$_4$CO, 2-FC$_6$H$_4$CO, 2-NptCO, CO$_2$Me, *t*-Bu, Ph, 4-MeC$_6$H$_4$, 4-CF$_3$C$_6$H$_4$, 4-ClC$_6$H$_4$, 3-thienyl

(6.6)

X=H, 5-F, 5-Cl, 5-Br, 6-Cl

Ar=Ph, 4-FC$_6$H$_4$, 4-ClC$_6$H$_4$, 4-MeC$_6$H$_4$, 4-MeOC$_6$H$_4$, 3-FC$_6$H$_4$, 3-ClC$_6$H$_4$, 3-MeC$_6$H$_4$, 3-MeOC$_6$H$_4$, 2-FC$_6$H$_4$, 2-ClC$_6$H$_4$, 2-MeC$_6$H$_4$, 3-MeOC$_6$H$_4$, 1-Npt, 2-furyl, (*E*)-PhCH=CH

(6.7)

Scheme 6.11.   Asymmetric haloamination of oxindoles.

AgNO$_3$ and NEt$_3$ allowed the synthesis of spiro-aziridine oxindole in 86% yield, >99:1 dr and 93% ee.

6.4. Sulfur

As in the previous sections, sulfur-substituted oxindoles can be obtained by sulfenylation of 3-alkyloxindoles or by alkylation of 3-sulfo-oxindoles. In particular, isatin-derived *N,S*-acetals have promising importance as drugs, and 3-CF$_3$S-oxindoles are interesting, because this substituent is highly lipophilic, so they generally increase the transmembrane permeation, that is the bioavailability. Owing to their promising potential as biopharmaceutics, oxindoles' dithioketal synthesis was also explored.

Although sulfur-based methines are ambident nucleophiles, *S*-nucleophilic attack on *S*-electrophiles has not been yet mentioned in literature. Table 6.3 reports many successful syntheses. Among them, the products obtained by Enders' group could be deprotected or oxidized at the thioether without any loss in enantioselectivity [32]. Li and co-workers envisaged hydrogen bond formation between the catalyst and the carbonyl of the Boc group, as already reported in other instances of this chapter, because indoles bearing non-amide protecting groups as well as *N*–H oxindole provided worse enantioselectivity [33]. Only the reaction set up by Feng's group worked on unprotected oxindoles [35], thus avoiding the protection and the deprotection steps. Moreover, it is also interesting that no undesired associations of the transition metal with sulfur are observed. Finally, carbonate is necessary for the formation of oxindole, because no reaction occurs without it. The substitution of the MeO with an *i*-PrO group on quinidine drastically increased enantioselectivity from 64% to 94% ee with 3-CF$_3$-oxindoles [38]. Furthermore, 3-CF$_3$-oxindoles significantly decomposed under these reaction conditions, and aliphatic *N*-(sulfanyl) phthalimides did not react. Yuan found that, under their conditions [41]:

(i) *N*-(alkylsulfanyl)succinimides gave only 61–90% yields with 50–67% ee in longer reaction times (13–14 days instead of less than a day),

(ii) free *N*–H indoles gave 75–78% yields with 50–83% ee,

(iii) *N*-(2-thienylsulfanyl)succinimide gave 72% yield with >99% ee only after 6 days.

Table 6.3. Sulfenylation reactions of prostereogenic oxindoles with *N*-(sulfanyl)imides.

R, R^1, X, PG	Catalyst	Yield (%)	ee (%)	Ref.
(a) Sulfenylating agent: 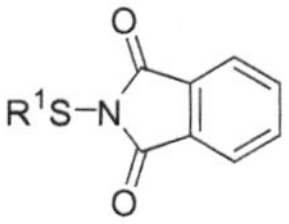				
R=Bn, Ph, Me, 4-MeC$_6$H$_4$CH$_2$, 3-FC$_6$H$_4$CH$_2$, (piperonyl)CH$_2$, 4-MeOC$_6$H$_4$, 4-FC$_6$H$_4$ R^1=Bn, Ph, 4-ClC$_6$H$_4$, 4-MeC$_6$H$_4$, 4-NO$_2$C$_6$H$_4$, 4-MeOC$_6$H$_4$, 3-NO$_2$C$_6$H$_4$ X=H, 5-Me PG=Boc	**TAK-III** (5 mol%)	86–98	85–96 (*R*)	[32]
R=Ph, 4-MeC$_6$H$_4$, 4-BuC$_6$H$_4$, 4-FC$_6$H$_4$, 4-MeOC$_6$H$_4$, 4-EtC$_6$H$_4$, 3-MeOC$_6$H$_4$, 3,5-Me$_2$C$_6$H$_3$ R^1=Ph, 4-FC$_6$H$_4$, 4-ClC$_6$H$_4$, 4-MeC$_6$H$_4$, 4-MeOC$_6$H$_4$ X=H, 5-MeO, 5-F, 5-Br, 4-MeO PG=Boc	**Quinidine** (10 mol%)	83–99^a	79–99 (*R*)	[33]
R=4-MeC$_6$H$_4$ R^1=3-FC$_6$H$_4$ X=H PG=Boc	**PHOS-VI** (1 mol%)	93	80 (*S*)	[34]
R=Bn, Me, Et, Pr, *i*-Pr, Bu, *i*-Bu, (3-Et)Bu, neopentyl, CH$_2$=CHCH$_2$, MeCO$_2$CH$_2$. 4-MeC$_6$H$_4$CH$_2$, 4-MeOC$_6$H$_4$CH$_2$, 4-ClC$_6$H$_4$CH$_2$, 4-FC$_6$H$_4$CH$_2$, 4-BrC$_6$H$_4$CH$_2$, 4-NO$_2$C$_6$H$_4$CH$_2$, 3-ClC$_6$H$_4$CH$_2$, 3-MeC$_6$H$_4$CH$_2$, 3-NO$_2$C$_6$H$_4$CH$_2$, 2-MeC$_6$H$_4$CH$_2$, 2-ClC$_6$H$_4$CH$_2$, (piperonyl) CH$_2$, 1-NptCH$_2$, 2-NptCH$_2$, 2-thienylCH$_2$, Ph, 4-MeOC$_6$H$_4$, 4-FC$_6$H$_4$, 4-ClC$_6$H$_4$, 4-MeC$_6$H$_4$, 3-MeC$_6$H$_4$, 2-Me C$_6$H$_4$, 1-Npt, 2-Npt R^1=Ph X=H, 5-Me, 5-MeO, 5-F, 5-Cl, 6-Br, 7-F PG=H	**NNO-Ia** (1 mol%) Sc(OTf)$_3$ (1.5 mol%) 4Å MS, Na$_2$CO$_3$	82–98^b	87–99 (*R*)	[35]

(Continued)

Table 6.3.   (*Continued*)

R, R^1, X, PG	Catalyst	Yield (%)	ee (%)	Ref.

R=Ph, 4-MeC$_6$H$_4$, 4-BuC$_6$H$_4$, 4-*t*-BuC$_6$H$_4$,
  4-PhC$_6$H$_4$, 4-EtOC$_6$H$_4$, 4-FC$_6$H$_4$,
  4-ClC$_6$H$_4$, 3-MeC$_6$H$_4$, 3-MeOC$_6$H$_4$,
  3-BrC$_6$H$_4$, 2-Npt
R^1=CF$_3$
X=H, 5-Me, 5-t-Bu, 5-MeO, 5-F, 5-Cl,
  5-Br
PG=Boc
	(DHQD)$_2$PYR
	(10 mol%)
		75–90	85–94 (*S*)	[36]

R=Ph, Me, 4-MeC$_6$H$_4$, 4-MeOC$_6$H$_4$, 4-FC$_6$H$_4$, 2-Npt R^1=Ph X=H, 5-Me, 5-MeO, 5-F, 7-F PG=Boc	**PHOS-Vb** (10 mol%)	92–99	72–90 (*S*)^c	[37]
R=Bn, *i*-Bu, 4-MeC$_6$H$_4$CH$_2$, 4-MeOC$_6$H$_4$CH$_2$, 4-FC$_6$H$_4$CH$_2$ R^1=Ph X=H, 5-Me, 5-MeO, 5-F, 7-F PG=Boc		93–99	66–90 (*R*)^c	

R=CF$_3$
R^1=Ph, 4-MeC$_6$H$_4$, 4-NO$_2$C$_6$H$_4$, 3-MeC$_6$H$_4$,
  2-MeC$_6$H$_4$, 2-pyridyl
X=H, 5-Me, 5-MeO, 6-Me, 6-MeO, 6-MeS,
  5,6-Me$_2$
PG=CO$_2$Me
	6-*i*-PrO-Quinidine
	(10 mol%)
		71–92	70–94 (*R*)	[38]

(*b*) Sulfenylating agent:

$$R^1S-N\left(\begin{array}{c}O\\ \\O\end{array}\right)$$

R, R^1, X, PG	Catalyst	Yield (%)	ee (%)	Ref.

R=Ph, 4-FC$_6$H$_4$, 4-MeC$_6$H$_4$, 4-MeOC$_6$H$_4$,
  3-FC$_6$H$_4$, 3-CF$_3$C$_6$H$_4$, 3-MeC$_6$H$_4$, 2-Npt
R^1=Bn, Ph, Et, Cy, 2-ClC$_6$H$_4$CH$_2$,
  2-BrC$_6$H$_4$CH$_2$, 4-BrC$_6$H$_4$CH$_2$,
  4-ClC$_6$H$_4$CH$_2$, 2-furylCH$_2$
X=H, 5-F, 5-Cl, 5-Br, 5-Me, 5-MeO, 6-Br,
  7-F
PG=Bn
	(DHQ)$_2$PHAL
	(5 mol%) or
	(DHQD)$_2$PHAL
	(10 mol%)
		68–98	85–97^d	[39]

(Continued)

Table 6.3.   (*Continued*)

R, R^1, X, PG	Catalyst	Yield (%)	ee (%)	Ref.
R=Me, Et, Bn, 4-FC$_6$H$_4$CH$_2$, 4-MeOC$_6$H$_4$CH$_2$, 3-FC$_6$H$_4$CH$_2$, 3-CF$_3$C$_6$H$_4$CH$_2$, 2-FC$_6$H$_4$CH$_2$2, 2-MeO C$_6$H$_4$ R^1=Pr, Cy, Bn, 4-FC$_6$H$_4$CH$_2$, 4-MeC$_6$H$_4$CH$_2$, 4-ClC$_6$H$_4$CH$_2$, 4-*t*-BuC$_6$H$_4$CH$_2$, 3-MeC$_6$H$_4$CH$_2$, 2-MeC$_6$H$_4$CH$_2$ X=H, 4-Me, 5-Me, 6-F, 7-F PG=Me	**GU** (20 mol%)	65–99	86–98 (*S*)	[40]
R=pyrrol-1-yl R^1=Ph, 4-ClC$_6$H$_4$, 4-MeC$_6$H$_4$, 4-MeOC$_6$H$_4$, 4-Br C$_6$H$_4$, 3-ClC$_6$H$_4$, 3-MeC C$_6$H$_4$, 1-Npt^e X=H, 5-F, 5-Cl, 5-Br, 5-Me, 6-Cl PG=Me, Et, Ph, Bn	**Cinchonidine** (10 mol%), 4Å MS	81–98^f	82–93 (*R*)	[41]
R=4-ClC$_6$H$_4$S, 2-FC$_6$H$_4$S, BnS, CH$_2$=CHCH$_2$S, 2-NptS R^1=Ph, Bn X=H, 5-MeO, 5-F, 5-Me, 6-MeO, 7-Cl, 5,7-Me2 PG=H, Me, Bn	**Dihydroquinine** (5 mol%), 13X MS	57–99%	81–96 (S)	[42]
R=4-ClC$_6$H$_4$S, R^1=CF$_3$, X=H, PG=Bn	**Dihydroquinine** (10 mol%), 13X MS	65	85 (+)^g	

Notes: ^{a}Performed also in 1 mmol scale. ^{b}(*R*)-3-PMP-3-(phenylthio)indolin-2-one was prepared in 94% yield with 97% ee in a gram-scale reaction. ^{c}Author did not explain the stereo-divergence of this reaction. The corresponding enantiomers were also obtained in comparable yield and selectivity with the enantiomeric iminophopshorane. ^{d}**(DHQ)$_2$PHAL** [(*R*)-isomer], **(DHQD)$_2$PHAL** [(*S*)-isomer]. ^{e}The reaction was also carried out with (phenylselanyl) succinimide and this is the only report about selenium-containing oxindoles. ^{f}Scaling up the reaction, 1.09 g of product was recovered without loss in the reactivity and enantioselectivity. ^{g}A positive α_D is reported, but not a Cahn–Ingold–Prelog descriptor.

Less common sulfenylation agents, such as trichloroisocyanuric acid and $AgSCF_3$ [43] or [2-(2-iodophenyl)propan-2-yloxy](trifluoromethyl)sulfane [44], have been employed for the trifluoromethylsulfenylation of oxindoles (Scheme 6.12, Eqs. (6.8) and (6.9), respectively). Tan and co-workers [43] recovered products with yield and stereochemistry comparable with those reported by Rueping [36], but the $^{+}SCF_3$ was generated *in situ* from readily available starting materials, instead of preparing the precursor in advance. Actually authors looked for, but they did not find convincing evidence of what true sulfenylating species is active under their experimental conditions. On the other hand, Shen and Lu firstly reported the trifluoromethylsulfenylation to 3-alkyloxindoles [44].

A particular synthesis of 3-CF_3S-3-alkyl/aryloxindoles was proposed by Shen *et al.*, who was able to transfer chirality from (1*S*)-(–)-*N*-trifluoromethylthio-2,10-camphorsultam to prochiral 3-alkyl/aryloxindoles, to give the products in good to high enantiomeric excesses (Scheme 6.13) [45]. Only a cyanomethyl substituent in the para-position of the 3-aryl group or 3-alkylated oxindoles occurred with moderate enantioselectivity. All papers describing trifluoromethylsulfenylation of oxindoles reported only (*S*)-configured 3-CF_3S-3-alkyl/aryloxindoles. However, Shen and Lu

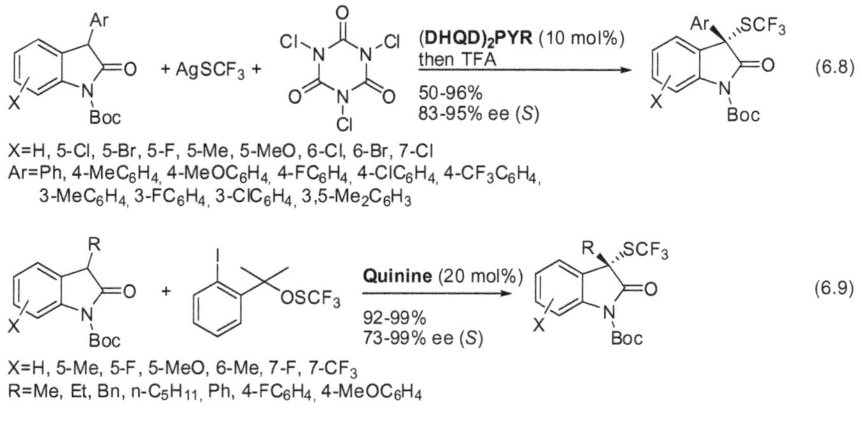

Ar
Cl
N
Cl
O
=O
+ AgSCF3 +
N
O
Boc
X
Cl
(DHQD)2PYR (10 mol%)
then TFA
50-96%
83-95% ee (S)
Ar
SCF3
(6.8)
X=H, 5-Cl, 5-Br, 5-F, 5-Me, 5-MeO, 6-Cl, 6-Br, 7-Cl
Ar=Ph, 4-MeC6H4, 4-MeOC6H4, 4-FC6H4, 4-ClC6H4, 4-CF3C6H4,
3-MeC6H4, 3-FC6H4, 3-ClC6H4, 3,5-Me2C6H3
R
I
OSCF3
Quinine (20 mol%)
92-99%
73-99% ee (S)
R
SCF3
(6.9)
X=H, 5-Me, 5-F, 5-MeO, 6-Me, 7-F, 7-CF3
R=Me, Et, Bn, n-C5H11, Ph, 4-FC6H4, 4-MeOC6H4

Scheme 6.12. Trifluoromethylthiolation of oxindoles.

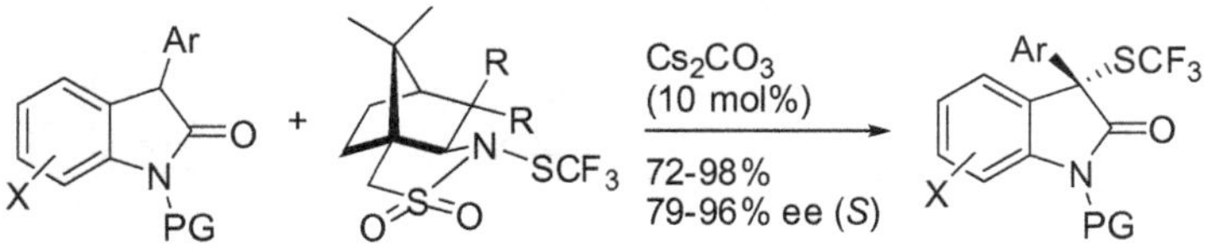

X=H, 7-F
PG=Me, Boc
R=Cl, OMe
Ar=Ph, Me, 4-MeOC$_6$H$_4$, 4-ClC$_6$H$_4$, 4-t-BuC$_6$H$_4$, 4-CNC$_6$H$_4$, 4-CO$_2$MeC$_6$H$_4$,
    4-MeCOC$_6$H$_4$, 4NCCH$_2$C$_6$H$_4$, 3-CNC$_6$H$_4$, 2-Npt

Scheme 6.13.   Synthesis of 3-CF$_3$S-3-alkyl/aryloxindoles by chiral transfer reaction.

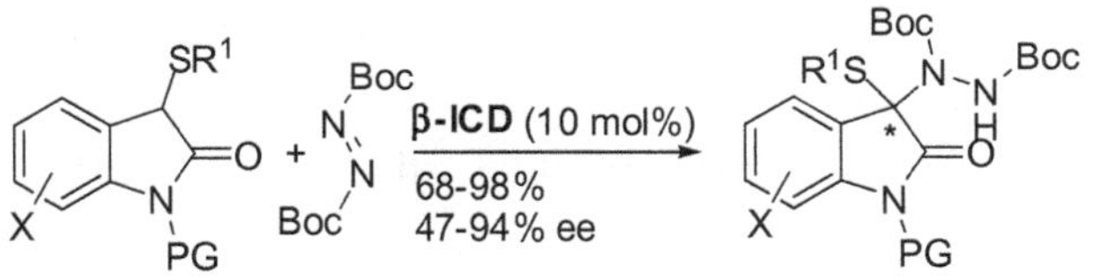

X=H, 5-Me, 5-F, 5-Cl, 5-Br, 5-I, 5-MeO, 6-MeO, 7-Me, 7-Cl, 5,7-Me$_2$, 6-Cl-7-Me
PG=H, Me Bn
R^1= 2-Npt, Bn, Ph, CH$_2$=CHCH$_2$

Scheme 6.14.   Asymmetric addition of 3-alkyl/arylthiooxindoles to DBAD.

used quinine as the catalyst, thus perhaps the synthesis of (*R*)-configured products could be performed by the use of the pseudo enantiomeric quinidine.

As mentioned above, the addition of electrophiles on 3-thiooxindoles may occur both on the C-3 or sulphur atom.

For instance, thiols were found to attack the sulfur atom to form a disulfane [46]. Therefore, the synthesis of 3-alkyl/arylthio-3-alkyl/aryloxindoles starting from 3-alkyl/arylthiooxindoles is scarcely studied. Among them, the first report using 3-thiooxindoles as nucleophiles was the addition of 3-alkyl/arylthiooxindoles to DBAD catalyzed by β-ICD (Scheme 6.14) [47]. Unfortunately, only a positive α_D value was reported because authors were unable to obtain an enantiopure single crystal.

Nakamura [48] and Enders [49] simultaneously wrote two papers describing the addition of thiols to isatin-derived ketimines, allowing a

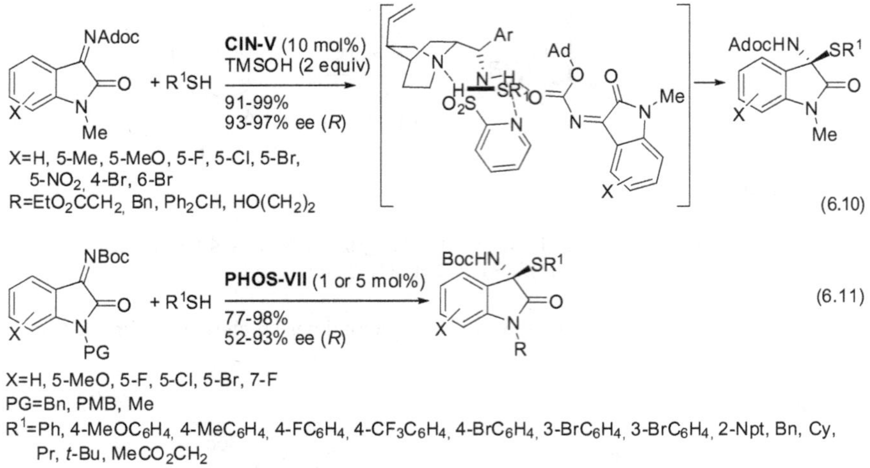

X=H, 5-Me, 5-MeO, 5-F, 5-Cl, 5-Br,
   5-NO$_2$, 4-Br, 6-Br
R=EtO$_2$CCH$_2$, Bn, Ph$_2$CH, HO(CH$_2$)$_2$

$$\text{CIN-V (10 mol\%), TMSOH (2 equiv), 91-99\%, 93-97\% ee }(R) \tag{6.10}$$

X=H, 5-MeO, 5-F, 5-Cl, 5-Br, 7-F
PG=Bn, PMB, Me
R^1=Ph, 4-MeOC$_6$H$_4$, 4-MeC$_6$H$_4$, 4-FC$_6$H$_4$, 4-CF$_3$C$_6$H$_4$, 4-BrC$_6$H$_4$, 3-BrC$_6$H$_4$, 3-BrC$_6$H$_4$, 2-Npt, Bn, Cy,
   Pr, t-Bu, MeCO$_2$CH$_2$

$$\text{PHOS-VII (1 or 5 mol\%), 77-98\%, 52-93\% ee }(R) \tag{6.11}$$

Scheme 6.15.   Asymmetric addition of thiols to isatin-derived ketimines.

synthesis of isatin-derived *N,S*-acetals (Scheme 6.15, Eqs. (6.10) and (6.11), respectively). Enders' group prepared only the (*R*)-configured enantiomer, while Nakamura's one both enantiomers, albeit the (*S*)-configured in low enantioselectivity (82–94% ee). Moreover, Nakamura did not report the synthesis of spirothiazolidines (a highly active growth inhibitor for different tumor cell lines, Appendix C), successfully attempted by Enders in three steps. Nakamura set up the one-pot synthesis from isatin (aza-Wittig reaction then thiol addition) in 90% yield with 95% ee, while Enders scaled up to 1.24 g of (*R*)-*t*-butyl-1-benzyl-2-oxo-3-(phenylthio) indolin-3-ylcarbamate in 92% yield with 93% ee. Finally, Nakamura envisaged the transition state depicted in Scheme 6.15, and Enders found that:

(i) less reactive alkyl thiols needed a higher catalyst loading,
(ii) provided a lower level of enantioselectivity,
(iii) non-protected ketimines or *N*-Boc, *N*-Ac protected as well as thiophene-2-thiol as substrate are unreactive.

Nitroalkenes were also employed as the electrophile in two papers one mainly devoted to the reaction of 3-benzylthiooxindoles [50], the other of

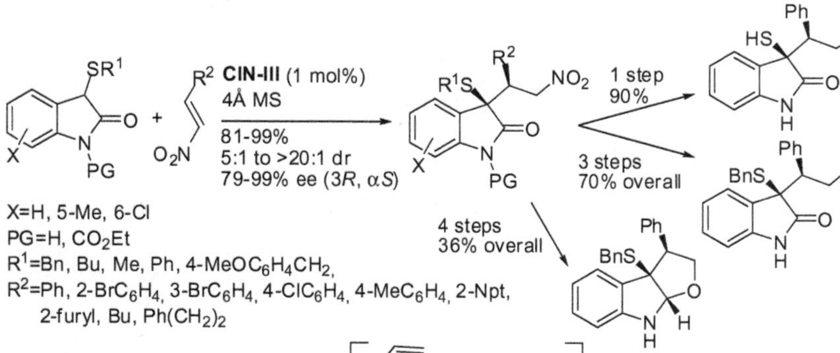
CIN-III (1 mol%)
4Å MS
81-99%
5:1 to >20:1 dr
79-99% ee (3R, αS)
1 step
90%
3 steps
70% overall
4 steps
36% overall
X=H, 5-Me, 6-Cl
PG=H, CO2Et
R1=Bn, Bu, Me, Ph, 4-MeOC6H4CH2,
R2=Ph, 2-BrC6H4, 3-BrC6H4, 4-ClC6H4, 4-MeC6H4, 2-Npt,
2-furyl, Bu, Ph(CH2)2

(6.12)

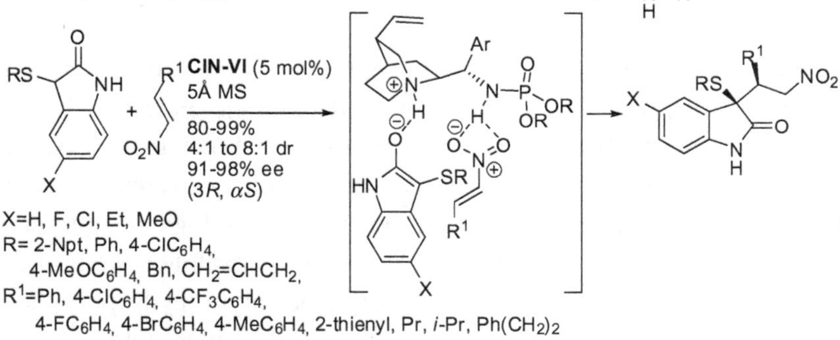
CIN-VI (5 mol%)
5Å MS
80-99%
4:1 to 8:1 dr
91-98% ee
(3R, αS)
X=H, F, Cl, Et, MeO
R= 2-Npt, Ph, 4-ClC6H4,
4-MeOC6H4, Bn, CH2=CHCH2,
R1=Ph, 4-ClC6H4, 4-CF3C6H4,
4-FC6H4, 4-BrC6H4, 4-MeC6H4, 2-thienyl, Pr, i-Pr, Ph(CH2)2

(6.13)

Scheme 6.16. Asymmetric addition of 3-arylthio- and 3-alkylthiooxindoles to nitroalkenes.

3-arylthiooxindoles (Scheme 6.16, Eqs. (6.12) and (6.13), respectively) [51]. However, both reported almost an entry on the other substrate. Yields and enantiomeric excesses are comparable, although the latter reaction is less diasteroselective. Lu and co-workers further manipulated their products and, in particular, they prepared a furoindoline, a core structure of some antitumor agents (Scheme 6.16, Eq. (6.12)) [50]. On the other hand, Wu and co-workers successfully attempted a gram-scale reaction (97% yield, 6:1 dr and 97% ee) only with 1.0 mol% of catalyst [51].

Finally, two asymmetric examples of the sulfenylation of 3-halooxindoles with sulfinate salts as nucleophiles had been reported by Yuan and co-workers (Scheme 6.17) [52]. The same products were obtained by Wang [26], but here the stereochemistry is introduced in the substitution reaction, whereas Wang introduces it in the formation of the halooxindole.

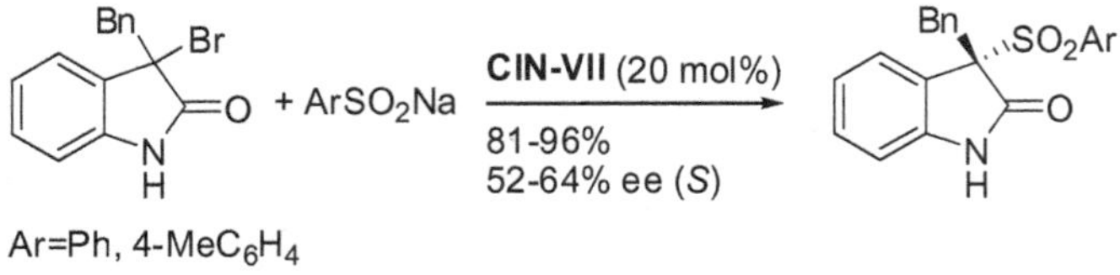

Ar=Ph, 4-MeC$_6$H$_4$

Scheme 6.17. Asymmetric sulfinate addition to 3-halooxindoles.

6.5. Phosphorus

Phosphonic esters play significant biological roles as metabolic intermediates, backbones for genetic information, and synthetic protein switches (for a review on enantioselective addition of P–H bond to C–C/C–X double bond, see Ref. [53]). In particular, α-amino-phosphonic acid derivatives are known to exhibit a broad spectrum of biological activities. In spite of this, only a few asymmetric additions of phosphite to ketimines have been reported. The first asymmetric organocatalytic hydrophosphonylation reaction of isatin imines was reported by Reddy *et al.* [54]. Then other reactions appeared in the literature (Table 6.4). It should be noted that, hydrophosphonylation reactions could occur only with a phosphite that is a nucleophile. However, under neutral conditions, phosphonate is the almost existing tautomer. Thus, all the catalysts must have a basic site to shift the phosphonate $\rightleftharpoons$ phosphite equilibrium towards the most nucleophilic phosphite. Then hydrogen bond interactions with the C=O and the C=N bonds of the ketimine isatins allow to assemble a transition state in which the *Re*-face of the oxindole is less hindered.

The reaction of dialkylphosphonates was attempted under cinchona-derived catalysts, but gave very poor results or did not work at all [54, 55], even with a titanium catalyst (15–85% yields with 40–72% ee) [56], while it was not attempted with a Takemoto-like catalyst [57].

Chimni and co-workers also attempted a one-pot sequential aza-Wittig and phospha-Mannich reactions from which products were obtained in 70–81% yield, with 70–85% ee [55]. The work of Kim had to be outlined for the low catalyst loading [57]. Khan and co-workers expanded the scope of their titanium complex to isatins, in order to prepare biologically important target compounds (Scheme 6.18, Eq. (6.13)) [57]. It should be

Table 6.4.   Hydrophosphonylation reactions of prostereogenic isatin imines.

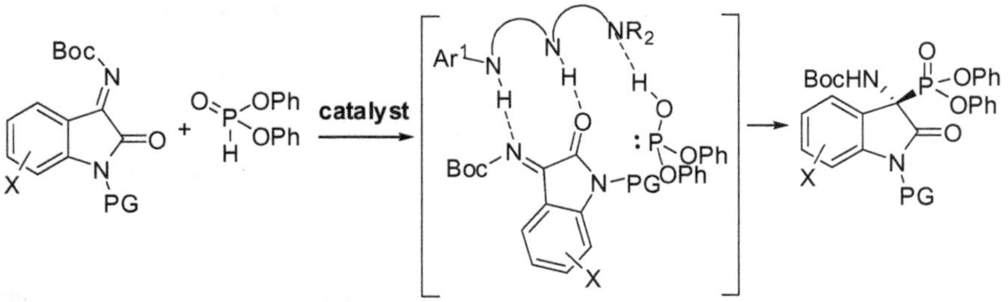

X, PG	Catalyst	Yield (%)	ee (%)	Ref.
X=H, 5-Cl, 5-Br, 5-Me, 5-MeO, 5-NO$_2$, 4-Br, 6-Br, 7-Me, 7-Cl PG=Bn, H, CH$_2$=CHCH$_2$, PMB	**CIN-IIb** (10 mol%)	85–96%	52–98 (*R*)	[54]
X=H, 5-Cl, 5-Br, 5-I, 5-MeO PG=CH$_2$=CHCH$_2$, CH$_2$=CMeCH$_2$, Bn, H	**CIN-Ib** (20 mol%) 4Å MS	71–88%	71–97 (*R*)	[55]
X=H, Me, Cl, F PG=Bn, Me, H	**TAK-IV** (2.5 mol%), Ti(O-*i*-Pr)$_4$ (5 mol%) 4Å MS	84–88%	46 to >99[b]	[56]
X=H, 5-F, 5-Cl, 5-Br, 5-MeO PG=CH$_2$=CHCH$_2$, CH$_2$=CMeCH$_2$, MeCH=CHCH$_2$, Bn, H	**TAK-IIb** (2.5 mol%) 4Å MS	70–94%[a]	73–99 (*R*)	[57]

Notes: [a]*N*-Boc protection of the indole nitrogen atom sharply decreased yield (45%) and selectivity (26% ee); 1.051 g of the desired product was recovered in 81% yield and 93% ee. [b]Unfortunately, neither the configuration nor any α_D were reported.

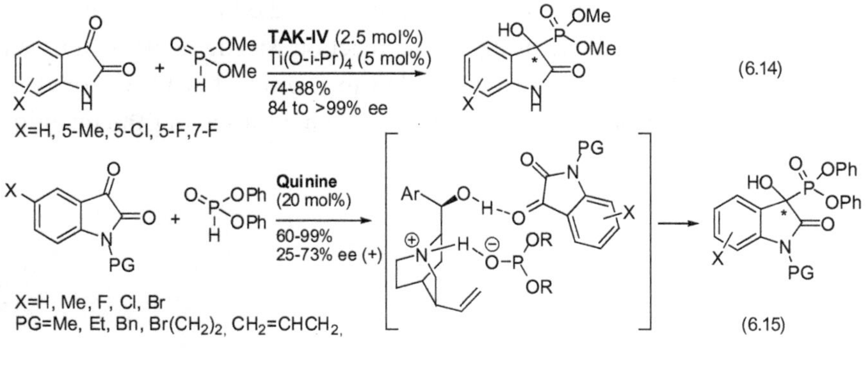

Scheme 6.18.   Asymmetric addition of phosphonates to isatin.

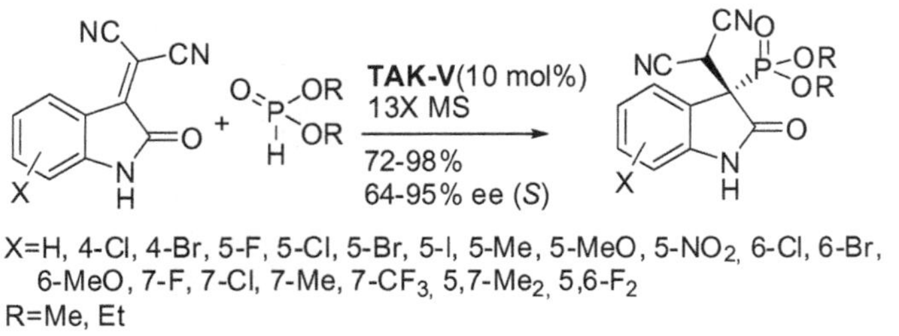

Scheme 6.19. Asymmetric Michael addition of dialkylphosphonates to isatylidene malononitriles.

noted that *N*-protected isatins always gave racemic products, in contrast to isatin imines, while *N*-protection isatins increased ee's. Moreover, the reaction was carried out with dimethylphosphonate and not with diphenylphosphonate.

Also Reddy and co-workers attempted the reaction of diphenylphosphonate and recovered a racemic mixture in 95% yield [54], but they did not attempt the reaction with isatin. Previously, Xu and Wang found that quinine allowed the enantioselective addition of diphenylphosphonate to *N*-protected isatin (Scheme 6.18, Eq. (6.15)) [58]. It should be noted that almost only racemic products (1% ee) were obtained with dibutyl and diisobutylphosphonate. The absolute stereochemistry was not determined in both papers.

Another phosphorylation reaction was the enantioselective Michael addition of dialkylphosphonates to isatylidene malononitriles (Scheme 6.19) [59]. Authors observed a significant drop in the enantiomeric excesses with *N*-protected derivatives and that diphenyl- and dibenzylphosphonates did not work. The synthetic utility of the reaction was demonstrated by an oxidative degradation reaction, which led to a product with carboxylate and phosphonate groups in 88% yield and without affecting the enantiomeric excess.

Appendix A. Catalysts Cited in this Chapter

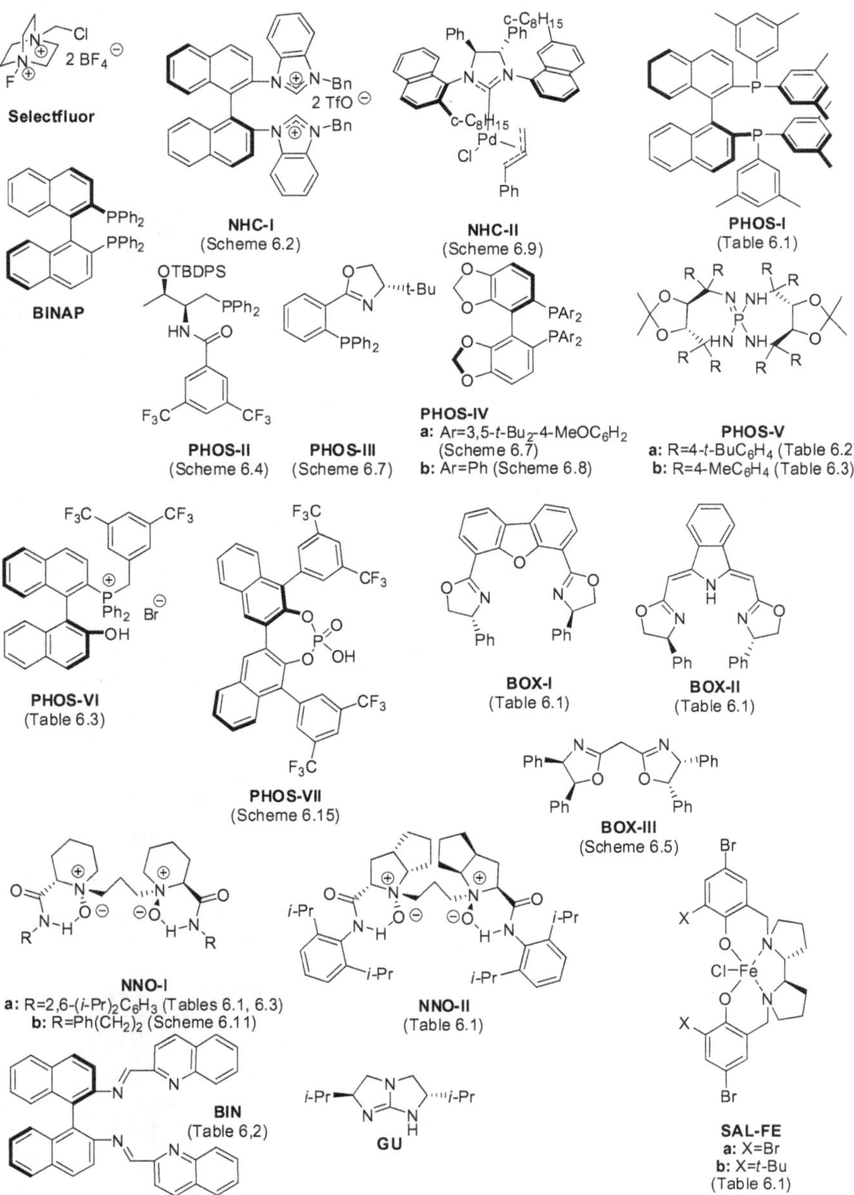

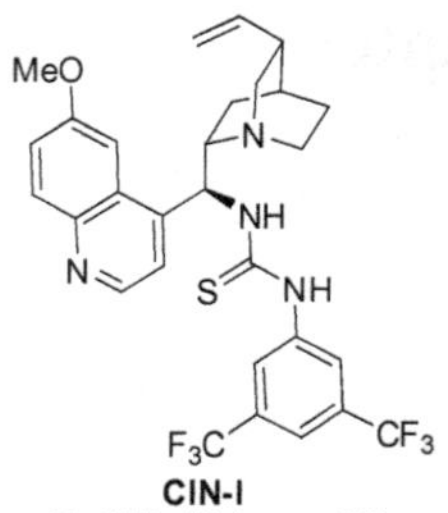

MeO
N
NH
S
NH
F₃C
CF₃
CIN-I
a: R=OMe (Scheme 6.3)
b: R=H (Table 6.4)

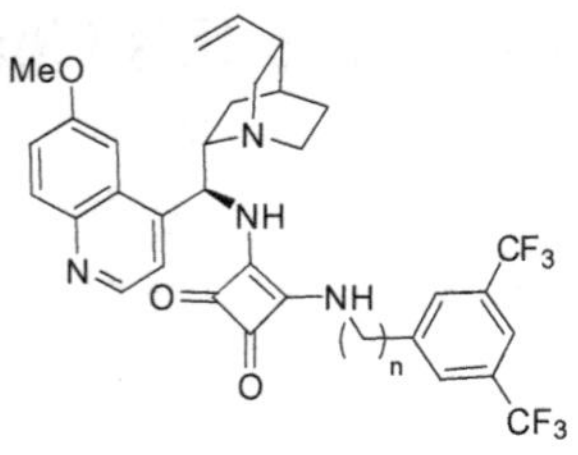

MeO
N
NH
O
O
NH
n
CF₃
CF₃
CIN-II
a: n=0 (Scheme 6.10)
b: n=1 (Table 6.4)

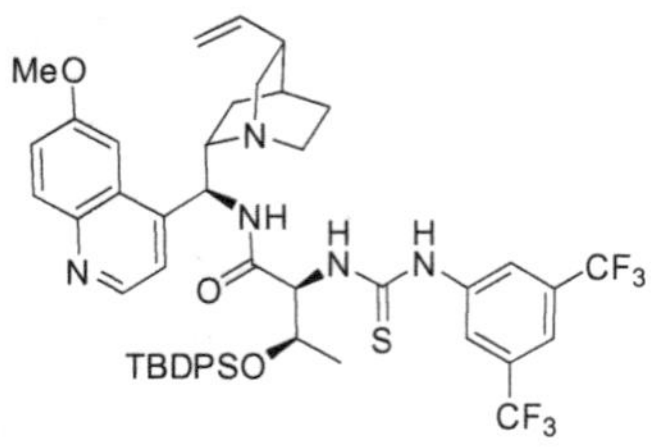

MeO
N
NH
O
N
H
S
N
H
CF₃
CF₃
TBDPSO
CIN-III
(Schemes 6.10, 6.16)

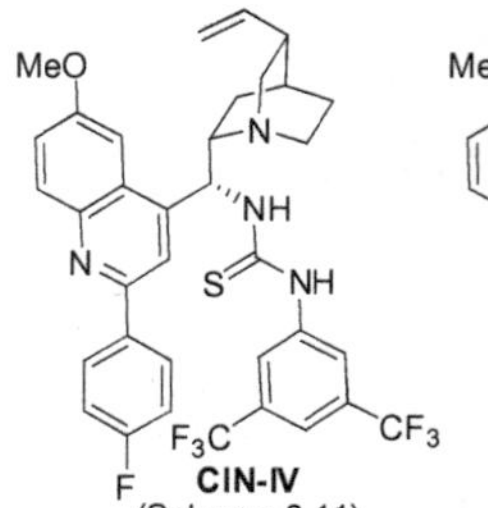

MeO
N
NH
S
NH
CF₃
CF₃
F
CIN-IV
(Scheme 6.11)

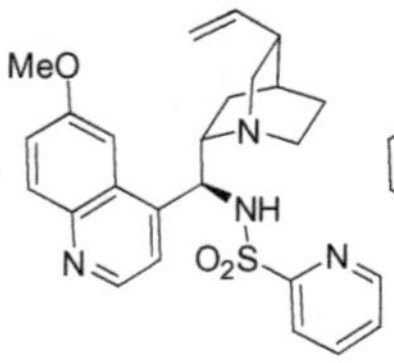

MeO
N
NH
O₂S
N
CIN-V
(Scheme 6.15)

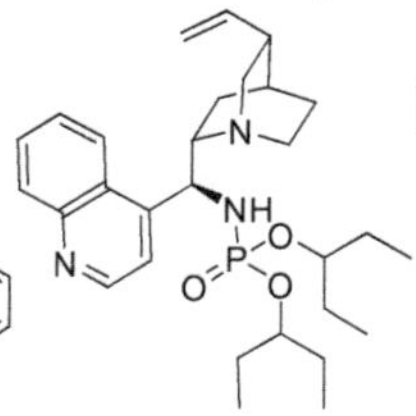

N
NH
P
O
O
O
CIN-VI
(Scheme 6.16)

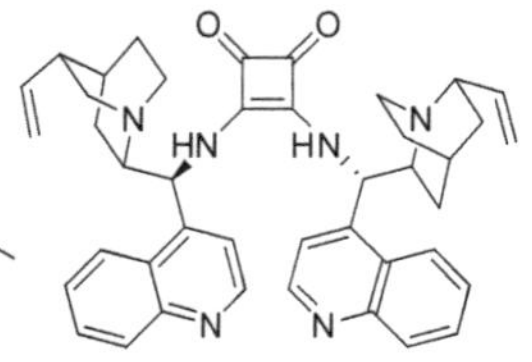

O
O
N
HN
HN
N
N
N
CIN-VII
(Scheme 6.17)

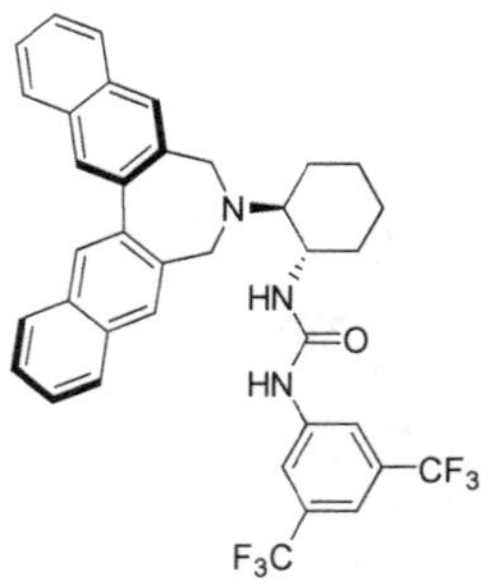

N
HN
O
HN
CF₃
F₃C
TAK-I
(Scheme 6.3)

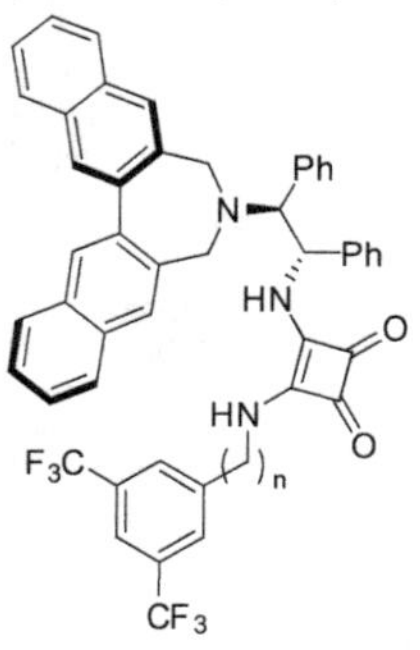

N
Ph
Ph
HN
O
HN
F₃C
n
CF₃
TAK-II
a: n=1 (Scheme 6.10)
b: n=0 (Table 6.4)

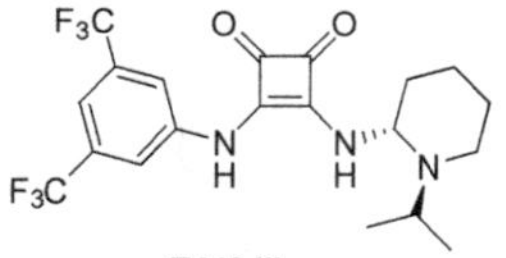

F₃C
O
O
NH
NH
N
F₃C
H
H
TAK-III
(Table 6.3)

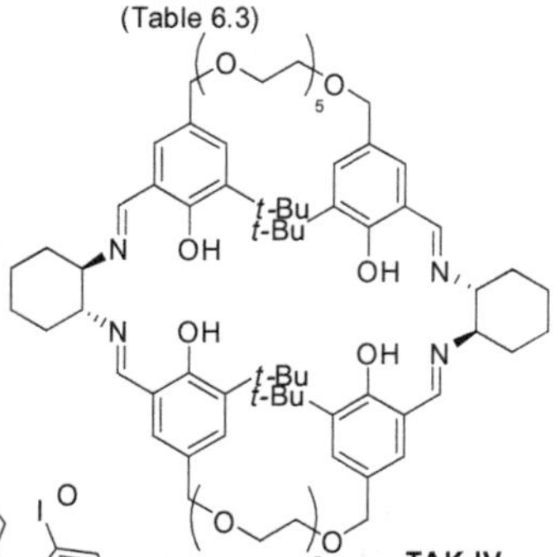

O
O
5
t-Bu
t-Bu
OH
N
N
OH
N
N
OH
OH
t-Bu
t-Bu
O
O
5
TAK-IV
(Table 6.4)

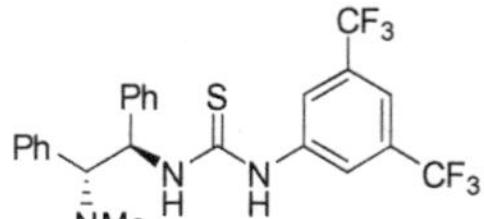

Ph
S
CF₃
Ph
N
H
N
H
CF₃
NMe₂
TAK-V
(Scheme 6.19)

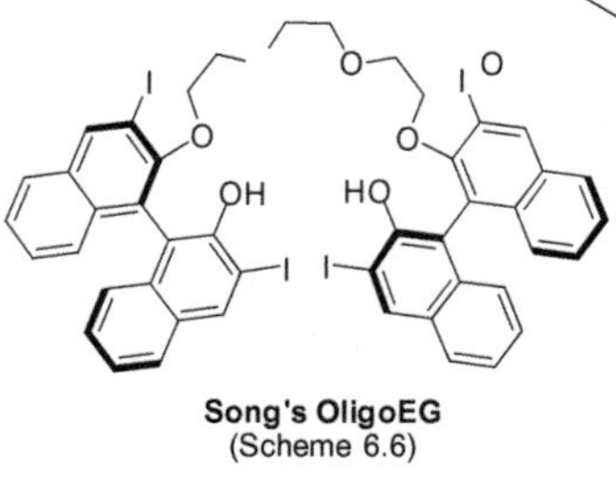

I
O
O
I
OH
HO
O
I
I
Song's OligoEG
(Scheme 6.6)

Appendix B. List of Abbreviation

Adoc	Adamantyloxycarbonyl
Acac	Acetylacetonate
BINAP	2,2′-*Bis*(diphenylphosphino)-1,1′-binaphthyl
BMIM	Butylmethylimidazolium
Bn	Benzyl (PhCH$_2$)
Boc	*tert*-Butoxycarbonyl
Bz	Benzoyl (PhCO)
dba	Dibenzylideneacetone
DBAD	Di-*tert*-butyl azodicarboxylate
(DHQ)$_2$AQN	Hydroquinine anthraquinone-1,4-diyl diether
(DHQ)$_2$PHAL	Hydroquinine phthalazine-1,4-diyl diether
(DHQD)$_2$PHAL	Hydroquinidine phthalazine-1,4-diyl diether
(DHQD)$_2$PYR	Hydroquinidine 2,5-diphenylpyrimidine-4,6-diyl diether
DIPEA	*N,N*-Diisopropylethaneamine
Fmoc	Fluorenylmethyloxycarbonyl
HDMS	Hexamethyldisilazane
HFIP	Hexafluoroisopropanol
ICD	Isocupreidine
MS	Molecular sieves
NFSI	*N*-Fluorobenzenesulfonimide
NHC	*N*-Heterocyclic carbene
Npt	Naphthyl
PG	Protecting group
PMB	4-Methoxybenzyl
PMP	4-Methoxyphenyl
TBDPS	*tert*-Butyldiphenylsilyl
TFA	Trifluoroacetic acid
TfO	Triflate (trifluoromethanesulfonate)
TMS	Trimethylsilyl
Tol	*o,m* or *p*-methylphenyl
VAPOL	2,2′-Diphenyl-(4-biphenanthrol)

Appendix C. Natural Products Cited in this Chapter

a: X=F, BMS-204352 or (*S*)-Maxipost
b: X=Cl, BMS-225113   a spiro-thiazolidine

References

1. J.-S. Yu, H.-M. Huang, P.-G. Ding, X.-S. Hu, F. Zhou, J. Zhou, *ACS Catal.* **2016**, *6*, 5319.

2. N. Shibata, E. Suzuki, T. Asahi, M. Shiro, *J. Am. Chem. Soc.* **2001**, *123*, 7001.

3. T. Ishimaru, N. Shibata, T. Horikawa, N. Yasuda, S. Nakamura, T. Toru, M. Shiro, *Angew. Chem. Int. Ed.* **2008**, *47*, 4157.

4. R. Zhang, D. Wang, Q. Xu, J. Jiang, M. Shi, *Chin. J. Chem.* **2012**, *30*, 1295.

5. N. Shibata, T. Ishimaru, E. Suzuki, K. L. Kirk, *J. Org. Chem.* **2003**, *68*, 2494.

6. L. Zoute, C. Audouard, J.-C. Plaquevent, D. Cahard, *Org. Biomol. Chem.* **2003**, *1*, 1833.

7. Y. Hamashima, T. Suzuki, H. Takano, Y. Shimura, M. Sodeoka *J. Am. Chem. Soc.* **2005**, *127*, 10164.

8. N. Shibata, J. Kohno, K. Takai, T. Ishimaru, S. Nakamura, T. Toru, S. Kanemasa, *Angew. Chem. Int. Ed.* **2005**, *44*, 4204.

9. Q.-H. Deng, H. Wadepohl, L. H. Gade, *Chem. Eur. J.* **2011**, *17*, 14922.

10. J. Li, Y. Cai, W. Chen, X. Liu, L. Lin, X. Feng, *J. Org. Chem.* **2012**, *77*, 9148.

11. X. Gu, Y. Zhang, Z.-J. Xu, C.-M. Che, *Chem. Commun.* **2014**, *50*, 7870.

12. Y. Zhang, X.-J. Yang, T. Xie, G.-L. Chen, W.-H. Zhu, X.-Q. Zhang, X.-Y. Yang, X.-Y. Wu, X.-P. He, H.-M. He, *Tetrahedron*, **2013**, *69*, 4933.

13. F. Wang, J. Li, Q. Hu, X.-J. Yang, X.-Y. Wu, H.-M. He, *Eur J. Org. Chem.* **2014**, 3607.

14. X. Dou, Y. Lu, *Org. Biomol. Chem.* **2013**, *11*, 5217.

15. Y. S. Kim, S. J. Kwon, D. Y. Kim, *Bull. Korean Chem. Soc.* **2015**, *36*, 1512.

16. T. Wang, D. L. Hoon, Y. Lu, *Chem. Commun.* **2015**, *51*, 10186.

17. L. Zhang, W. Zhang, H. Mei, J. Han, V. A. Soloshonok, Y. Pan, *Org. Biomol. Chem.* **2017**, *15*, 311.

18. X. Chen, Y. Li, J. Zhao, B. Zheng, Q. Lu, X. Ren *Adv. Synth. Catal.* **2017**, *359*, 3057.

19. S. Paladhi, S. Y. Park, J. W. Yang, C. E. Song, *Org. Lett.* **2017**, *19*, 5336.

20. K. Balaraman, C. Wolf, *Angew. Chem. Int. Ed.* **2017**, *56*, 1390.

21. Y. Jin, M. Chen, S. Ge, J. F. Hartwig, *Org. Lett.* **2017**, *19*, 1390.

22. L. Wu, L. Falivene, E. Drinkel, S. Grant, A. Linden, L. Cavallo, R. Dorta, *Angew. Chem. Int. Ed.* **2012**, *51*, 2870.

23. M.-X. Zhao, Z.-W. Zhang, M.-X. Chen, W.-H. Tang, M. Shi, *Eur. J. Org. Chem.* **2011**, 3001.

24. W. Zheng, Z. Zhang, M. J. Kaplan, J. C. Antilla, *J. Am. Chem. Soc.* **2011**, *133*, 3339.

25. D. Wang, J.-J. Jiang, R. Zhang, M. Shi, *Tetrahedron: Asymmetry* **2011**, *22*, 1133.

26. X. Gao, J. Han, L. Wang, *Org. Lett.* **2015**, *17*, 4596.

27. A. Noole, I. Järving, F. Werner, M. Lopp, A. Malkov, T. Kanger, *Org. Lett.* **2012**, *14*, 4922.

28. H. J. Jeong, S. J. Kwon, D. Y. Kim, *Bull. Korean Chem. Soc.* **2015**, *36*, 1516.

29. X. Dou, W. Yao, B. Zhou, Y. Lu, *Chem. Commun.* **2013**, *49*, 9224.

30. Y. Cai, X. Liu, P. Zhou, Y. Kuang, L. Lin, X. Feng, *Chem. Commun.* **2013**, *49*, 8054.

31. J. Li, T. Du, G. Zhang, Y. Peng, *Chem. Commun.* **2013**, *49*, 1330.

32. C. Wang, X. Yang, C. C. J. Loh, G. Raabe, D. Enders, *Chem.-Eur. J.* **2012**, *18*, 11531.

33. X. Li, C. Liu, X.-S. Xue, J.-P. Cheng, *Org. Lett.* **2012**, *14*, 4374.

34. S. Shirakawa, A. Kasai, T. Tokuda, K. Maruoka *Chem. Sci.* **2013**, *4*, 2248.

35. Y. Cai, J. Li, W. Chen, M. Xie, X. Liu, L. Lin, X. Feng *Org. Lett.* **2012**, *14*, 2726.

36. M. Rueping, X. Liu, T. Bootwicha, R. Pluta, C. Merkens, *Chem. Commun.* **2014**, *50*, 2508.

37. X. Gao, J. Han, L. Wang, *Synthesis*, **2016**, *48*, 2603.

38. Y. E, T. Yuan, L. Yin, Y. Xu, *Tetrahedron Lett.* **2017**, *58*, 2521.

39. Z. Han, W. Chen, S. Dong, C. Yang, H. Liu, Y. Pan, L. Yan, Z. Jiang, *Org. Lett.* **2012**, *14*, 4670.

40. L. Huang, J. Li, Y. Zhao, X. Ye, Y. Liu, L. Yan, C.-H. Tan, H. Liu, Z. Jiang, *J. Org. Chem.* **2015**, *80*, 8933.

41. Y. You, Z.-J. Wu, Z.-H. Wang, X.-Y. Xu, X.-M. Zhang, W.-C. Yuan, *J. Org. Chem.* **2015**, *80*, 8470.

42. K. Liao, F. Zhou, J.-S. Yu, W.-M. Gao, J. Zhou, *Chem. Commun.* **2015**, *51*, 16255.

43. X.-L. Zhu, J.-H. Xu, D.-J. Cheng, L.-J. Zhao, X.-Y. Liu, B. Tan, *Org. Lett.* **2014**, *16*, 2192.

44. T. Yang, Q. Shen, L. Lu, *Chin. J. Chem.* **2014**, *32*, 678.

45. H. Zhang, X. Leng, X. Wan, Q. Shen, *Org. Chem. Front.* **2017**, *4*, 1051.

46. F. Zhu, F. Zhou, Z.-Y. Cao, C. Wang, Y.-X. Zhang, C.-H. Wang, J. Zhou, *Synthesis*, **2012**, *44*, 3129.

47. F. Zhou, X.-P. Zeng, C. Wang, X.-L. Zhao, J. Zhou, *Chem. Commun.* **2013**, *49*, 2022.

48. S. Nakamura, S. Takahashi, D. Nakane, H. Masuda, *Org. Lett.* **2015**, *17*, 106.

49. C. Beceño, P, Chauhan, A. Rembiak, A. Wang, D. Enders, *Adv. Synth. Catal.* **2015**, *357*, 672.

50. X. Dou, B. Zhou, W. Yao, F. Zhong, C. Jiang, Y. Lu, *Org. Lett.* **2013**, *15*, 4920.

51. W.-M. Gao, J.-S. Yu, Y.-L. Zhao, Y.-L. Liu, F. Zhou, H.-H. Wu, J. Zhou, *Chem. Commun.* **2014**, *50*, 15179.

52. J. Zuo, Z.-J. Wu, J.-Q. Zhao, M.-Q. Zhou, X.-Y. Xu, X.-M. Zhang, W.-C. Yuan, *J. Org. Chem.* **2015**, *80*, 634.

53. D. Zhao, R. Wang, *Chem. Soc. Rev.* **2012**, *41*, 2095.

54. J. George, B. Sridhar, B. V. S. Reddy, *Org. Biomol. Chem.* **2014**, *12*, 1595.

55. A. Kumar, V. Sharma, J. Kaur, V. Kumar, S. Mahajan, N. Kumar, S. S. Chimni, *Tetrahedron* **2014**, *70*, 7044.

56. M. Nazisha, A. Jakhara, N. H. Khana, S. Vermaa, R. I. Kureshya, S. H. R. Abdia, H. C. Bajaj, *Appl. Catal. A: General.* **2016**, *515*, 116.

57. H. S. Jang, Y. Kim, D. Y. Kim, *Beilstein J. Org. Chem.* **2016**, *12*, 1551.

58. L. Peng, L.-L. Wang, J.-F. Bai, L.-N. Jia, Q.-C. Yang, Q.-C. Huang, X.-Y. Xu, L.-X. Wang, *Tetrahedron Lett.* **2011**, *52*, 1157.

59. Z.-M. Liu, N.-K. Li, X.-F. Huang, B. Wu, N. Li, C.-Y. Kwok, Y. Wang, X.-W. Wang, *Tetrahedron*, **2014**, *70*, 2406.

Index

CPSIA information can be obtained
at www.ICGtesting.com
Printed in the USA
BVHW050131090919
557609BV00001B/3/P